Electrochemical Capacitors
Theory, Materials and Applications

Edited by
Inamuddin
Mohammad Faraz Ahmer
Abdullah M. Asiri
Sadaf Zaidi

Electrochemical capacitors are most important for the development of future energy storage systems and sustainable power sources. New superior hybrid supercapacitors are based on binary and ternary thin film nanocomposites involving carbon, metal oxides and polymeric materials. The synthesis of materials and fabrication of electrodes for supercapacitor applications is discussed in detail. The book also presents the fundamental theory and a thorough literature review of supercapacitors.

Keywords: Energy Storage, Electrochemical Capacitors, Nanocomposites, Hybrid Supercapacitors, Carbon/Metal Oxide Composites, Metal Oxides/Hydroxides Composites, Polymer Type Capacitors, Nanoscience, Hydrothermal Synthesis, Graphene-based Composites, Ultrasonic Assisted Synthesis

Electrochemical Capacitors
Theory, Materials and Applications

Edited by

Inamuddin[1,2], Mohammad Faraz Ahmer[3],
Abdullah M. Asiri[1] and Sadaf Zaidi[4]

[1] Chemistry Department, Faculty of Science, King Abdulaziz University, Jeddah 21589, Saudi Arabia

[2] Department of Applied Chemistry, Faculty of Engineering and Technology, Aligarh Muslim University, Aligarh-202 002, India

[3]Department of Electrical Engineering, Mewat College of Engineering and Technology, Mewat-122103, India

[4]Department of Chemical Engineering, Faculty of Engineering and Technology, Aligarh Muslim University, Aligarh 202002, India

Published by **Materials Research Forum LLC**
Millersville, PA 17551, USA

Published as part of the book series
Materials Research Foundations
Volume 26 (2018)
ISSN 2471-8890 (Print)
ISSN 2471-8904 (Online)

Print ISBN 978-1-945291-56-2
ePDF ISBN 978-1-945291-57-9

This book contains information obtained from authentic and highly regarded sources. Reasonable efforts have been made to publish reliable data and information, but the author and publisher cannot assume responsibility for the validity of all materials or the consequences of their use. The authors and publishers have attempted to trace the copyright holders of all material reproduced in this publication and apologize to copyright holders if permission to publish in this form has not been obtained. If any copyright material has not been acknowledged please write and let us know so we may rectify in any future reprint.

Distributed worldwide by

Materials Research Forum LLC
105 Springdale Lane
Millersville, PA 17551
USA
http://www.mrforum.com

Manufactured in the United States of America
10 9 8 7 6 5 4 3 2 1

Table of Contents

Preface

The book on **Electrochemical Capacitors: Theory, Materials and Applications** provides up-to-date knowledge related to the experimental background and state-of-the-art survey of composite thin films for electrochemical capacitors: the most important innovation for energy storage, which is imperative for future energy storage. The broad-spectrum coverage of this book will provide a premise for innovative work of electrochemical capacitors in the coming decades. This edition covers the theory, fundamental and literature review of supercapacitors. The synthesis of materials and fabrication of electrodes for supercapacitor applications are discussed in details. The reader will appreciate the contextual investigations extending from carbon, metal oxides and polymeric materials based binary and ternary composites energy storage systems for superior hybrid supercapacitors. This book is also covering the whole range of nanocomposite science and technology and innovation for sustainable power sources.

This book is the consequence of the precarious accountability of experts from various interdisciplinary science fields. It extensively explores the most plenteous, top to bottom, and innovative research and reviews. We are grateful to all the contributing authors and their co-author for their technically sound and in-depth contributions. We would also like to thank all distributors, authors, and other who conceded consent to utilize their figures, tables, and schemes.

Inamuddin[1,2], Mohammad Faraz Ahmer[3], Abdullah M. Asiri[1], Sadaf Zaidi[4]

[1] Chemistry Department, Faculty of Science, King Abdulaziz University, Jeddah 21589, Saudi Arabia

[2] Department of Applied Chemistry, Faculty of Engineering and Technology, Aligarh Muslim University, Aligarh-202 002, India

[3] Department of Electrical Engineering, Mewat College of Engineering and Technology, Mewat-122103, India

[4] Department of Chemical Engineering, Faculty of Engineering and Technology, Aligarh Muslim University, Aligarh 202002, India

Chapter 1

Theory, Fundamentals and Application of Supercapacitors

B. Saravanakumar [a], G. Muralidharan[b], S.Vadivel *[c], D. Maruthamani *[c], M.Kumaravel [c]

[a] Department of Physics, Dr. Mahalinggam College of Engineering and Technology, Pollachi, 642003, Tamilnadu, India

[b]Department of Physics, Gandhigram Rural Institute - Deemed university, Gandhigram, Tamilnadu – 624302, India

[c] Department of Chemistry, PSG College of Technology, Coimbatore-641004, India

vlvelu7@gmail.com, saravanakumar123@gmail.com

Abstract

Supercapacitors or electrochemical capacitors are electrochemical energy storage devices which have drawn huge attention from the scientific community in recent years due to their compactness and long cyclic stability. Great efforts have been paid to improve the energy density of supercapacitor electrodes. Recent research focuses on synthesis of advanced carbon metals, polymers, and metal oxide based composites which can be used as electrode materials for supercapacitor applications. In this chapter, we focus on the history, recent developments concerning supercapacitors electrode materials, electrolytes and separators that have been widely used in supercapacitor devices. The basic parameters and electrochemical properties of supercapacitors are also summarized. Finally, to achieve high specific capacitance in supercapacitors some perspectives and outlook are proposed.

Keywords

Supercapacitors, Energy Storage, Electrodes, Carbon Materials

Contents

1.   Introduction

In the past few decades, the fast advancements made in various areas such as industrialization have resulted in increased demands for energy storage devices. At the meantime, the global increase in population has also led to these growing energy requirements. These factors have combined to cause a major problem on the existing power infrastructure and pose serious troubles for the future of mankind [1]. Until now, petroleum fuels have been largely exploited for the power needs of our society. However, with limited resources of petroleum products, there is a requirement for an alternate energy sources. In this context, notable efforts have been devoted to developing more efficient energy storage devices and technologies. An example of such an efficient device is the electrochemical capacitors, often called as a supercapacitor. Supercapacitor technology has gained considerable research interest among the scientific community in recent years due to its unique properties. They exhibit higher capacitances that are much higher than traditional capacitors [2]. Supercapacitors can be used in various energy storage devices, either stand-alone or in combination with batteries. Supercapacitors have significant advantages over conventional capacitors in terms of their large charging-discharging stability and high power capacity. However, the energy density of supercapacitors is much lower than that of rechargeable batteries. This necessitates coupling with batteries for applications requiring a surplus energy supply for longer periods of time. Therefore, there is a huge interest in increasing the energy density of supercapacitors [3]. This is one of the major obstacles for the development of advanced electrode materials in supercapacitor technology. This chapter describes the fundamentals, electrode materials and their properties, and electrode fabrication and analysis techniques for supercapacitors.

2.   Energy needs and energy storage devices

The challenge associated with the production of renewable energy using the available natural resources is of serious concern for researchers throughout the world. Furthermore, the burning of fossil fuel produces large amounts of CO_2 emissions which lead to adverse effects such as global warming and ocean acidification.

There is no single solution to solve these challenges related to energy and environment problems. Furthermore, the growth of novel technologies and devices with a large interest in the protection of nature and environment is utmost importance. On the other hand, we cannot imagine our modern day life without electronic portable devices such as laptops, cell phones etc., also without the transportation vehicles like airplanes, buses, cars, and many other innovations made our life more sophisticated [1-2]. Though they provide a high degree of comfort, they are the major cause for environmental issues.

The last decade was a witness to more significant research efforts towards the development of various renewable energy systems like solar cells and fuel cells. However, these alternative energy sources still suffer on account of some limitations. Among the energy storage devices, batteries and supercapacitors (SCs) are viewed with greater attention due to their superior electrochemical properties. Batteries provide higher energy density but suffer from poor power density and cycle life. In contrast, the SCs can store and release a huge amount of power within a short time [3]. SCs are appealing as a modern energy storage system that possess most fascinating features such as high power density, long cycle life, environment safety [4]. SCs provide best solutions for electrical energy requirement in our day to day life. For instance, many electric vehicle companies are utilizing SCs for greater power delivery during acceleration. Apart from this, SCs are used in digital cameras, power back up systems and various electronic devices [5]. The cyclic performance of a supercapacitor is determined by the electrode, which can be designed through novel routes with nano or microstructures to improve the charge storage capability. Selection of an appropriate electrolyte might also improve the energy density of a supercapacitor system [6,7].

3.   Breakthrough in supercapacitor research

Howard Becker from General electric company first reported electrical double-layer charge storage in 1957. After Becker's initiations, Robert A. Rightmire (1966) and Donald L. Boos (1970) from (SOHIO) fabricated a new capacitor with higher capacitance value which is now commonly available.

In 1978, Nippon Electric Company (NEC) commercialized supercapacitors under license from SOHIO. After commercialization, SCs have been produced with several designs. Initially, these capacitors were termed as a power back up systems for many consumer electronic devices. Nippon made great efforts towards the commercialization of various new series of supercapacitors [8]. More significantly supported by organic electrolytes; in 2001 they introduced spiral-wound, thin-format series capacitors. Further, they changed the product name as NEC-Tokin in the year 2002. Nippon ChemiCon of Japan started mass production of their supercapacitors with higher performance (3000F, 2.5 V) in 1998.

The gold cap double layer capacitors were marketed by Panasonic in the year of 1978. Use of non-aqueous electrolytes and non pasted electrodes is the major deviation from the Nippon design. Due to these differences, the Panasonic capacitors gained higher operating voltage [9,10].

Daewoo group of Korea developed supercapacitors named as Nesscap supercapacitor in 1998. Nesscap operated with the help of organic electrolyte with a spiral wound prismatic cell construction. NessCap presently supplies supercapacitors from a few farads to 5,000F at 2.7V [8]. In 1993, ESMA a Russian company developed supercapacitors of capacity ranging from 3,000F to100,000F.

In recent years many research groups are working towards the fabrication of novel materials as an electrode for supercapacitors. Few of the inspiring works are discussed here. P.W. Ruch et al. investigated the characteristics of single-wall carbon nanotubes (SWCNTs) as supercapacitors [12]. For the first time, Ruch exploited the SWCNT for supercapacitor electrode applications. Graphene-based SCs were first reported by M. D. Stolleret al. [13]. This advanced electrode active material exhibited high surface area (2630 m^2g^{-1}) and with the maximum specific capacitance of 99 Fg^{-1} and 135 Fg^{-1} in organic and aqueous electrolytes respectively. Recently Liu et al. reported graphene-based SC energy density of (86 $Whkg^{-1}$ at room temperature and 136$Whkg^{-1}$at 80°C) [14]. El-Kady et al. reported a facile method of fabricating supercapacitor devices on

DVD discs [15]. Choudhary et al.core/shell nanowire supercapacitor using two-dimensional (2D) transition-metal chalcogenides which shows a superior charge retention of over 30,000 cycles. This work may be considered as a recent breakthrough in supercapacitor research.

Recently, the new concept of supercapattery has been introduced to understand the hybrid systems in which the charge storage principles of supercapacitor and battery are integrated into a single device. In this device, one electrode acts as capacitor like characteristics and other one shows battery like properties. However, there is a possibility of higher influence of the battery electrode over the capacitor electrode and this hybrid device will behave as a battery. Such device is termed as a super battery. If the capacitor property is dominant in the hybrid device then it is termed as supercapattery[17].

4.   Energy storage principles: EDLC *Vs* pseudocapacitance

Famous German physicist Helmholtz was the first person to demonstrate the idea of the electric double layer in the year of 1853 [18]. Generally an electrochemical capacitor has two electrodes with a separator. The electrodes are electrically connected by the aqueous electrolyte. The electrodes immersed in an electrolyte solution attract the counter ions by the application of an external potential. The voltage on this setup produces two layers of polarized ions. The first one is created on the surface of the electrode the second one from the opposite polarity is separated by a layer of solvated molecules. This plane separates the oppositely polarized ions and acts as a molecular dielectric.

In 1913, Gouy and Chapman proposed the "diffuse double layer" theory to explain the double layer capacitance [19-21]. In this theory, the solvated ions in the electrolyte are assumed as point charges contained within a single diffuse layer. Otto Stern in 1924 suggested a realistic proof to explain the physical situation at the electrode/ electrolyte interface [22]. He combined both Helmholtz and Gouy-Chapman theories. He used different electrolytes and the ionic radius of the associated ions to arrive at the capacity of the double layer capacitor. Stern model proposed that $q_{solution}$ (charge from solution side) to be a combination of $q_{Helmholtz}$ (Helmholtz charge) and q_{Gouy} (charges from the diffuse layer).

$q_{solution} = q_{Helmholtz} + q_{Gouy}$

Then the total capacity at the interface "C" is denoted as,

$$\frac{1}{C} = \frac{1}{C_H} + \frac{1}{C_G}$$

In this equation, C_H is the capacity due to Helmholtz layer and C_G the capacity due to the diffuse layer. Graham from America proposed a model to further demonstrates the double layer [23]. According to Graham concept, some of the ionic species can penetrate the Stern layer and approach the solvated ion molecules. Bockris et al. proposed the BDM model which we are commonly using for double layer capacitance [24]. In this model solvent molecules have a fixed alignment with the electrode. Most specifically adsorbed ions could appear in this layer. The electrical energy is stored by means of Faradic (redox process) reactions on the surface of the electrode with an electric double layer. Basically, pseudocapacitance (redox process) arises from charge transfer between the electrode and electrolytes. Only the charge transfer (or) change in oxidation process takes place.

5. Design of Supercapacitor

In General, a supercapacitor consists of three main components namely electrode, electrolyte and separator. The electrochemical properties of SCs mainly depend on electrode and electrolyte. Nevertheless, the electrode is the primary component for storing energy and delivery. The following sections elaborate on the individual components of a supercapacitor system.

5.1 Electrode materials

During past few decades, the invention of novel electrode materials with excellent specific capacitance has enabled the rapid growth in the supercapacitor field. The electrode material is the most important component in determining the electrochemical property and stability. There are three different types of electrode materials which are usually used for supercapacitor devices. These include carbon-based materials and metal oxides and conducting polymers.

5.1.1 Carbon and allied materials

In practice, Carbon with large surface area is widely used for supercapacitor electrodes. Generally, carbon materials exhibit an advantageous combination of physical and chemical properties. The superior characteristics of low cost, high specific surface area, high conductivity, porous nature, corrosion resistance, processability and being eco-friendly make carbon-based materials an attractive candidate for supercapacitor electrode applications. Activated carbon (AC), carbon aerogels (CA), carbon nanofibers (CNF), carbon nanotubes (CNT) and graphene are the various forms of carbon which are exploited as electrode materials for SCs.

The carbon-based materials derived from natural resources like wood and other biomasses like plants contain impurities like ash, cause a reduction in the supercapacitor performance. Further, the pore sizes also vary from source to source. Therefore the method of improving the surface area as well as the porosity of carbons by thermal or chemical process is called as "activation" which converts carbon into activated carbon (AC). The ACs has a porous structure composed of micro, meso and macro pores [25-28]. Generally, the presence of a large number of mesopores (2-50nm) in the AC electrode materials provides maximum capacitance values. The ACs possesses higher surface area but the limited availability of mesopores limits the specific capacitance values due to lower ion accessibility [29]. The carbon aerogels are other forms of carbon with excellent conductivity, high surface area, and tunable porous structure. These attractive properties of carbon aerogels are well suited for electrode materials for SCs with long cyclic stability [30-37]. Since 1990's major research interest has been renowned to carbon nanotubes (CNTs) as a superior electrode material for supercapacitor applications. It holds many profitable properties such as narrow pore size distribution, higher surface area, and high structural stability [38]. These profitable features of CNTs have made them an alternative material for supercapacitor [39-43]. However, due to high cost, CNTs are not viable for large-scale production. But at the same time, they are widely used as conductivity booster and replace activated carbons in supercapacitor electrodes.

Recently Graphene, the King of carbons with 2D lattices are considered as alternative electrode materials for SCs due to their high intrinsic surface area and excellent electrical conductivity. In addition, graphene possesses outstanding intrinsic strength, good chemical stability, and excellent thermal conductivity. Owing to these unique features, graphene provides a better platform for storage of ions and electrons. Graphene has been extensively studied for supercapacitor electrode applications by many researchers [44-47].

5.1.2 Conducting polymers

The conducting polymers are a new class of electrode materials for supercapacitors owing to their three-dimensional (3D) porous structure, high capacitance value, scalable synthesis and lightweight. The most commonly applicable conducting polymers for SCs include polypyrrole, polyaniline, polythiophene, and poly[3,4-ethylenedioxythiophene] (PEDOT) and their derivatives. [48-51]. However, some of the key issues like poor cyclic stability, bulging of the electrolyte and oxidative degradation of active material limits the usage of conducting polymers in supercapacitor electrodes [52].

5.1.3 Metal oxides

Owing to its high redox chemistry, the metal oxides (TMOs) are exploited as smart candidates for supercapacitor electrodes. More specifically, nanostructured ruthenium oxide (RuO_2), manganese oxide (MnO_2), nickel oxide (NiO), cobalt oxide (CO_3O_4) and bismuth oxide(Bi_2O_3) and some metal oxides have been examined as supercapacitor electrodes. In addition, mixed metal oxides, metal oxides with carbon and conducting polymers also have been examined towards enhancing the electrochemical performance of the supercapacitor.

The RuO_2 is a suitable material for supercapacitor due to its various oxidation states. It showed maximum capacitance of 1450 Fg^{-1} with low electrical resistivity. This attractive material has been studied extensively towards supercapacitor electrode applications [53-55]. Nevertheless, the higher cost and environmental issues restrict its usage in commercialization. Worldwide many research groups devoted their research work towards an alternative material for RuO_2 in energy storage applications. The MnO_2 is considered as a better alternate material for RuO_2 in supercapacitor electrodes. MnO_2 with the peculiar properties like large specific capacitance, cost effectiveness, environmental friendliness have gained huge attention in constructing supercapacitors. Many researchers have reported the use of MnO_2 as an alternative electrode material for SCs [56-61]. However poor specific surface area and dissolution of manganese ions in aqueous media restrict its commercial usage.

Co_3O_4 nanostructures are also emerging in supercapacitor electrodes due to their higher specific capacitance and cost-effective nature. Thus, Co_3O_4 shows good structural stability against charge-discharge process [62-67].

The NiO is one among the most extensively studied metal oxide for supercapacitor electrode applications. This material also exhibits high theoretical specific capacitance in a wide potential range of 0.5 V, low cost, high chemical stability. Owing to these unparalleled characteristics NiO is used for supercapacitor electrodes nowadays [68-70].

Due to less toxicity, magnetic separation, and natural abundance, Fe_2O_3 has attracted the attention of many researchers. Apart from that Fe_2O_3 is incombustible and eco-friendly. Recently many researchers reported performance of Fe_2O_3 nanostructures for supercapacitor electrode applications [71-74]. Apart from NiO and Co_3O_4 nanostructures $Ni(OH)_2$ and $Co(OH)_2$ nanostructures are also extensively studied for supercapacitor applications [75-78].

5.1.4 Hybrid electrode

The use of Transition metal oxide as a supercapacitor electrode material is emerging recently due to its better ion transportation kinetics and high electrical conductivity. Furthermore, the conductivity of transition metal oxide could be improved by incorporating carbon-based materials like activated carbon; carbon nanotubes (CNTs) and graphene. Over the past decade variety of hybrid nanostructures have been synthesized and examined for supercapacitor electrode applications [79-83].

5.1.5 Mixed metal oxide electrodes

Recently mixed metal oxide composites are considered as most popular in terms of achieving high electrochemical performance in SCs. The blending of two different transition metal oxides to form a heterostructure greatly reduces the inactive sites present in the material and it is possible to take full advantage of both kinds of active materials. This kind of synergetic architectures exhibits better results in SCs [84-88].

5.2   Electrolyte

Electrolytes play a major role in SC devices, which connect two electrodes and establish the charge transfer. In General, there are three different categories of electrolytes used, aqueous electrolytes, organic electrolytes, and ionic liquids. All these electrolytes are associated with their own advantages and disadvantages. For instance, SCs utilizing aqueous electrolytes show higher conductivity and specific capacitance. But its working voltage is restricted to only (1.2V). In contrast, organic electrolytes operate at a still higher voltage but suffer from poor ionic conductivity. Though the best operating voltage is obtained in room temperature ionic liquids, the prohibitive cost makes them unsuitable. For supercapacitors the selection of electrolytes mainly depends upon the stability of electrolyte at an applied potential window and its ionic conductivity.

5.2.1 Aqueous electrolytes

Aqueous electrolytes have been used extensively for studying the supercapacitor performance. For example, nearly 90% of published work in 2014 descripted the use od aqueous electrolytes [89]. Further; cost-effectiveness and easy handling of aqueous electrolytes simplify the device fabrication process. Neutral aqueous electrolytes for SCs show high ionic conductivity and low cost. The aqueous electrolytes possess smaller sized solvated ions and high dielectric constants values. Both of these features significantly enhance the specific capacitance of a supercapacitor [90,91]. However, the low value of the operating window (~1.2V) limits its usage. Increase in the electrolyte voltage window improves the energy density of supercapacitor.

5.2.2 Organic electrolytes

Nowadays organic electrolyte based commercial SC devices influencing market due to improved voltage performance. The increased cell voltage increases the electrochemical properties like specific capacitance and energy density. In addition, the organic electrolytes permit the use of inexpensive materials as electrodes and packaging. There are many efforts focused to design stable organic electrolytes with higher conductivity and wide potential window. However, the high cost, lower ionic conductivity, flammability, and toxicity are the important issues to be seriously considered before employing as an electrolyte for SC applications.

5.2.3 Ionic liquid electrolytes

The ionic liquids are termed as solvent-free electrolytes are another class of electrolytes for SCs. Use of ionic liquids as electrolytes pushes the potential window to over 3V [92]. Fluoromethane sulphonyl imide is mainly employed as the ionic liquid electrolyte for supercapacitor. However, the ionic conductivity of these ionic liquids is very low at room temperature. But the energy storage systems are used in the temperature range from − 30°C to + 60°C. The ionic liquids still need improvements to increase their stability and conductivity at room temperature. Though the ionic liquids are effective, their use is restricted by the high cost of these materials.

5.3 Separator

The separator is also an important component of a supercapacitor system. The separators are ion-permeable and prevent the electrical and physical contact between two electrodes and allow the ionic charge transfer from electrolytes. To achieve high performance in SC's the separators must possess higher ionic conductivity and smaller thickness. Generally, nonwoven paper and polymer separators are utilized with aqueous electrolytes for SCs.

5.4 Different analytical techniques used for the analysis of supercapacitors

Cyclic voltammetry, galvanostatic charge-discharge, and impedance techniques are important characterization techniques usually employed to analyze the properties of supercapacitors.

5.4.1 Cyclic voltammetry (CV)

The CV analysis was recorded using three electrode systems consisting of working, counter and reference electrode with suitable electrolyte at various scan rates. An aqueous electrolyte is usually added to the sample to increase its conductivity. The commonly used materials for the working electrode are glassy carbon, platinum, and gold. The specific capacitance performance in appropriate electrolytes was carried out in their corresponding potential range and recorded at various scan rates (mVs^{-1}). In general scan rate increases, the specific capacitance value usually decreases. The following equation has been used to calculate the specific capacitance value by using cyclic voltammetry method

$$SC = \frac{1}{V \times m \ (Va - Vc)} \int_{Va}^{Vc} IV.dV$$

The SC values were calculated by integrating the area and then dividing by the sweep rate (mVs^{-1}), the mass of the active material (m), and the potential window (Va to Vc).

5.4.2 Chronopotentiometry

The galvanostatic charge/discharge behavior of the supercapacitor electrode was evaluated using chronopotentiometry. From the charge/discharge method, it was observed that the charging curve was almost similar to the discharging curve, but slightly differs from the initial charging time. The specific capacitance at a particular current density of the electrode materials can be evaluated from the following Equation

$$SC = \frac{I \times \Delta t}{\Delta V \times m} Fg^{-1}$$

Where I is the discharge current, Δt is the discharging time, ΔV is the discharging voltage, and m is the mass of the electroactive material. When compare to cyclic voltammetry the specific capacitance values calculated from the charge-discharge technique was almost reliable. This technique also has been employed to evaluate the stability of electrode materials. From charging and discharging curves, the SC values were calculated.

5.4.3 Power density (P) and energy density (E)

The energy density and power densities are two important parameters for the investigation of the electrochemical performance of supercapacitors. The charge-discharge method is the most suitable technique to evaluate the power density and energy density values of the supercapacitor. The energy density of supercapacitor decreases with increase in power density. Energy density (E) and power density (P) were calculated from the following Equations

$$E = \frac{1}{2} C (\Delta V)^2$$

$$P = \frac{E}{t}$$

Where C is specific capacitance (F g^{-1}), ΔV is a potential window (V), t is discharge time (s), E is energy density (Wh kg^{-1}) and P is power density (KW kg^{-1}).

Conclusions

With the invention of the supercapacitors, an alternative energy storage device emerged TO offer high power density and good stability. As it was reported supercapacitors have the great quality that makes them suited for various applications including batteries, hybrid power systems, and emergency power supplies. As it can be seen from the review presented here, there is room for improvement in the fabrication of supercapacitor electrodes to be used in more applications needing high energy density.

References

[1]   J.R. Miller, P. Simon, Electrochemical capacitors for energy management, Science 321 (2008) 651-652. https://doi.org/10.1126/science.1158736

[2]   P.G. Bruce, B. Scrosati, J.M. Tarascon, Nanomaterials for rechargeable lithium batteries, Angew. Chem. Int. Ed. 47 (2008) 2930-2946. https://doi.org/10.1002/anie.200702505

[3]   B.G. Choi, J. Hong, W.H. Hong, P.T. Hammond, H. Park, Facilitated ion transport in all-solid-state flexible supercapacitors, ACS Nano. 5 (2011) 7205-7213. https://doi.org/10.1021/nn202020w

[4]   P. Simon, Y. Gogotsi, Materials for electrochemical capacitors, Nat. Mater. 7 (2008) 845-854. https://doi.org/10.1038/nmat2297

[5]   P.J. Hall, M. Mirzaeian, S.I. Fletcher, F.B. Sillars, A.J.R. Rennie, G.O. Shitta-Bey, G. Wilson, A. Cruden, R. Carter, Energy storage in electrochemical capacitors: designing functional materials to improve performance, Energy Environ. Sci. 3 (2010) 1238-1251. https://doi.org/10.1039/c0ee00004c

[6]   Z. Yang, J. Zhang, M. C.W. Kintner-Meyer, X. Lu, D. Choi, J.P. Lemmon, J. Liu, Electrochemical Energy Storage for Green Grid, Chem. Rev. 111 (2011) 3577-3613. https://doi.org/10.1021/cr100290v

[7] D. Wang, R. Kou, D. Choi, Z. Yang, Z. Nie, J. Li, L.V Saraf, D. Hu, J. Zhang, G.L. Graff, J. Liu, M.A. Pope, I.A. Aksay, Ternary self-assembly of ordered metal oxide-graphene nanocomposites for electrochemical energy storage, ACS Nano 4 (2010) 1587-1595. https://doi.org/10.1021/nn901819n

[8] B. Zhu, S. Tang, S. Vongehr, H. Xie, X. Meng, Hierarchically MnO2–nanosheet covered Submicrometer-FeCo2O4-tube forest as binder-free electrodes for high energy density all-solid-state supercapacitors, ACS Appl. Mater. 8 (2016) 4762–4770. https://doi.org/10.1021/acsami.5b11367

[9] D. Gong, J. Zhu, B. Lu, RuO2@Co3O4 heterogeneous nanofibers: a high-performance electrode material for supercapacitors, RSC Adv. 6 (2016) 49173–49178. https://doi.org/10.1039/C6RA04884F

[10] H. Aruga, K. Hiratsuka, T. Morimoto, Y. Sanada, U. S. Patent 4, 725, 927 A (1988).

[11] S. Razoumov, A. Klementov, S. Litvinenko, A. Beliakov, U.S. Patent 6, 222,723 B1 (2001).

[12] Y. Wang, Y. Song, Y. Xia, Electrochemical capacitors: mechanism, materials, systems, characterization and applications, Chem. Soc. Rev. 45 (2016) 5925–5950. https://doi.org/10.1039/C5CS00580A

[13] M.D. Stoller, S. Park, Y. Zhu, J. An, R.S. Ruoff, Graphene-based ultracapacitors, Nano. Lett. 10 (2008) 3498-3502. https://doi.org/10.1021/nl802558y

[14] C. Liu, Z. Yu, D. Neff, A. Zhamu, B.Z. Jang, Graphene-based supercapacitor with an ultrahigh energy density, Nano Lett. 10 (2010) 4863-4868. https://doi.org/10.1021/nl102661q

[15] M.F. El-Kady, R.B. Kaner, Scalable fabrication of high-power graphene micro-supercapacitors for flexible and on-chip energy storage, Nat. Commun. (2013) 1475. https://doi.org/10.1038/ncomms2446

[16] N. Choudhary, C. Li, H.K. Chung, J. Moore, J. Thomas, Y. Jung, High-performance one-body core/shell nanowire supercapacitor enabled by conformal growth of capacitive 2D WS2 layers, ACS Nano. 10 (2016) 10726-10735. https://doi.org/10.1021/acsnano.6b06111

[17] B. Akinwolemiwa, C. Peng, George Z. Chen. Redox electrolytes in supercapacitors, J. Electrochem. Society. 162 (2015) 5054-5059. https://doi.org/10.1149/2.0111505jes

[18] Z. Zeng, D. Wang, J. Zhu, F. Xiao, Y. Li, X. Zhu, NiCo2S4
      nanoparticles//activated balsam pear pulp for asymmetric hybrid capacitors, Cryst.
      Eng. Comm. 18 (2016) 2363–2374. https://doi.org/10.1039/C6CE00319B

[19] G. Gouy, Constitution of the electric charge at the surface of an electrolyte, J.
      Phys. Radium. 9 (1910) 457-467.

[20] G. Gouy, constitution of the electric charge at the surface of an electrolyte, Compt.
      Rend. 149 (1910) 654-657.

[21] D.L. Chapman, LI. A contribution to the theory of electrocapillarity, Phil. Mag. 25
      (1913) 475-481. https://doi.org/10.1080/14786440408634187

[22] H. O. Stern, Zur theorie der elektrolytischen doppelschicht, Z. Anorg. Allg. Chem.
      30 (1924) 508-516.

[23] D.C. Grahame, The electrical double layer and the theory of electrocapillarity,
      Chem. Rev. 41 (1947) 441-501. https://doi.org/10.1021/cr60130a002

[24] J. O. M. Bockris, M. A. V. Devanathan, K. Muller, On the structure of charged
      interfaces, Proc. Roy. Soc. A 274 (1963) 55.
      https://doi.org/10.1098/rspa.1963.0114

[25] A. Laheaar, S. Delpeux-Ouldriane, E. Lust, F. Beguin, Ammonia treatment of
      activated carbon powders for supercapacitor electrode application, J. Electrochem.
      Soc. 161 (2014) A568-A575. https://doi.org/10.1149/2.051404jes

[26] G. Xu, C. Zheng, Q. Zhang, J. Huang, M. Zhao, J. Nie, X. Wang, F, Wei, Binder-
      free activated carbon/carbon nanotube paper electrodes for use in supercapacitors,
      Nano Res. 4 (2011) 870-881. https://doi.org/10.1007/s12274-011-0143-8

[27] X. Li, C.Han, X. Chen, C. Shi, Preparation and performance of straw based
      activated carbon for supercapacitor in non-aqueous electrolytes, Micropor and
      Mesopor. Mater. 131 (2010) 303-309.
      https://doi.org/10.1016/j.micromeso.2010.01.007

[28] A. Alonso, V. Ruiz, C. Blanco, R. Santamaría, M. Granda, R. Menéndez, S.G.E.
      de Jager, Activated carbon produced from Sasol-Lurgigasifier pitch and its
      application as electrodes in supercapacitors, Carbon 44 (2006) 441-446.
      https://doi.org/10.1016/j.carbon.2005.09.008

[29] T. Chen, L. Dai, Carbon nanomaterials for high-performance supercapacitors,
      Nanotoday 16 (2013) 272-280. https://doi.org/10.1016/j.mattod.2013.07.002

[30] A.G. Gomez, P. Miles, T.A. Centeno, J.M. Rojo, Uniaxially oriented carbon monoliths as supercapacitor electrodes, Electro. Chim. Acta 55 (2010) 8539-8544. https://doi.org/10.1016/j.electacta.2010.07.072

[31] E.G. Calvo, C.O. Ania, L. Zubizarreta, J.A. Menendez, A. Arenillas, Exploring new routes in the synthesis of carbon xerogels for their application in electric double-layer capacitors, Energy & Fuels 24 (2010) 3334-3339. https://doi.org/10.1021/ef901465j

[32] S. He, W. Chen, 3D graphene nanomaterials for binder-free supercapacitors: scientific design for enhanced performance, Nanoscale 7 (2015) 6957–6990. https://doi.org/10.1039/C4NR05895J

[33] M. Inagaki, H. Konno, O. Tanaike, Carbon materials for electrochemical capacitors, J. Power Sources. 195 (2010) 7880-7903. https://doi.org/10.1016/j.jpowsour.2010.06.036

[34] J. Li, X. Wang, Q. Huang, S. Gamboa, P.J. Sebastian, Studies on preparation and performances of carbon aerogel electrodes for the application of supercapacitor, J. Power Sources 158 (2006) 784-788. https://doi.org/10.1016/j.jpowsour.2005.09.045

[35] J. Biener, M. Stadermann, M. Suss, M.A. Worsley, M.M. Biener, K.A. Rose, T.F. Baumann, Advanced carbon aerogels for energy applications, Energy Environ. Sci. 4 (2011) 656-667. https://doi.org/10.1039/c0ee00627k

[36] L. Wang, D. Wang, X.Y. Dong, Z.J. Zhang, X.F. Pei, X.J. Chen, B. Chen, J. Jin, Layered assembly of graphene oxide and Co-Al layered double hydroxide nanosheets as electrode materials for supercapacitors, Chem. Commun. 47 (2011) 3556–3558. https://doi.org/10.1039/c0cc05420h

[37] S. Sepehri , B.B. Garcia, Q. Zhang, G. Cao, Enhanced electrochemical and structural properties of carbon cryogels by surface chemistry alteration with boron and nitrogen, Carbon 47 (2009) 1436-1443. https://doi.org/10.1016/j.carbon.2009.01.034

[38] P.J. Hall, M. Mirzaeian, S.I. Fletcher, F.B. Sillars, A.J.R. Rennie, G.O. Shitta-Bey, G. Wilson, A. Cruden, R. Carter, Energy storage in electrochemical capacitors: Designing functional materials to improve performance, Energy Environ. Sci. 3 (2010) 1238-1251. https://doi.org/10.1039/c0ee00004c

[39] J. Yu, W. Lu, S. Pei, K. Gong, L. Wang, L. Meng, Y. Huang, J.P. Smith, K.S. Booksh, Q. Li, J.H. Byun, Y. Oh, Y. Yan, T.W. Chou, Omni directionally

stretchable high-performance supercapacitor based on isotropic buckled carbon nanotube films, ACS Nano 10 (2016)5204-5211. https://doi.org/10.1021/acsnano.6b00752

[40] A.I. Najafabadi, S. Yasuda, K. Kobashi, T. Yamada, D.N. Futaba, H. Hatori, M. Yumura, S. Iijima, K. Hata, Extracting the full potential of single-walled carbon nanotubes as durable supercapacitor electrodes operable at 4 V with high power and energy density, Adv. Mater. 22 (2010) 235-241. https://doi.org/10.1002/adma.200904349

[41] R.N.A.R. Seman, M.A. Azam, A.A. Mohamad, Systematic gap analysis of carbon nanotube-based lithium-ion batteries and electrochemical capacitors, Renew. Sustain. Energ. Rev. 75 (2016) 644-659. https://doi.org/10.1016/j.rser.2016.10.078

[42] S.K. Simotwo, C. DelRe, V. Kalra, Supercapacitor electrodes based on high-purity electrospunpolyaniline and polyaniline–carbon nanotube nanofibers, ACS Appl. Mater. Interfaces 8 (2016) 21261-21269. https://doi.org/10.1021/acsami.6b03463

[43] M.V. Kiamahalleh, S.H.S. Zein, G. Najafpour, S. Buniran, Multiwalled carbon nanotubes based nanocomposites for supercapacitors: a review of electrode materials, NANO: Brief Rep. Rev. 7 (2012) 1230002. https://doi.org/10.1142/S1793292012300022

[44] W.K. Chee, H.N. Lim, Z. Zainal, N.M. Huang, I. Harrison, Y. Andou, Flexible graphene-based supercapacitors: a review, J. Phys. Chem. C. 120 (2016) 4153-4172. https://doi.org/10.1021/acs.jpcc.5b10187

[45] J. Yu, J. Wu, H. Wang, A. Zhou, C. Huang, H. Bai, L. Li, Metallic fabrics as the current collector for high-performance graphene-based flexible solid-state supercapacitor, ACS App. Mater. Interfaces 8 (2016) 4724-4729. https://doi.org/10.1021/acsami.5b12180

[46] C. Zhu, T. Liu, F. Qian, T.Y.J. Han, E.B. Duoss, J.D. Kuntz, C.M. Spadaccini, M.A. Worsley, Y. Li, Supercapacitors based on three-dimensional hierarchical graphene aerogels with periodic macropores, Nano Lett. 16 (2016) 3448-3456. https://doi.org/10.1021/acs.nanolett.5b04965

[47] E.D. Walsh, X. Han, S.D. Lacey, J.W. Kim, J.W. Connell, L. Hu, Y. Lin, Dry-processed, binder-free holey graphene electrodes for supercapacitors with ultrahigh areal loadings, ACS App. Mater. Interfaces 43 (2016) 29478-29485. https://doi.org/10.1021/acsami.6b09951

[48] Y.Shi, L. Pan, B. Liu, Y. Wang, Y. Cui, Z. Bao, G. Yu, Nanostructured conductive polypyrrole hydrogels as high-performance, flexible supercapacitor electrodes, J. Mater. Chem. A. 2 (2014) 6086-6091. https://doi.org/10.1039/C4TA00484A

[49] P. R. Deshmukh, S.N. Pusawale, N.M. Shinde, C.D. Lokhande, Growth of polyaniline nanofibers for supercapacitor applications using successive ionic layer adsorption and reaction (SILAR) method, J. Korean. Phy. Soc. 65 (2014) 80-86. https://doi.org/10.3938/jkps.65.80

[50] T. Liu, L. Finn, M. Yu, H. Wang, T. Zhai, X. Lu, Y. Tong, Y. Li, Polyaniline and polypyrrole pseudocapacitor electrodes with excellent cycling stability, Nano Lett. 14 (2014) 2522-2527. https://doi.org/10.1021/nl500255v

[51] N. Kurra, J. Park, H.N. Alshareef, A conducting polymer nucleation scheme for efficient solid-state supercapacitors on paper, J. Mater. Chem. A 2 (2014) 17058-17065. https://doi.org/10.1039/C4TA03603D

[52] T. Kobayashi, H. Yoneyama, H. Tamura, Electrochemical reactions concerned with electrochromism of polyaniline film-coated electrodes, J. Electroanal. Chem. 177 (1984) 281-291. https://doi.org/10.1016/0022-0728(84)80229-6

[53] Z. Peng, X. Liu, H. Meng, Z. Li, B. Li, Z. Liu, S. Liu, Design and tailoring of the 3D macroporous hydrous RuO2 hierarchical architectures with a hard-template method for high-performance supercapacitors, ACS Appl. Mater. Interfaces 9 (2017) 4577-4586. https://doi.org/10.1021/acsami.6b12532

[54] C.C. Hu, C.W. Wang, K.H. Chang, M.G. Chen, Anodic composite deposition of RuO2/reduced graphene oxide/carbon nanotube for advanced supercapacitors, Nanotechnology 26 (2015) 274004. https://doi.org/10.1088/0957-4484/26/27/274004

[55] C. Liu, C. Li, K. Ahmed, W. Wang, I. Lee, F. Zaera, C.S. Ozkan, M. Ozkan, Scalable, Binderless, and carbonless hierarchical Ni nanodendrite foam decorated with hydrous ruthenium dioxide for 1.6 V symmetric supercapacitors, Adv. Mater. Interfaces 3 (2016) 1500503. https://doi.org/10.1002/admi.201500503

[56] J. Li, Z. Ren, S. Wang, Y. Ren, Y. Qiu, J. Yu, MnO2 nanosheets grown on internal surface of macroporous carbon with enhanced electrochemical performance for supercapacitors, ACS Sustainable Chem. Eng. 4 (2016) 3641-3648. https://doi.org/10.1021/acssuschemeng.6b00092

[57]   K. Xu, X. Zhu, P. She, Y. Shang, H. Sun, Z. Liu, Macroscopic porous MnO2 aerogels for supercapacitor electrodes, Inorg. Chem. Front. 3 (2016) 1043-1047. https://doi.org/10.1039/C6QI00110F

[58]   Y. Huang, Y. Huang, W. Meng, M. Zhu, H. Xue, C.S. Lee, C . Zhi, Enhanced tolerance to stretch-induced performance degradation of stretchable MnO2-based supercapacitors, ACS Appl. Mater. Interfaces 7 (2015) 2569-2574. https://doi.org/10.1021/am507588p

[59]   R.B. Rakhi, B. Ahmed, D. Anjum, H.N. Alshareef, Direct chemical synthesis of MnO2 nanowhiskers on transition-metal carbide surfaces for supercapacitor applications, ACS Appl. Mater. Interfaces 8 (2016) 18806-18814. https://doi.org/10.1021/acsami.6b04481

[60]   M.L.Thai, G. T. Chandran, R.K. Dutta, X. Li, R.M. Penner, 100k cycles and beyond: extraordinary cycle stability for MnO2 nanowires imparted by a gel electrolyte, ACS Energy Lett. 1 (2016) 57-63. https://doi.org/10.1021/acsenergylett.6b00029

[61]   S. Nagamuthu, S. Vijayakumar, G. Muralidharan, Biopolymer-assisted synthesis of λ-MnO2 nanoparticles as an electrode material for aqueous symmetric supercapacitor devices, Ind. Eng. Chem. Res. 52 (2013) 18262-18268. https://doi.org/10.1021/ie402661p

[62]   G. Godillot, P.L. Taberna, B. Daffos, P. Simon, C. Delmas, L.G. Demourgues, High power density aqueous hybrid supercapacitor combining activated carbon and highly conductive spinel cobalt oxide, J. Power Sources 331 (2016) 277-284. https://doi.org/10.1016/j.jpowsour.2016.09.035

[63]   S. Kong, F. Yang, K. Cheng, T. Ouyang, K. Ye, G. Wang, D. Cao, In-situ growth of cobalt oxide nanoflakes from cobalt nanosheet on nickel foam for battery-type supercapacitors with high specific capacity, J. Electroanal. Chem. 785 (2017) 103-108. https://doi.org/10.1016/j.jelechem.2016.12.002

[64]   Y. Xu, L. Wang, P. Cao, C. Cai, Y. Fu, M. Xiaohua, Mesoporous composite nickel cobalt oxide/graphene oxide synthesized via a template-assistant co-precipitation route as electrode material for supercapacitors, J. Power Sources 306 (2016) 742-752. https://doi.org/10.1016/j.jpowsour.2015.12.106

[65]   Y. Ding, W. Bai, J. Sun, Y. Wu, M. A. Memon, C. Wang, C. Liu, Y. Huang, J. Geng, Cellulose tailored anatase TiO2 nanospindles in three-dimensional graphene composites for high-performance supercapacitors, ACS Appl. Mater. Interfaces 8 (2016) 12165-12175. https://doi.org/10.1021/acsami.6b02164

[66] C. Yu, Y. Wang, J. Zhang, X. Shu, J. Cui, Y. Qin, H. Zheng, J. Liu, Y. Zhang, Y. Wu, Integration of mesoporous nickel cobalt oxide nanosheets with ultrathin layer carbon wrapped TiO2 nanotube arrays for high-performance supercapacitors, New J. Chem. 40 (2016) 6881-6889. https://doi.org/10.1039/C6NJ00359A

[67] L.S Aravinda, K.K. Nagaraja, H.S. Nagaraja, K. UdayaBhat, B.R. Bhat, Fabrication and performance evaluation of hybrid supercapacitor electrodes based on carbon nanotubes and sputtered TiO2, Nanotechnology 27 (2016) 314001. https://doi.org/10.1088/0957-4484/27/31/314001

[68] K.K. Purushothaman, I. ManoharaBabu, B. Sethuraman, G. Muralidharan, Nanosheet-assembled NiO microstructures for high-performance supercapacitors, ACS Appl. Mater. Interfaces 5 (2013) 10767-10773. https://doi.org/10.1021/am402869p

[69] S. Vijayakumar, S. Nagamuthu, G. Muralidharan, Supercapacitor studies on NiO nanoflakes synthesized through a microwave route, ACS Appl. Mater. Interfaces 5 (2013) 2188-2196. https://doi.org/10.1021/am400012h

[70] F. Cao, G.X. Pan, X.H. Xia, P.S. Tang, H.F. Chen, Synthesis of hierarchical porous NiO nanotube arrays for supercapacitor application, J. Power Sources 264 (2014) 161-167. https://doi.org/10.1016/j.jpowsour.2014.04.103

[71] P.M. Kulal, D.P. Dubal, C.D. Lokhande, V.J. Fulari, Chemical synthesis of Fe2O3 thin films for supercapacitor application, J. Alloys. Compd. 509 (2011) 2567-2571. https://doi.org/10.1016/j.jallcom.2010.11.091

[72] G. Binitha, M.S. Soumya, A.A. Madhavan, P. Praveen, A. Balakrishnan, K.R.V. Subramanian, M.V. Reddy, Shantikumar V. Nair, A. Sreekumaran Nair, N. Sivakumar. Electrospun α-Fe2O3 nanostructures for supercapacitor applications, J. Mater. Chem. A 1 (2013) 11698-11704. https://doi.org/10.1039/c3ta12352a

[73] J. Huang, S.Yang, Y. Xu, X. Zhou, X. Jiang, N. Shi, D. Cao, J. Yin, G. Wang, Fe2O3 sheets grown on nickel foam as electrode material for electrochemical capacitors, J. Electroanal. Chem. 713 (2014) 98-102. https://doi.org/10.1016/j.jelechem.2013.12.009

[74] B. Sethuraman, K.K. Purushothaman, G. Muralidharan, Synthesis of mesh-like Fe2O3/C nanocomposite via greener route for high performance supercapacitors, RSC Adv. 4 (2014) 4631-4637. https://doi.org/10.1039/C3RA45025B

[75] Z. Lu, Z. Chang, W. Zhu, X. Sun, Beta-phased Ni(OH)2 nanowall film with reversible capacitance higher than theoretical Faradic capacitance, Chem. Commun. 47 (2011) 9651-9653. https://doi.org/10.1039/c1cc13796d

[76] M. Aghazadeh, A.N. Golikand, M. Ghaemi, Synthesis, characterization, and electrochemical properties of ultrafine β-Ni (OH)2 nanoparticles, Int. J. Hydrogen Energy 36 (2011) 8674-8679. https://doi.org/10.1016/j.ijhydene.2011.03.144

[77] T. Xue, J.M. Lee, Capacitive behavior of mesoporous Co(OH)2 nanowires, J. Power Sources 245 (2014) 194-202. https://doi.org/10.1016/j.jpowsour.2013.06.135

[78] J. Tang, D. Liu, Y. Zheng, X. Li, X. Wang, D. He, Effect of Zn-substitution on cycling performance of α-Co(OH)2 nanosheet electrode for supercapacitors, J. Mater. Chem. A 2 (2014) 2585-2591. https://doi.org/10.1039/c3ta14042c

[79] P. Chen, H. Chen, J. Qiu, C. Zhou. Inkjet printing of single-walled carbon nanotube/RuO2 nanowire supercapacitors on cloth fabrics and flexible substrates, Nano Res. 3 (2010) 594-603. https://doi.org/10.1007/s12274-010-0020-x

[80] G. Xiong, K.P.S.S. Hembram, R.G. Reifenberger, T.S. Fisher, MnO2-coated graphitic petals for supercapacitor electrodes, J. Power Sources 227 (2013) 254-259. https://doi.org/10.1016/j.jpowsour.2012.11.040

[81] W. Tang, Y.Y. Hou, X.J. Wang, Y. Bai, Y.S. Zhu, H. Sun, Y.B. Yue, Y.P. Wu, K. Zhu, R. Holze, A hybrid of MnO2 nanowires and MWCNTs as cathode of excellent rate capability for supercapacitors, J. Power Sources 197 (2012) 330-333. https://doi.org/10.1016/j.jpowsour.2011.09.050

[82] G. Yu, L. Hu, M. Vosgueritchian, H. Wang, X. Xie. J.R. McDonough, X. Cui, Y. Cui, ZhenanBao, Nano Lett. 11 (2011) 2905-2911. https://doi.org/10.1021/nl2013828

[83] G. Yu, L. Hu, N. Liu, H. Wang, M. Vosgueritchian, Y. Yang, Y. Cui, Z. Bao, Solution-processed graphene/MnO2 nanostructured textiles for high-performance electrochemical capacitors, Nano Lett. 11 (2011) 2905-2911. https://doi.org/10.1021/nl2013828

[84] R.B. Rakhi, W. Chen, M.N. Hedhili, D. Cha, H.N. Alshareef, enhanced rate performance of mesoporous Co3O4 nanosheet supercapacitor electrodes by hydrous RuO2 nanoparticle decoration, ACS Appl. Mater. Interfaces 6 (2014) 4196-4206. https://doi.org/10.1021/am405849n

[85]   J.H. Kim, K.Z.Y. Yan, C.L. Perkins, A.J. Frank, Microstructure and pseudocapacitive properties of electrodes constructed of oriented NiO-TiO2 nanotube arrays, Nano Lett. 10 (2010) 4099-4104. https://doi.org/10.1021/nl102203s

[86]   Y. Yang, D. Kim, M. Yang, P. Schmuki, Vertically aligned mixed V2O5–TiO2 nanotube arrays for supercapacitor applications, Chem. Commun. 47 (2011) 7746-7748. https://doi.org/10.1039/c1cc11811k

[87]   S. Hou, G. Zhang , W. Zeng, J. Zhu, F. Gong, F. Li, H. Duan, Hierarchical Core–shell structure of ZnO nanorod@NiO/MoO2 composite nanosheet arrays for high-performance supercapacitors, ACS Appl. Mater. Interfaces 6 (2014) 13564-13570. https://doi.org/10.1021/am5028154

[88]   D. Cai, H. Huang, D. Wang, B. Liu, L. Wang, Y. Liu, Q. Li, T. Wang, High-performance supercapacitor electrode based on the unique ZnO@Co3O4 core/shell heterostructures on nickel foam, ACS Appl. Mater. Interfaces 6 (2014) 15905-15912. https://doi.org/10.1021/am5035494

[89]   C. Zhong, Y. Deng, W. Hu, J. Qiao, L. Zhang, J. Zhang. A review of electrolyte materials and compositions for electrochemical supercapacitors, Chem. Soc. Rev. 44 (2015) 7484-7539. https://doi.org/10.1039/C5CS00303B

[90]   F. Béguin, V. Presser, A. Balducci, E. Frackowiak, Carbons and electrolytes for advanced supercapacitors, Adv. Mater. 26 (2014) 2219-2251. https://doi.org/10.1002/adma.201304137

[91]   Q.T. Qu, P. Zhang, B. Wang, Y.H. Chen, S. Tian, Y.P. Wu, R.J. Holze, Electrochemical performance of MnO2 nanorods in neutral aqueous electrolytes as a cathode for asymmetric supercapacitors, J. Phys. Chem. C 113 (2009) 14020-14027. https://doi.org/10.1021/jp8113094

[92]   T. Tsuda, C.L. Hussey, Electrochemical applications of room-temperature ionic liquids, Electrochem. Soc. Inter. 16 (2007) 42-49.

Chapter 2

Metal Oxides/Hydroxides Composite Electrodes for Supercapacitors

Rajendran Ramachandran[1,2], Murugan Saranya[3], Fei Wang [1,2,4,*]

[1]Department of Electronic and Electrical Engineering, Southern University of Science and Technology, Shenzhen, 518055, China.

[2] Shenzhen Key Laboratory of 3rd Generation Semiconductor Devices, Shenzhen 518055, China

[3]Platinum Retail Ltd, Chorleywood Road, Rickmansworth, UK.

[4] State Key Laboratories of Transducer Technology, Shanghai 200050, China

ramnano2009@gmail.com, msaran19@gmail.com, *wangf@sustc.edu.cn

Abstract

Supercapacitors (Electrochemical capacitors) are promising energy storage devices and have attracted significant attention as 'bridges' for the energy/power gap between traditional capacitors and batteries. The performance of supercapacitors is essentially determined by its electrode material. Among various supercapacitor electrodes, transition metal oxides/hydroxides usually exhibit high specific capacitance and energy densities. This chapter discusses the advantages and disadvantages of the supercapacitor and the supercapacitor performance of various metal oxide/hydroxides. Also, this review focus on the development and challenges of metal oxide/hydroxide based electrode materials.

Keywords

Specific Capacitance, Metal Oxides/Hydroxides, Current Density, Ruthenium Oxide

Contents

1.   Introduction

With the rapid improvement of the worldwide economy, the consumption of fossil fuels, and increasing environmental pollution there is an urgent need for efficient, clean, and scale up sustainable sources of energy, as well as new technologies related to energy conversion and storage [1,2]. Most renewable clean energy sources are highly dependent on the day time and regional weather conditions. Evolution of related energy conversion and energy storage devices is therefore required in order to effectively harvest these intermittent energy sources [2]. Among different energy storage and conversion technologies, electrochemical ones such as batteries, fuel cells, and electrochemical supercapacitors (ESs) are the three kinds of most important electrochemical energy storage/conversion devices [3]. The ES, also known as a supercapacitor or ultracapacitor, has attracted considerable interest in both academia and industry applications. Because supercapacitors are devices capable of managing high power rates compared to batteries. i.e, it has some distinct advantages such as long cyclic life, higher power density and bridging function for energy gap between conventional capacitors and batteries/fuel cells [4].

The Ragone plot (Fig. 1) shows the comparison of specific power and specific energy for different electrical energy storage devices. This plot indicates that the supercapacitors are able to store more energy density than conventional capacitors and are capable of delivering more power than batteries [5]. Basically, with regards to applications the ones that need quicker discharge rate go for capacitor while the slower ones go for batteries. From fig.1, it can be apparent that batteries are capable of attaining up to 150 Wh/kg of energy density, around 10 times higher than that of electrochemical capacitors. In terms of power density, batteries don't have the capability of reaching the values of electrochemical capacitors. Electrochemical capacitors hardly reach 20 times higher than that of batteries power density (200 W/kg). Due to fast charge-discharge cycles or cold environmental temperature, batteries are experiencing weaknesses like a rapid decrease in electrochemical performances, they are expensive to maintain and have a limited lifetime [6]. Due to the characteristic differences between supercapacitors and batteries with respect to energy storage mechanism and electrode materials, the characteristic supercapacitor performance makes it unique and apart from batteries. Summary of such differences between batteries and supercapacitors as well as the conventional electrolytic capacitor are listed in Table 1. [7,8].

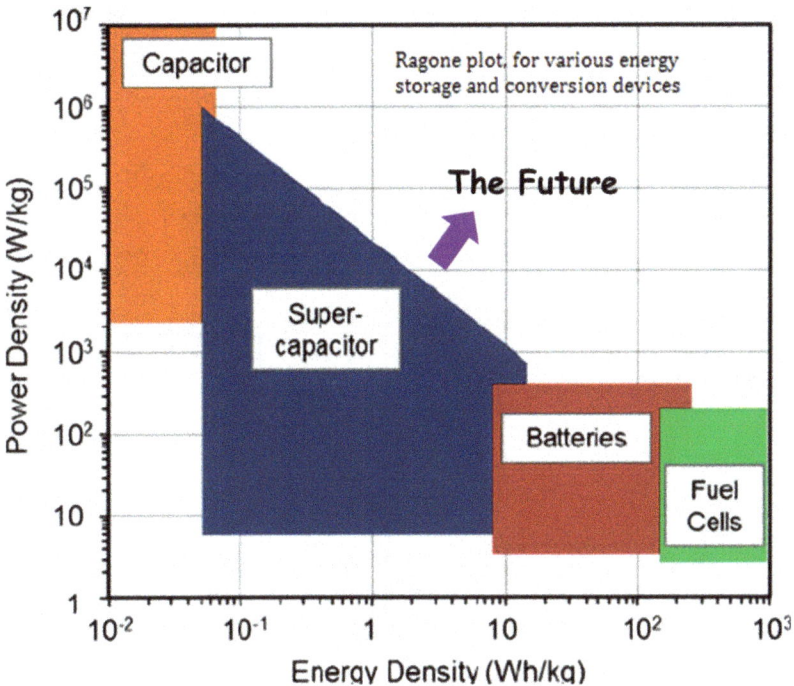

Fig. 1. Ragone plot of specific power and specific energy for different electrical energy storage devices.

Table1 Difference between batteries, conventional capacitors, and supercapacitors.

Parameters	Electrolytic capacitors	Supercapacitor	Battery
Storage mechanism	Physical	Physical	Chemical
Charging time	10^{-6} -10^{-3} Sec	1-30 Sec	1-5 hrs
Discharging time	10^{-6} -10^{-3} Sec	1-30 Sec	0.3-3 hrs
Energy density (Wh/kg)	<0.1	1-10	20-100
Power density (Wh/kg)	~10	5-10	0.5-1
Cycle life	Infinite	>500000	500-2000
Coulombic efficiency (%)	About 100	85-98	70-85

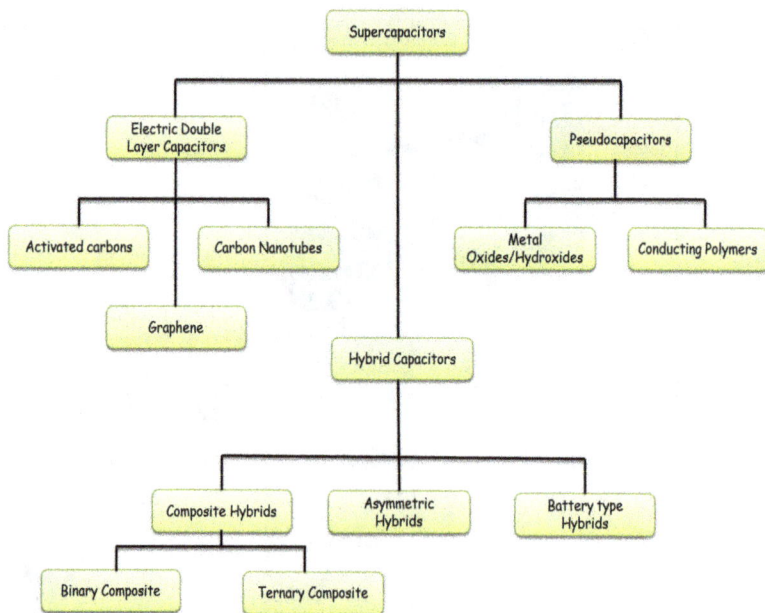

Fig. 2. Classification of supercapacitors.

2. Classification of supercapacitors

Based on energy storage and distribution of ions coming from the electrolyte to the electrode surface, the mechanism of the supercapacitors is classified into three categories. 1. Electric Double- Layer Capacitors (EDLC's) 2. Pseudocapacitors and 3. Hybrid capacitors as shown in Fig. 2.

2.1 Electric double-layer capacitors

The charge storage mechanism of EDLCs is based on the electrostatic charge accumulation at the electrode/electrolyte interface [9]. The material of the electrode plays a crucial role for the final supercapacitor performance. For the supercapacitor modern world, the most common electrode material for the practical application is activated carbon. It is cheap, has large surface area and is easy to process. Other carbon-based materials such as carbon nanotubes, carbon aerogels and recently graphene are also utilized as electrode materials in EDLC's. The good electronic conductivity, high surface area and large chemical stability of these carbon materials provide high power density with excellent cyclic stability. The lower value of the specific capacitance of these carbon electrodes is the major drawback of EDLC's. The energy density of a supercapacitor depending on the specific capacitance (C) and the operating voltage (V) can be expressed as follows [10].

$$E = \frac{1}{2}CV^2$$

(1)

To improve the energy density (E), the specific capacitance and operating voltage should be enhanced in supercapacitors. Since the delivered specific capacitance in carbon-based electrode materials is less and hence achieving a high energy density has become a difficult task in EDLC's. Thanks to pseudocapacitor for achieving this enhancement.

2.2 Pseudocapacitors

Pseudocapacitors store charge electrostatically compared to EDLC's. The charge storage of the pseudocapacitance is achieved by redox reactions. i.e, electrical energy stored via a Faradic process which involves the charge transfer between electrode and electrolyte [11]. The greater specific capacitance and energy densities could be achieved due to the Faradic process involved in pseudocapacitors compared to EDLC's. Metal oxides/hydroxides, metal sulfides and conducting polymers are mainly used as pseudocapacitor electrodes. The specific capacitance of the pseudocapacitors is usually greater than that of EDLC's. The main drawbacks of the pseudocapacitors are low power density and lack of stability during cycling process due to the Faradic nature, involving reduction- oxidation reaction just like in the case of batteries [12].

2.3   Hybrid capacitors

EDLC's deliver good cyclic stability and large power performance while pseudocapacitors offer greater specific capacitance and energy densities. Novel hybrid supercapacitors offer a combination of fast charging rate of supercapacitor electrodes and the high energy density of battery like electrodes in the same cell [13,14]. The combination of two types of different electrodes results in more energy storage due to the much wider operating voltage window of the organic electrolyte and the good specific capacity of the battery type electrode. To improve the electrochemical performance of the hybrid capacitors, several combinations have been tested with both positive and negative electrodes in aqueous and inorganic electrolytes [15]. Three types of hybrid supercapacitors have identified by researchers, which can be classified on the basis of configurations of electrodes like (a) composite, (b) asymmetric and (c) battery-types.

2.3.1 Composite

The combination of carbon electrodes with either metal oxides or conducting polymers in a single electrode will have both chemical and physical charge storage mechanisms. The capacitance of the composite has increased through the Faradic reactions of pseudocapacitive materials. Binary and ternary composites are the two different types of composites [16]. Two different electrode materials are involved in the binary composite, while three different electrode materials involved in the case of the ternary composite to form a single electrode.

2.3.2 Asymmetric hybrids

In asymmetric hybrid supercapacitors, the carbon-based materials are used as a negative electrode while either metal oxides or conducting polymers act as positive electrodes. Thus, it combines Non-Faradic process from EDLC and Faradic process from pseudocapacitive materials [16].

2.3.3 Battery type

Battery type supercapacitors are alike of asymmetric hybrids. But, these supercapacitors are made up by supercapacitor electrode with battery electrode. Both supercapacitors, as well as battery properties, can be utilized in a one cell configuration [16].

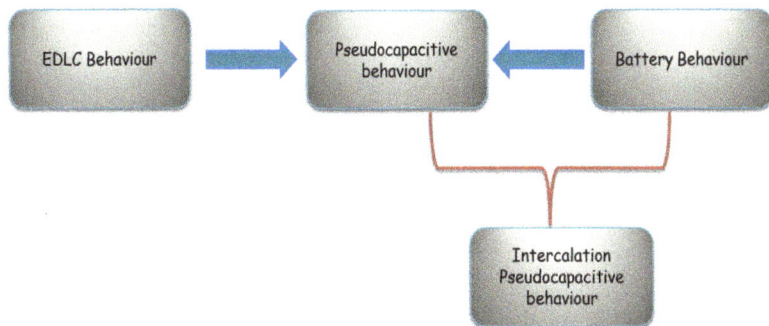

Fig. 3. EDLC, pseudocapacitance and battery type relationship.

Fig. 3 demonstrates the relationship among EDLC, pseudocapacitive and battery type supercapacitors. It can be noticed that the pseudocapacitive behavior bridge the gap between EDLC's and battery type behaviors. In recent years, more attention has been intercalation pseudocapacitive behavior [17]. It fills the gap between pseudocapacitive behavior and the battery type behavior. The concept of "pseudocapacitive behavior" and "intercalation pseudocapacitive behavior" is almost similar. B.E. Convey explained the theory of the electrochemical behavior of RuO_2 [1]. The electrochemical behavior of RuO_2 bridges the gap between the EDLC's behavior and battery behavior. i.e, "transition from 'supercapacitor (EDLC's)' to 'battery' behavior in electrochemical energy storage system". Because the electrochemical profile of RuO_2 (clear rectangular shape of cyclic voltammetry curve) and the fast kinetics of the electrodes are all close to that of EDLCs, the RuO_2 electrode was considered as pseudocapacitive behavior.

The rechargeable batteries current capacity mainly depends on the intercalation/de-intercalation of cations (H^+ or Li^+) within the crystalline structure of the electrode materials, coupled with the redox reactions of metal ions within the crystalline structure [18]. From Fig. 4 it is clear that the batteries charge storage is controlled by cation diffusion within a crystalline framework. It should be noted that besides the intercalation reaction, the "phase-transformation" and/or insertion of Li into the crystal structure of the electrodes mechanisms are also involved in all rechargeable batteries. Like rechargeable batteries, pseudocapacitors also store charges via redox reactions of the metal ions in the electrode materials [19]. The only difference is that the rechargeable batteries are limited by cation diffusion within the framework of the active electrode material, whereas the pseudocapacitance is not controlled by the diffusion process. A linear dependence of the charge stored with the charging potential within the window of interest displayed by pseudocapacitive electrodes, but the origin of the charge storage is from the electron-transfer mechanism, not like relying on the accumulation of ions in the electric double layer. The cyclic voltammetry (CV) curve is an essential investigation to differentiate between pseudocapacitance and battery type behaviors [20,21]. A rectangular CV shape appeared in the former case (like above- mentioned RuO_2 electrode material) which is close to that of the capacitive performance of carbon electrode, whereas the battery type displays clear redox peaks in CV curves.

The charge storage mechanism of intercalation pseudocapacitance depends on the cations intercalation/de-intercalation (Na^+, K^+, H^+ and Li^+) in the bulk of active materials, but is not limited by the cations diffusion within the crystalline framework of electrode material (Fig.4). In the case of "interaction pseudocapacitive behaviour" similar kinetics as like typical pseudocapacitive behaviour (i.e, a linearly proportional voltammetric response) [22], but it displays the electrochemistry of same as like battery-type electrode behaviour (i.e, the interaction of cations in the crystalline framework of the electrode materials involved redox reactions), in which charge storage occurs in a very narrow potential window [19].

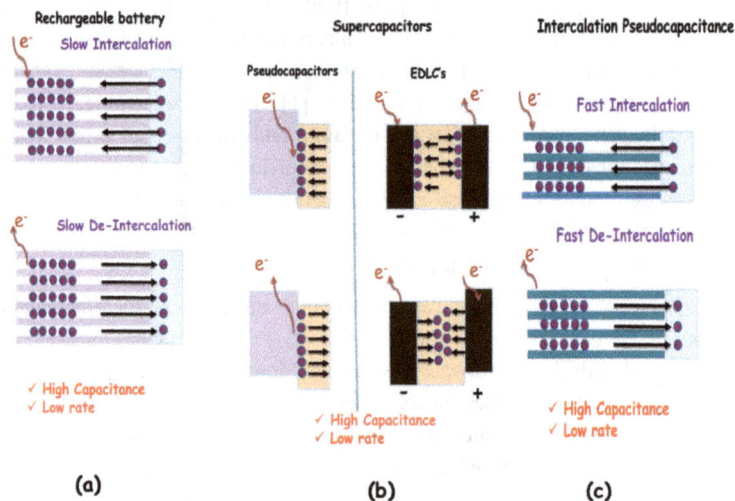

Fig. 4. Schematic illustration of different charge storage mechanisms. (a) charge storage mechanism of the rechargeable battery. (b) charge storage mechanism of the supercapacitor. (c) charge storage mechanism of intercalation pseudocapacitance [19].

3. Advantages and challenges of supercapacitors

3.1 High power density

It is clear that supercapacitors are capable of much higher power delivery than batteries. As evidenced from Fig. 1 supercapacitors can be able to deliver much higher power density (1-10 kW kg^{-1}) than lithium ion batteries (150W kg^{-1}). The charging and discharging rates in supercapacitors are much faster than that of electrochemical redox reactions inside batteries. Since the charge-discharge reactions in supercapacitors are not necessarily limited by ionic conduction into the electrode bulk and hence the supercapacitors are capable of delivering high power density. For example, approximately within 30 sec, supercapacitors can be fully charged and discharged. Also, the maximum duration of energy taken from it within 0.1 s [23]. The charging time in batteries is few hours, which is shown in Table 1.

3.2 Long cycle life

The electrochemical energy stored in batteries is caused via Faradic reactions, which undergo irreversible phase changes and irreversible interconversion of the chemical electrode reagents. But, there are no chemical charge transfer reactions and phase changes involved during the charge-discharge process when electrochemical energy is stored in supercapacitors. Thus, supercapacitors had infinite cyclability and no need of any maintenance during their lifetimes. Moreover, it can withstand up to 10^6 charge-discharge cycles without any changes in their characteristics. Such charge-discharge cycles are impossible for batteries even small depth discharge. Even though the charge-discharge processes are involved fast redox reactions, supercapacitors are estimated to have a life expectancy of up to 30 years, which is much higher than batteries [24].

3.3 Long shelf life

Another advantage of the supercapacitor is having long shelf life. If batteries are unused for months, they will degrade and cannot be used due to self-discharge and corrosion. But, supercapacitors maintain their capacitance and can being recharged to their original condition even after several months. Burke reported that unused supercapacitors for several years can be capable of retaining their original condition [25].

3.4 Wide range of operating temperature

Supercapacitors can be operated from -40 to 70°C temperature range, this advantage makes them promising candidates for applications in the military, where stable energy storage is required under high and low temperatures.

3.5 High efficiency

During each cycle of supercapacitors, the energy loss to heat is relatively small and negligible. So that, much cycle efficiency (around 95%) could be achieved even when operating at rates above 1 kW kg^{-1} [26].

4. Disadvantages of supercapacitors

Though supercapacitors have many advantages over batteries and other energy storage systems, at the current stage they face some challenges like low energy density, high cost, high rate of self-discharging and industrial standards for commercialization.

4.1 Low energy density

Low energy density is the major drawback of supercapacitors in short and medium terms. The commercially available electrochemical capacitors can deliver energy densities from 3-5 Whkg^{-1} only. This energy density value of supercapacitors is lower and relatively small when compared to batteries (> 50Wh kg^{-1}). Larger supercapacitors are to be constructed for higher energy density requiring applications.

4.2 High cost

Another drawback of the supercapacitor is the cost of raw materials and manufacturing cost. At the current stage, for commercial purpose carbon and RuO_2 are the two common electrodes used in EC. Though carbon based materials have high surface area and provide good performance, the cost of these carbon electrodes is quite high.

4.3 High self-discharging rate

Supercapacitors have a self- discharging rate of 10-40 % per day [27]. This leads to a major problem in some practical applications.

4.4 Industrial standards for commercialization

Though currently available commercial carbon/carbon supercapacitors have to deliver capacitance of 50-5000 F, it is necessary to establish some general industrial standards such as performance, the structure of electrodes, the thickness of electrode and porosity. Due to limited commercially available supercapacitor products, they have limited applications.

5.   Electrode materials for supercapacitors

The charge storage and capacitance of supercapacitors depend on the type of electrode materials used. Therefore, further developing of new electrode materials with high capacitance is most important to overcome the above-discussed challenges. The specific capacitance and energy density of supercapacitors heavily depend on the specific surface area of the electrode materials. During the electrochemical reactions, when the electrode material is in contact with an electrolyte, the only limited specific surface area of the electrode involves electrochemically with the electrolyte. Therefore, for various materials, the measured capacitance does not increase linearly with increasing surface area [28]. According to the surface area and charge storage mechanism, the electrode materials of supercapacitors can be categorized into four types:

(1) carbon materials, (2) conducting polymers, (3) metal oxides/hydroxides and (4) transition metal sulfides.

6.   Strategies to enhance the performance of supercapacitor

As discussed in early sections, achieving a high energy and power density of supercapacitors are still a challenging and difficult task. Therefore, researchers from all over the world are trying to develop electrode materials of high capacitance. There are two possible ways to enhance the performance of the supercapacitor electrodes.

1. Producing the electrode materials of the supercapacitor with different nano structure morphology and 2. Combining two different electrodes into a single electrode material. Since the specific surface area and electrochemically active sites depend on the structures of the electrode material, the capacitance of supercapacitors can be improved by different nanostructures. Similarly, combining two electrode materials also improved the performance of supercapacitors because of synergistic effects. Generally, the composites have good electrical conductivity rather than their individual forms. Different nanostructures and composite electrodes provide more ion adsorption/active sites for the charge transfer reactions and transfer pathways of the electrolyte ions (Fig. 5)

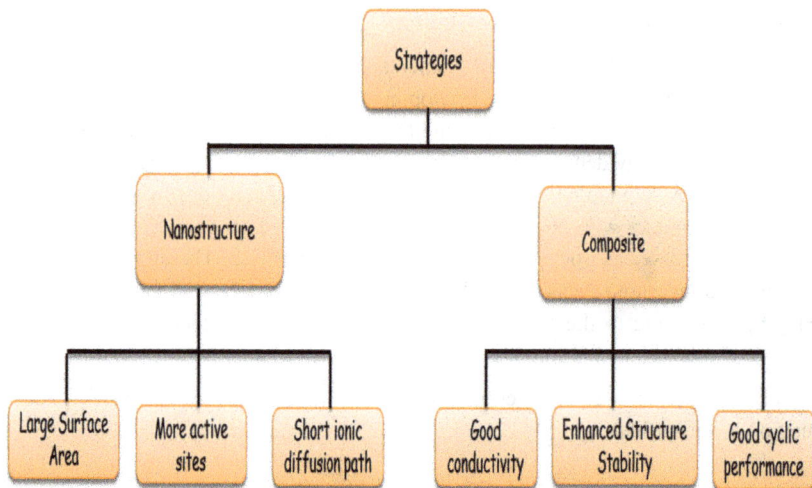

Fig. 5. Strategies to enhance the performance of supercapacitor electrodes.

7. Factors affecting the performance of metal oxides/hydroxide based supercapacitor electrodes

7.1 Crystallinity

The crystallinity of electrode materials is one of the main factors which affect the pseudocapacitance of metal oxide and hydroxide materials. In general, the amorphous structure of materials with highly porous nature is beneficial for ion and cation diffusion into the electrolyte to the electrode. Also, the porous structures in amorphous electrode materials can support more redox reactions to enhance the specific capacitance due to their higher specific surface area [29]. Thus, amorphous structure materials exhibit better electrochemical performance towards supercapacitor applications than a well-crystallized structure metal oxides/hydroxides. Nevertheless, poorly or partially crystallized metal oxide/hydroxide leads to a lower electrical conductivity. Thus, it is necessary and important to explore the appropriate crystallinity of material and optimal conductivity and ionic transport.

7.2 Crystal structure

The different crystal structures such as α, β, γ and δ play a crucial role in determining the cation interactions into electrode/electrolyte interface and significant influence on the metal oxide/hydroxide pseudocapacitance [30]. Brousse et al., [31] reported that the MnO_2 with α crystal structure delivered a relatively high specific capacitance of 110 F/g, because α-MnO_2 has a large tunnel (1D (2x2)) and K^+ cations could easily transport to these tunnels. On the other hand, the tunnel size of β-MnO_2 electrode is narrow and lesser transport of K^+ cations leads to a low specific capacitance of 110 μF cm^{-2} only. Thus, the limited electrochemical performance is mainly due to the surface/crystal structure of MnO_2 electrode. It is well known that the crystal structures of metal oxide/hydroxide can effectively modulate the charge storage process in supercapacitors.

7.3 Specific surface area

The redox reactions of pseudocapacitance of metal oxides and hydroxides depend on the specific surface area. The higher specific surface area material can deliver higher specific capacitance due to more active sites are involved in redox reactions [32]. To explore a higher specific surface area of metal oxides/hydroxide is an effective method to achieve high energy density supercapacitors. There are several ways to increase the specific surface area including decreasing particle size, morphology changes and combining metal oxide/hydroxide with carbon-based materials such as CNT, graphene, etc.

7.4 Particle size

Another notable factor for affecting the electrochemical performance of metal oxides is particle size. Small-sized particles have short diffusion and transport pathways of electrolyte ions. As a result, the more electrochemical active material can be utilized during redox reactions and rate of charge/discharge capability of the supercapacitors will be improved. The ion diffusion time constant (τ) can be described as follows [33],

$$\tau = \frac{L^2}{2D} \tag{2}$$

where L is the transport length and D is the ion transport coefficient. When the particle size is decreased, the ion diffusion coefficient time also decreased simultaneously. So, it is necessary to use the smaller size of particles in supercapacitors to achieve better electrochemical performance [34].

7.5   Morphology

Morphology of electrode materials is very closely related to the specific surface area, surface to volume ratios and ionic transport in the electrolyte. Thus, the specific capacitance of metal oxide/hydroxide based supercapacitors can be tuned via different morphologies such as nanoparticles, nanorods, nanowire, nanotube, nanoflowers and thin films. In general, one-dimensional (1D) nanostructured metal oxides/hydroxide can improve the supercapacitor performance due to the large specific surface area and short diffusion path length for both ions and electrons. For example, Co_3O_4 nanowires on Ni foam have been synthesized via template-free method by Gao et al. [35], and the 1D nanowire with 205 nm diameter exhibited a maximum specific capacitance of 746 F/g. Similarly, NiO nanorod with a diameter of less than 20 nm displayed specific capacitance of 2018 F/g as reported by Lu et. al [36]. The smaller diameter of NiO nanorods offers more active sites as well as large specific surface area and leads to better electrochemical behavior toward supercapacitive properties. Next to nanowire and nanorods structures, hollow spheres morphology is a notable nanostructure to enhance the specific capacitance of electrode materials. Du et al., reported hollow sphere structure of Co_3O_4 by a facile carbonaceous microsphere template with a high surface area of $60 m^2 g^{-1}$ which greatly improved the specific capacitance and excellent cyclic stability of supercapacitors [37]. It has now become necessary to design morphologies with high specific surface area and porous structure for improving the electrochemical performance of metal oxides/hydroxides electrode.

7.6   Electrical conductivity

It is well-known that the supercapacitor performance of electrodes highly depends on their electrical conductivity. Unfortunately, the electrical conductivity of most metal oxides/hydroxides is poor which affects the ion and electron transport in systems. For example, the theoretical specific capacitance of MnO_2 electrode material is ~1370 F/g. But, due to its measured lower electrical conductivity (~10^{-5} to 10^{-6} S cm^{-1}) can deliver the specific capacitance up to 350 F/g only [38]. Like MnO_2, other metal oxides and hydroxide such as NiO, $Ni(OH)_2$, Co_3O_4, $Co(OH)_2$ and V_2O_5 exhibit lower conductivity. Thus, it is very important to improve the electrical conductivity of metal oxide/hydroxide electrodes to get superior supercapacitor performance. Compositing with carbon materials/conducting polymers or doping with metal elements to metal oxides/hydroxides is the effective approach to improve the electrical conductivity of these materials.

7.7   Electrode mass loading

The specific capacitance, energy and power density of supercapacitor depend on the quantity of active materials loading on the substrate. A small quantity of mass loading leads to better transport paths for the diffusion of ions into electrode systems and hence, high specific capacitance could be achieved. On the other hand, the large mass loading can cause higher series resistance and lower electrical conductivity indicating that only partial active materials on the surface of electrode involve the redox reactions. Yang et al., [39] reported that increased mass loading from 6 to 10 mg cm^{-2} of MnO_2 thin film, results in decreased specific capacitance from 203 to 155 F/g. Therefore, still, it is one of the drawbacks for metal oxides/hydroxides to achieve both excellent specific capacitance and high mass loading.

8.   Metal oxide/hydroxides based supercapacitor electrodes

In supercapacitor devices, the electrode materials are considered key component for its electrochemical performance. Therefore, selecting and designing efficient electrode materials towards high energy density supercapacitor is a great challenge. Generally, metal oxides/hydroxides delivered more specific capacitance than carbon materials because of their pseudocapacitance behavior arose from redox reactions.

8.1   Ruthenium oxide based supercapacitors

RuO_2 is the well-known electrode material for commercial supercapacitor applications. Ruthenium oxide is still a recognized electrode material due to its high theoretical specific capacitance (~2000 F/g), long cycle life, high electrical conductivity and good electrochemical reversibility [40]. It is in either crystalline or amorphous hydrous form that exhibits larger electrochemical capacitance. RuO_2 has a high specific capacitance which is ten times higher than the specific capacitance of carbon electrodes. This high value of specific capacitance is due to the pseudocapacitance from the redox reactions between Ru ions and cations of the electrolyte. The charge storage process of RuO_2 can be described by the following equation [41],

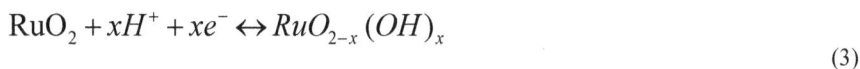

$$RuO_2 + xH^+ + xe^- \leftrightarrow RuO_{2-x}(OH)_x$$

(3)

The Ru oxidation states might change from (II) to (IV) in the acidic electrolyte. It is reported that the RuO_2 electrode material will be oxidized to RuO_4^{2-}, RuO_4^- and RuO_4 during the charging process and these high state compounds reduced to RuO_2 during the discharge process [42]. RuO_2 supercapacitive performance depends on the following three factors.

1. Specific surface area 2.Combined water in RuO_x and 3.Crystallinity of $RuO_2.xH_2O$. There are several approaches to increase the specific surface area of RuO_2 such as depositing RuO_2 films on rough surface substrates, coating RuO_2thin films on high surface area materials like graphene and synthesizing nano-sized oxide materials, etc. [43]. From the equation (3), it is clear that RuO_2 in acidic electrolyte involves the insertion and extraction of H^+. Hydrous RuO_2 ($RuO_2. nH_2O$) is a good proton conductor [44]. So, these water molecules which attached to RuO_2 can accelerate the diffusion of H^+ in the electrode and improve the electrochemical performance. It was reported that the diffusion of cations in hydrated electrodes could achieve through hopping of alkaline ions and H^+ ions between water molecules of RuO_2 and OH^- sites in the electrolyte, indicating that the hydrogen atoms were relatively mobile in $RuO_2.xH_2O$ sample than a pristine form of RuO_2. The hydrous form of ruthenium oxide electrode having fast ionic conduction and good proton conductivity (H^+ diffusion coefficient reaches from 10^{-8} to 10^{-12} cm^2 s^{-1}) [45] leads to enhance the capacitive behavior. It is notable that a number of combined water molecules in RuO_2 also play a crucial role in the supercapacitor performance. The specific capacitance was decreased from 720 to 19.2 F/g for $RuO_2 \cdot 0.5H_2O$ and $RuO_2 \cdot 0.03H_2O$, respectively [46]. Zheng et al. proposed four steps to determine the charge storage performance of $RuO_2.nH_2O$.

(i) Electron hopping within $RuO_2.nH_2O$ particles.

(ii) Electron hopping between the particles.

(iii) Electron hopping between the electrode and current collector.

(iv) Proton diffusion within $RuO_2.nH_2O$ particles [47].

As discussed in the early section, the amorphous nature of RuO_2 has been responsible for its high electrical conductivity and good electrochemical reversibility than crystalline RuO_2. It is very difficult to expand and contract a well-crystallized structure of RuO_2, so it blocks protons transporting the bulk materials and restricts the diffusion process. The capacitance of well crystalline RuO_2 comes from only the surface reaction, whereas the pseudocapacitance of amorphous RuO_2 not only from the surface but also in bulk of the powder. Amorphous RuO_2 exhibited a maximum specific capacitance of 768 F/g which is two times greater than that of crystalline RuO_2 [47]. Another advantage of amorphous ruthenium oxide is that the maximum potential range is about 1.35 V which is greater than that of a measured potential range of crystalline RuO_2 (1.05 V) in aqueous electrolyte. Up to now, various nanostructures based RuO_2 supercapacitors have been fabricated such as nanotubes, nanosheets, and nanorods.

In spite of all the advantages, due to its high cost and scare availability, RuO_2 is still not suitable for commercial application with supercapacitors. The production cost of RuO_2 based supercapacitors can be reduced in two ways.

1. Composites of RuO_2 with cheap metal oxides such as MnO_2, NiO, TiO_2, and SnO. [48,49] 2. Deposition of RuO_2 into carbon materials like carbon nanotubes, graphene and conducting polymers such as polyaniline and polypyrrole [50].

8.1.1 Ruthenium oxide based composites

In order to reduce the cost of precious metal use, many studies have been done on combining RuO with cheap metal oxides such as SnO_2, MnO_2, NiO, TiO_2, to form composite oxide electrodes. The specific capacitance of RuO_2 has increased to 710 F/g in a KOH electrolyte when RuO_2 was deposited onto SnO_2 by an incipient wetness precipitation method [42]. The enhanced specific capacitive behavior was found by doping vanadium into RuO_2 electrode [51]. The Vanadium introduction into $Ru_{0.36}V_{0.64}O_2$ not only extended the electrochemical window from 1.0 to 1.1V but also increased the utilization of Ru species as well as the electrode's electrochemical stability [52]. Various amounts loading of amorphous $Ru_{1-y}Cr_yO_2$ on TiO_2 nanotubes via reduction of aqueous $Ru_{1-y}Cr_yO_2$ with $RuCl_3$ have been investigated for supercapacitor electrode material. The proper amount of $Ru_{1-y}Cr_yO_2$ loading with TiO_2 nanotubes enhanced specific capacitance up to 1272.5 F/g. TiO_2 nanotubes in $Ru_{1-y}Cr_yO_2$ not only help the transportation of ions but also provide double-layer charge [53].

8.1.1.1 Ruthenium oxide /Carbon based composites

In recent years carbon materials like the carbon nanotube [54], graphene [55], and carbon Nano-onions [56] have been used with RuO_2 to enhance the supercapacitor properties at low cost. The specific capacitance of the composites depends on the amount of RuO_2 in carbon electrodes such as 256 F/g with 14% of Ru in a carbon aerogel [57], 647 F/g with 60% RuO_2 in Ketjenblack [58] and 850 F/g for carbon- RuO_2 composites [59]. It should be noted that many investigations have utilized a very high temperature to synthesis RuO_2/carbon composites [57]. These high-temperature annealing processes will lead to higher crystallinity of RuO_2 and hence the supercapacitor performance can be increased as mentioned previously. In RuO_2/carbon composites, the presence of carbon plays the following activities.

(i) Carbon material in composite seems to favor the dispersion of amorphous $RuO_2.xH_2O$ particles. A strong interaction between RuO_2 and the carbon surface carboxyl groups contributed to the high dispersion of RuO_2 nanoparticles. These high dispersed RuO_2 nanoparticles in carbon exhibited capacitance enhancement, because of the ions were able to access the inner part of RuO_2 also.

(ii) Ions and electrons transfer can be facilitated by a carbon support in the electrode layer. The porous morphology of carbon can allow excellent electrolyte access in the three-dimensional network when a thin layer of RuO_2 was coated on CNT [60].

8.1.1.2   Ruthenium oxide /Polymer composites

Like carbon, polymers have been recently incorporated into RuO_2 to enhance the supercapacitor performance. A maximum specific capacitance of 302 F/g has been reported for polypyrrole/ RuO_2 composite [61]. Liu et al., prepared RuO_2 composites with poly(3,4-ethylenedioxythiophene) and reported a high specific capacitance of 1217 F/g and power density of 20 kW kg^{-1}[62]. The special hollow nanotube structure of the polymer allowed penetration of ions into the composite easily and provided a short diffusion distance for the ion transport.

Polymers in RuO_2 serve several roles, (a) due to steric and electrostatic stabilization mechanism, polymers can prevent the aggregation of $RuO_2.xH_2O$ particles. (b) Uniformly distributing $RuO_2.xH_2O$ particles in the polymer matrix (c) Polymer matrix can increase the active surface area of $RuO_2x.H_2O$ for redox reactions and (d) Improving the adhesion of R $RuO_2x.H_2O$ to the current collector.

Although RuO based electrodes can provide extremely high specific capacitance and power density but the high cost and environmental harmfulness are their drawbacks. Thus, researchers have put significant effort into finding cheaper and environmentally friendly materials towards supercapacitor applications.

Table 2. Comparison of RuO$_2$ based supercapacitors.

Electrode	Synthesis method	Cs (F/g)	Scan rate/Current density	Potential window (V)	Electrolyte	Ref
RuO$_2$/carbon dot	Sol-gel method	460	50 A/g	0 to 1.0	1M Na$_2$SO$_4$	63
RuO$_2$-rGO	Sol-gel method	500	1 A/g	0 to 1.5	1M H$_2$SO$_4$	64
RuO$_2$/quasi graphene	Hydrothermal method	453.7	1 A/g	-0.9 to 0.8	1M KOH	50
RuO$_2$/quasi graphene	Hydrothermal method	415.7	1 A/g	-0.9 to 0.8	1M H$_2$SO$_4$	50
RuO$_2$/quasi graphene	Hydrothermal method	287.5	1 A/g	-0.9 to 0.8	1M Na$_2$SO$_4$	50
RuO$_2$@C	Chemical route	107	1 A/g	0 to 3.5	EMIM-BF4 IL	65
RuO$_2$/MnO$_2$	Electrodeposition	793	2 mV/s	0 to 0.8	1 M Na$_2$SO$_4$	66
RuO$_2$@Co$_3$O$_4$	Electrospinning	1103.6	10 A/g	0 to 0.5	2 M KOH	67
RuO$_2$.xH$_2$O/ FMWCNT	Polymer-assited technique	1474	10 mV/s	0 to 1.0	1 M H$_2$SO$_4$	68
RuO$_2$/PANI	Chemical bath deposition	830	1 A/g	0 to 1.0	1 M H$_2$SO$_4$	69
RuO$_2$/PEDOT nanotubes	Step-wise template synthesis	1217	-	0 to 1.0	1 M H$_2$SO$_4$	62

8.2 Manganese Oxide based supercapacitors

Manganese oxide provides a replacement for ruthenium oxide based supercapacitors. The theoretical capacitance of MnO$_2$ ranging from 1100 to 1300 F/g which makes it a good potential candidate for supercapacitor electrodes [70]. The capacitance of MnO$_2$ comes from mainly pseudocapacitance which attributed to reversible redox reactions. There are several oxidation states, including Mn(0), Mn(II), Mn(III), Mn(IV), Mn(V), Mn(VI), and

Mn(VII), for manganese oxides which involving in redox reactions. Generally, the insertion of electrolyte cations such as H^+, Li^+, Na^+ and K^+ into the bulk material and the surface adsorption of the electrolyte cations on the MnO_2 electrode control the charge storage behavior of MnO_2 [71].

$$MnO_2 + C^+ + e^- \leftrightarrow MnOOC \qquad (4)$$

$$(MnO_2)_{surface} + C^+ e^- \leftrightarrow (MnOOC)_{surface} \qquad (5)$$

As discussed in previous sections, different crystal structures such as α, β, γ, δ of MnO_2 play a crucial role in the supercapacitive performance. Different synthesis conditions can lead to various MnO_2 structures. Several types of synthesis methods reported in the literature like thermal decomposition, co-precipitation,sol–gel processes, electrodeposition, mechanical milling processes, and hydrothermal synthesis for MnO_2 preparations [32]. Various structures of MnO_2 like nanowhiskers, nanoplates and nanorods can be controlled via the hydrothermal method among other techniques. Wei et al. synthesized a novel 2D β- MnO_2 network with the long-range order through a one-pot hydrothermal reaction which exhibited a maximum specific capacitance of 453 F/g at a current density of 0.5 A/g [71]. Recently, amorphous MnO_2 films have been prepared by the sol-gel method followed by the an annealing process. The resultant MnO_2 films exhibited a maximum specific capacitance of 360 F/g at a current of 0.82 A/g and a large density of 48.8 W h /kg at a power density of 1.16 W /kg [72]. Urchin-like nanostructures of γ- MnO_2 and α- MnO_2 have been prepared by the microwave-assisted method within 5 min under neutral and acidic conditions, respectively. A maximum specific capacitance of 311 F/g was achieved at a current density of 0.2A/g for γ- MnO_2 nanoparticles, which is higher than that of α- MnO_2 nanostructure [73]. It is very difficult to control the morphologies and phases in microwave synthesis. In recent years, an emerging attractive concept is to combine the microwave method with the hydrothermal method, which can dramatically reduce the reaction time from 10 h or even several days down to 30 min, and also keep the ability to control the morphology and size distribution of the materials [74]. Birnessite-type MnO_2 nanospheres prepared by the microwave-hydrothermal method at 75°C within 30 min under low pressure have exhibited a specific capacitance of 210 F/g at 0.2 A/g current density [75].

8.2.1 Challenges for manganese oxide electrodes

8.2.1.1 Dissolution problem

The major challenge of MnO_2 electrodes in supercapacitors is its tendency of dissolution. The partial dissolution of MnO_2 in the electrolyte during cycling process has been the problem. The capacitance degradation also occurred in MnO_2 electrodes [76]. This can be expressed as follows,

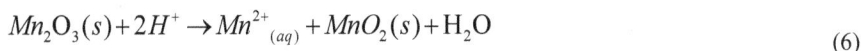

$$Mn_2O_3(s) + 2H^+ \rightarrow Mn^{2+}_{(aq)} + MnO_2(s) + H_2O \qquad (6)$$

There are two approaches studied to prevent MnO_2 from dissolution during cycling.

1. Developing a new electrolyte salt to avoid forming acidic species in solutions.

2. Applying a protective shell to the MnO_2 surface.

Babakhani and Ivey prepared MnO_2/CP (conducting polymer) coaxial core/shell electrode for supercapacitor applications [77]. The prepared product exhibited a specific capacitance of 285 F/g with 92% retention after 250 cycles in 0.5 M Na_2SO_4 at 20 mV/s. The conducting polymer in the MnO_2 suppresses the dissolution of MnO_2 and improves its resistance to failure, leading to both enhanced capacitance and high cycling rate capability.

8.2.1.2. Poor electrical conductivity and low surface area

Another concern is the low surface areas and poor electronic conductivity of MnO materials. This problem can be improved by introducing some transition metal elements. Doping of transition metals like ruthenium, nickel, and Mo into MnO_x has proven an effective way to increase the surface area and electrical conductivity. Doping of Ni into MnO_2 has increased the surface area of MnO_2 by about 46% and the specific capacitance by 37% [78]. Like transition metals, doping of highly conductive supports such as active carbon, CNT, graphene and conducting polymers can increase the conductivity of MnO_2. Li et al., have reported that when MnO_2 nanoparticles were partially coated on MCNT's surface via hydrothermal route, the specific capacitance was increased up to 550 F/g [79]

8.2.2 Manganese oxide based composites

The introduction of other materials into MnO enhances the working potential range, improves the electron conductivity of the electrode, and provides an effective utilization of more active surface area of MnO_2. As a result, the composite electrodes can display higher specific capacitances, higher energy, and higher power densities.

For example, graphene- MnO_2 composite which was prepared by microwave irradiation showed a maximum specific capacitance of 310 F/g at 2 mV/s. The pure graphene and birnessite-type of MnO_2 displayed specific capacitance only 104 and 103 F/g, three times lesser than that of the composite. The presence of graphene has increased the electrical conductivity of electrode as well as the interfacial area between MnO_2 and the electrolyte [80]. A novel hybrid material of graphene nanoribbons and MnO_2 nanoparticles (GNR-MnO_2) has been successfully synthesized via one-step method by Liu et al. [81]. The high-performance asymmetric supercapacitor of a widened operating potential window (2.0V) was fabricated with the GNS- MnO_2 hybrid material and pure GNS as the positive and negative electrodes, respectively and the device exhibited a specific capacitance of 212 F/g. MnO_2/porous carbon microspheres with a partially graphitic structure were fabricated based on a hydrothermal emulsion polymerization and common activation process for supercapacitor electrodes. As a result, the high specific capacitance of 459 F/g exhibited at 1.0 A/g current density [82]. Carbon nanofiber is the promising backbone for the conformal coating of MnO_2.Kang et al., have synthesized CNFs/ MnO_2 nanocomposite for freestanding supercapacitor electrode materials. This freestanding electrode displayed high power density of 13.5 kW/kg and energy density 20.9Wh/kg with long term cyclic stability (94% retention) [83].

8.2.3 Manganese Oxide /Polymer composites

The conductivity and pseudocapacitive properties of MnO_2 can be improved by incorporation of PANI into MnO_2 nanostructures. A hybrid nanocomposite of MnO_2 with PANI (α-MnO_2-PANI) has been prepared through *in-situ* polymerization process by Jafri et al. [84], and the prepared nanocomposite displayed the maximum specific capacitance of 626 F/g at a current density of 2A/g in the potential range of 0-0.7 V. Another research group has synthesized the ultrathin MnO_2 nanorods on the surface of PANI nanofibers. The amount of MnO_2 nanorods loading can be controlled by $KMnO_4$ concentration during synthesis. The prepared MnO_2-PANI composite has substantially increased specific capacitance up to 417 F/g[85]. Due to the high stability, excellent conductivity and good capacitive behavior, the poly(3,4-ethylene dioxythiophene) (PEDOT) and MnO_2-PEDOT have received much attention in supercapacitor electrodes. Tang et al., have been developed MnO_2-PEDOT nanocomposite by step-by-step anodic deposition on Ni foam and the optimized PEDOT/ MnO_2/PEDOT sandwich electrode showed excellent capacitive behavior with the high specific capacitance of 487.5 F/g [86]. Like PANI, polypyrrole also provides a large specific area, good stability and high conductivity for MnO_2. The electrochemical polymerization of pyrrole deposited on γ-MnO_2 enhanced the specific capacitance to 141.6 F/g compared with 73.7 F/g for pure MnO_2 before pyrrole coating [87]. Liu et al. have reported the preparation of MnO_2 nanosheets-PPy composite film by chemical polymerization process and the presence of PPy promoted the charge transfer in the MnO_2 nanofilm. An excellent rate performance with the real capacitance of 25.9 mF/cm was achieved for symmetric supercapacitor system [88].

Table 3 Comparison of MnO₂ based supercapacitors.

Electrode	Synthesis method	Cs(F/g)	Scan rate/Current density	Potential window (V)	Electrolyte	Ref
γ- MnO_2	Microwave assisted reflux	311	0.2 A/g	-0.1 to 0.8	1 M Na_2SO_4	73
α- MnO_2	Microwave assisted reflux	163	0.2 A/g	-0.1 to 0.8	1 M Na_2SO_4	73
MnO_2	Hydrothermal method	461	5 mV/s	0 to 0.9	1 M Na_2SO_4	89
MnO_2 hollow sphere	Solvothermal method	147	0.2 A/g	0 to 2.5	EMIM-BF4 IL	90
MnO_2 nanosheets	Spray coating	414.28	0.35 A/g	-0.5 to 2.0	-	91
MnO_2/CNT	Electrodeposition	199	1 A/g	0 to 1.0	1 M Na_2SO_4	92
MnO_2 nanowire/Graphene oxide	Hydrothermal route	360.3	0.5 A/g	0 to 1.0	1 M Na_2SO_4	93
MnO_2 nanowire	Hydrothermal route	128	0.5 A/g	0 to 1.0	1 M Na_2SO_4	93
β- MnO_2/rGO	Hydrothermal route	362	1.0 A/g	0 to 1.0	6 M KOH	94
MnO_2/3D porous carbon	Electrodeposition	416	1mV/s	0 to 1.0	1 M Na_2SO_4	95
MnO_2/PANI	Polymerization	383	0.5 A/g	-0.1 to 0.9	1 M Na_2SO_4	96
MnO_2/PEDOT	Electrodeposition	359	0.5A/g	0 to 1.0	1 M Na_2SO_4	97
MnO_2/PPy	Chemical oxidation	352.8	8mA/cm^2	-0.2 to 0.8	0.5 M Na_2SO_4	98
MnO_2/Ni(OH)/ rGO	One step deposition	1985	2 A/g	0 to 0.5	1 M KOH	99

8.3 Nickel oxide based supercapacitors

Nickel oxide (NiO) and its hydroxides (Ni(OH)₂ are the materials suitable for pseudocapacitor electrode applications owing to their high theoretical specific capacitance (3750 F/g), low cost, and high chemical and thermal stabilities [100]. In alkaline electrolyte, the redox reactions of NiO and Ni(OH)₂ electrodes can be expressed as follows [101],

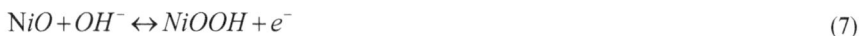

$$NiO + OH^- \leftrightarrow NiOOH + e^- \tag{7}$$

$$Ni(OH)_2 + OH^- \leftrightarrow NiOOH + H_2O + e^- \tag{8}$$

The electrochemical surface reactivity of NiO/ Ni(OH)$_2$ strongly depends on its crystallinity and the temperature can affect the crystallinity of NiO. Cheng et al., have reported that the maximum specific capacitance of 696 F/g can be reached for NiO annealed at 250°C, but the capacitance decreased significantly for 300°C annealed NiO electrode [102]. Nanostructured NiO materials can offer fast reaction kinetics due to large surface area, short diffusion, and transport pathways. There are various NiO nanostructures such as nanobelts, nanowires, nanorods, and nanoflowers which have been fabricated for supercapacitor electrodes. Wu et al. reported the preparation of porous NiO with macropores by electrophoresis and electrodeposition which have the larger specific capacitance (351 F/g) than bare NiO film [103]. The high porosity of this porous architecture can increase the ion transport into the electrolyte and hence the specific capacitance is increased. Different morphologies of the cubic NiO films were prepared by hydrothermal and electrodeposition methods on Indium Tin oxide (ITO) glass, the maximum specific capacitance has achieved 148 F/g at 100 mV/s [104]. The electrochemical behavior of NiO also depends on the preparation method. It is reported that the NiO with cubic structure, synthesized via chemical process displayed maximum specific capacitance of 167 F/g, whereas the specific capacitance of NiO synthesized by a sol-gel method was 250 F/g [105]. Ni(OH)$_2$ is a hexagonal layered structure and has two polymorphs such as α and β- Ni(OH)$_2$. The practical specific capacitance of Ni(OH)$_2$ is much higher than NiO. The specific capacitance of powdered Ni(OH)$_2$ materials usually found in the range from 500-600 F/g, while for Ni(OH)$_2$ films, the specific capacitance was 3000 F/g which is closer to its theoretical specific capacitance. Electrodeposited Ni(OH)$_2$ on Ni foam exhibited a very high specific capacitance of 3152 F/g at a current density of 4 A/g. The loose crystal of Ni(OH)$_2$ film, contributes to the easy insertion/extraction of ions into electrolyte during the electrochemical reaction [106]. There are two major problems with NiO electrode materials in supercapacitors: (a) Poor cycle performance and (b) high resistivity. The conductivity of NiO/Ni(OH)$_2$ can be improved by introducing carbon materials or cobalt ions into the NiO matrix. For example, Co-Ni/Co-Ni oxides exhibited a maximum specific capacitance of 331 F/g as well as high cyclic stability [107]

8.4   Cobalt oxides/hydroxides based supercapacitors

Cobalt oxide (Co$_3$O$_4$) has a cubic structure that exhibits excellent reversible redox behavior, high electrical conductivity, and long-term stability. Thus, it has been considered as an alternative electrode material for supercapacitors. The redox reactions in alkaline electrolyte can be described as follows [108],

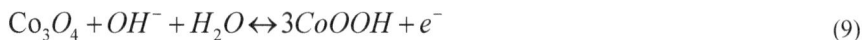

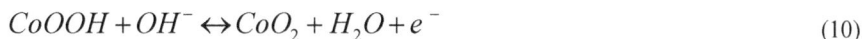

$$Co_3O_4 + OH^- + H_2O \leftrightarrow 3CoOOH + e^- \qquad (9)$$

$$CoOOH + OH^- \leftrightarrow CoO_2 + H_2O + e^- \qquad (10)$$

Co_3O_4 microspheres with a mesoporous crater-like structure which synthesized from mesoporous silica material as a template could provide a specific capacitance of 102 F/g with smaller inner resistance (0.4Ω) [109]. Xia et al. have synthesized a single crystalline Co_3O_4 nanowire array via the hydrothermal route and the prepared nanowires show specific capacitance of 754 F/g at a current density of 2 A/g with long cycle life. The one-dimensional structure of Co_3O_4 provides a fast ionic diffusion path [110]. Composite of Co_3O_4 can significantly enhance the supercapacitor performance. Co_3O_4 nanowire grown on CVD graphene has delivered a maximum specific capacitance of 1100 F/g at 10 A/g [111]

Due to layered structure and large interlayer spacing of $Co(OH)_2$ based materials. They provide a faster ion insertion/desertion rate than Co_3O_4. $Co(OH)_2$ has similar properties like $Ni(OH)_2$. The redox reaction of cobalt hydroxide can be shown as follows [112],

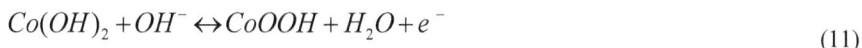

$$Co(OH)_2 + OH^- \leftrightarrow CoOOH + H_2O + e^- \qquad (11)$$

The theoretical specific capacitance of $Co(OH)_2$ is ~3700 F/g and it can reach a high value in practical applications. Mesoporous $Co(OH)_2$ deposited on Ni foam by electrodeposition method, delivered a maximum specific capacitance of 2646 F/g [113]. It should be noted that the specific capacitance of Co based materials is higher than RuO_2, due to the low potential window, which limits the materials in practical applications.

Table 3 Comparison of Ni, Co and V$_2$-oxidesbased Supercapacitors

Electrode	Synthesis method	Cs (F/g)	Scan rate/Current density	Potential window (V)	Electrolyte	Ref
Porous NiO film	ammonia-evaporation method	232	2 A/g	0 to 0.7	1M KOH	119

NiO	MOF derived	324	1 A/g	0.1 to 0.55	6M KOH	1120
Ni(OH)₂/Graphene	Two-step oxidation	1335	2.8 A/g	0 to 0.5	1M KOH	121
Co₃O₄/Co(OH)₂	Hydrothermal	1164	1.2 A/g	0 to 0.45	3M KOH	122
Co₃O₄/graphite	Co-precipitation	395.4	0.5 A/g	-0.3 to 0.3	6M KOH	123
CoNiO	Co-precipitation	156.3	0.5 A/g	1.0 to 4.0	1M KOH	124
V₂O₅/graphene@Ni	CVD + hydrothermal	1235	2 A/g	0 to 0.8	3 MKCl	125
Mo-V₂O₅	Thermal evaporation	175	1 mA/g	-0.1 to -0.6	1 M KCl	126
V₂O₅/CNT	Hydrothermal	553.33	5 mV/s	0 to 1.0	1 M Na₂SO₄	127

8.5 Other metal oxides/hydroxides

Other than RuO_2, NiO, MnO and Co_3O_4 electrodes, vanadium oxide, zinc oxide (ZnO), tin oxide (SnO_2) and iron oxides (Fe_2O_3 and Fe_3O_4) have been studied for supercapacitor electrode materials. Due to variable oxidation states, V_2O_5 can be used for high-performance supercapacitors. Lee et al., have reported the preparation of amorphous V_2O_5 by quenching V_2O_5 powder at 950°C and the prepared electrode exhibited a maximum specific capacitance of 350 F/g at 5 mV/s 1 M KCl electrolyte [114]. To improve the conductivity of V_2O_5, Jayalakshmi et al. combined CNT with V_2O_5 and demonstrated that the addition of CNT could increase the capacitance to almost three times compared to pure V_2O_5 material [115]. Amorphous SnO_2 nanostructures have been synthesized via the electrochemical deposition method and the electrodes showed excellent supercapacitor behavior with a specific capacitance of 285 F/g at a scan rate of 10mV/s [116]. Iron oxides such as Fe_2O_3 and Fe_3O_4 have relatively high electrical conductivities ~200 S cm^{-1} compared to other metal oxides. However, the low specific capacitance values of iron oxides limit their commercial applications. The incorporation of MWCNT into Fe_2O_3 nanospheres enhanced the energy density to 50Wh/kg at a specific power density of 1000 W/kg. But, due to the poor electron transport, the specific capacitance dropped at higher current density and scan rates [117]. Recently, a Fe_2O_3 film has been prepared via a hydrothermal method which exhibited a specific capacitance of 118.2 F/g in 1 M Na₂SO₄solution between -1 to 0.1 V potential window and 88.8% of capacitance retention achieved after 500 cycles of charging/discharging [118].

Conclusion

Since supercapacitors are one of the electrochemical energy devices, they play an important role in our society. In particular, supercapacitors have been used widely in hybrid power sources, backup power sources and burst-power generation in electronic devices. However, the limited energy density is one of the key challenges of supercapacitors, which restricts their wider applications in the energy storage field. The performance of the supercapacitors depends on the type of electrode material. We have reviewed the comparison of energy storage devices like supercapacitors, batteries, fuel cell and discussed the advantages and disadvantages of supercapacitors. Various transition metal oxides/hydroxides have been used for supercapacitor electrodes, due to their larger surface area, high conductivity, and good stability. The short diffusion path of ions and electrons of metal oxides/hydroxides leads to high specific capacitance and good rate capability. Thus, the developing nanostructured metal oxides/hydroxides can efficiently enhance the supercapacitor performance. However, the specific capacitance of the metal oxides/hydroxides is high but the poor electrical conductivity makes the limited usage in practical applications. Introducing of carbon-based materials such as graphene, CNT, etc. and metal doping into the transition metal oxides/hydroxides is used to enhance the electrochemical behavior of the electrodes. Hence, high performance of supercapacitors could be achieved by metal oxide/hydroxides electrodes with composite materials.

Acknowledgement

This work was supported by National Natural Science Foundation of China (Project No. 51505209) and Shenzhen Science and Technology Innovation Committee (Projects No. JCYJ20170412154426330). Fei Wang is also supported by Guangdong Natural Science Funds for Distinguished Young Scholar (Project No.: 2016A030306042). This chapter is partly supported by the State Key Laboratories of Transducer Technology, Shanghai, China.

References

[1]   B. E. Conway, Electrochemical Supercapacitors: Scientific fundamentals and technological applications, Kluwer Academic/Plenum, New York, 1999. https://doi.org/10.1007/978-1-4757-3058-6

[2]   C. Zhong, Y. Deng, W. Hu, J. Qiao, L. Zheng, J. Zhang, A review of electrolyte materials and compositions for electrochemical supercapacitors, Chem. Soc. Rev. 44 (2015) 7484-7539. https://doi.org/10.1039/C5CS00303B

[3]   R. Ramachandran, M. Saranya, A. Grace, F. Wang, MnS nanocomposites based on doped graphene: simple synthesis by a wet chemical route and improved electrochemical properties as an electrode material for supercapacitors, RSC Adv. 7 (2017) 2249-257. https://doi.org/10.1039/C6RA25457H

[4]   Z. Zhou and X. F. Wu, Graphene-beaded carbon nanofibers for use in supercapacitor electrodes: Synthesis and electrochemical characterization, J. Power Sources 222 (2013) 410–416. https://doi.org/10.1016/j.jpowsour.2012.09.004

[5]   C. Meng, O.Z. Gall, P.P. Irazoqui, A flexible super-capacitive solid-state power supply for miniature implantable medical devices, Biomedical Microdevices, 15 (2013) 973-983. https://doi.org/10.1007/s10544-013-9789-1

[6]   A. Schneuwly, R. Gallay, Properties and applications of supercapacitors from the state-of-the-art to future trends, Proceedings PCIM (2000) 1-10.

[7]   A.N. Grace, R. Ramachandran, Advanced materials for supercapacitors, eco-friendly nano-hybrid materials for advanced engineering applications. (Apple Academic Press) (2016) 99 -128.

[8]   A. Pandolfo, A. Hollenkamp, Carbon properties and their role in supercapacitors, J Power Sources 157 (2016) 11–27. https://doi.org/10.1016/j.jpowsour.2006.02.065

[9]   M. Jayalakshmi, K. Balasubramanian, Single step solution combustion synthesis of zno/carbon composite and its electrochemical characterization for supercapacitor application, Int. J. Electrochem. Sci. 3 (2008) 96-103.

[10]   L. Zhang, X. S. Zhao, Carbon-based materials as supercapacitor electrodes, Chem. Soc. Rev. 38 (2009) 2520-2531. https://doi.org/10.1039/b813846j

[11]   R. Ramachandran, M. Saranya, V. Velmurugan, B.P.C. Ragupathy, S.K. Jeong, A.N. Grace, Effect of reducing agent on graphene synthesis and its influence on charge storage towards supercapacitor applications, Appl. Energy 153 (2015) 22–31 https://doi.org/10.1016/j.apenergy.2015.02.091

[12]   R. Ramachandran, M. Saranya, P. Kollu, B.P.C. Ragupathy, S.K. Jeong, A.N. Grace, Solvothermal synthesis of Zinc sulfide decorated Graphene (ZnS/G) nanocomposites for novel Supercapacitor electrodes, Electrochimica Acta 178 (2015) 647–657. https://doi.org/10.1016/j.electacta.2015.08.010

[13]   A. Burke, R&D considerations for the performance and application of electrochemical capacitors, Electrochem. Acta 53 (2007) 1083-1091. https://doi.org/10.1016/j.electacta.2007.01.011

[14] F. Zhang, T. Zhang, X. Yang, L. Zhang, K. Leng, Y. Huang, Y. Chen, A high-performance supercapacitor-battery hybrid energy storage device based on graphene-enhanced electrode materials with ultrahigh energy density, Energy Environ. Sci. 6 (2013) 1623-1632. https://doi.org/10.1039/c3ee40509e

[15] P. Simon, K. Naoi, new materials and new configurations for advanced electrochemical capacitors, Electrochem. Soc. (2008) 34-38.

[16] Z.S. Iro, C. Subramani, S.S. Dash, A brief review on electrode materials for supercapacitor, Int. J. Electrochem. Sci. 11 (2016) 10628 – 10643. https://doi.org/10.20964/2016.12.50

[17] V. Augustyn, J. Come, M.A. Lowe, J.W. Kim, P.L. Taberna, S.H. Tolbert, H.D. Abruna, P. Simon, B. Dunn, High-rate electrochemical energy storage through Li+ intercalation pseudocapacitance, Nat. Mater. 12 (2013) 518–522. https://doi.org/10.1038/nmat3601

[18] C. Daniel, J. O. Besenhard, Handbook of battery materials, Wiley-VCH Verlag GmbH & Co. KGaA, 2nd edn, 2011 ch. 5, print-ISBN: 9783527326952, online-ISBN: 9783527637188.

[19] Y. Wang, Y. Song, Y. Xia, Electrochemical capacitors: mechanism, materials,systems, characterization and applications, Chem. Soc. Rev. 45 (2016) 5925-5950. https://doi.org/10.1039/C5CS00580A

[20] H. Lindstrom, S. Sodergren, A. Solbrand, H. Rensmo, J. Hjelm, A. Hagfeldt, S. E. Lindquist, Lithium intercalation in nanoporous anatase TiO2 studied with XPS, J. Phys. Chem. B 101 (1997) 7717–7722

[21] A. J. Bard, L. R. Faulkner, Electrochemical Methods Fundamentals and Applications, John Wiley, Inc., New York, 2nd edn, 6 (2001) 233, 235.

[22] T. Brezesinski, J. Wang, S. H. Tolbert, B. Dunn, Ordered mesoporous α-MoO3 with iso-oriented nanocrystalline walls for thin-film pseudocapacitors, Nat. Mater. 9 (2010) 146–151. https://doi.org/10.1038/nmat2612

[23] M. Uzunoglu, M. S. Alam, Modeling and Analysis of an FC/UC Hybrid Vehicular Power System Using a Novel-Wavelet-Based Load Sharing Algorithm IEEE Trans. Energy Convers., 23 (2008) 263-272. https://doi.org/10.1109/TEC.2007.908366

[24] Y. Zhang, H. Feng, X. Wu, L. Wang, A. Zhang, T. Xia, H. Dong, X. Li, L. Zhang, Progress of electrochemical capacitor electrode materials: A review, Int. J.

Hydrogen Energy 34 (2009) 4889-4899.
https://doi.org/10.1016/j.ijhydene.2009.04.005

[25] A. Burke, Ultracapacitors: why, how, and where is the technology J. Power
      Sources 91 (2000) 37-50. https://doi.org/10.1016/S0378-7753(00)00485-7

[26] J. R. Miller, A. F. Burke, Electrochemical capacitors: Challenges and opportunities
      for real-world applications, Electrochem. Soc. Interface, 17 (2008) 53-57.

[27] H. Chen, T. N. Cong, W. Yang, C. Tan, Y. Li, Y. Ding, Progress in electrical
      energy storage system: A critical review, Prog. Nat. Sci. 19 (2009) 291-312.
      https://doi.org/10.1016/j.pnsc.2008.07.014

[28] A. S. Arico, P. Bruce, B. Scrosati, J. Tarascon, W. V. Chalkwijk, Nanostructured
      materials for advanced energy conversion and storage devices, Nat. Mater. 4
      (2005) 366-377. https://doi.org/10.1038/nmat1368

[29] J. Zheng, P. Cygan, T. Jow, Hydrous ruthenium oxide as an electrode material for
      electrochemical capacitors. J. Electrochem. Soc. 142 (1995) 2699–2703.
      https://doi.org/10.1149/1.2050077

[30] S.L. Brock, N. Duan, Z.R. Tian, O. Giraldo, H. Zhou, S.L. Suib, A review of
      porous manganese oxide materials. Chem. Mater. 10 (1998) 2619–2628.
      https://doi.org/10.1021/cm980227h

[31] T. Brousse, M. Toupin, R. Dugas, L. Athouel, O.Crosnier, D. Belanger,
      Crystalline MnO2 as possible alternatives to amorphous compounds in
      electrochemical supercapacitors. J. Electrochem. Soc. 153 (2006) A2171-A2180.
      https://doi.org/10.1149/1.2352197

[32] G. Wang, L. Zhang, J. Zhang, A review of electrode materials for electrochemical
      supercapacitors. Chem. Soc. Rev. 41 (2012) 797–828.
      https://doi.org/10.1039/C1CS15060J

[33] C. Liu, F. Li, L.P. Ma, H.M. Cheng, Advanced materials for energy storage. Adv.
      Mater. 22 (2010) E28–E62. https://doi.org/10.1002/adma.200903328

[34] W. Sugimoto, K. Yokoshima, Y. Murakami, Y. Takasu, Charge storage
      mechanism of nanostructured anhydrous and hydrous ruthenium-based oxides,
      Electrochim. Acta 52 (2006) 1742–1748.
      https://doi.org/10.1016/j.electacta.2006.02.054

[35] Y. Gao, S. Chen, D. Cao, G. Wang, J. Yin, Electrochemical capacitance of Co3O4
      nanowire arrays supported on nickel foam. J. Power Sources 195 (2010) 1757–
      1760. https://doi.org/10.1016/j.jpowsour.2009.09.048

[36] Z. Lu, Z. Chen, J. Liu, X. Sun, Stable ultra high specific capacitance of NiO nanorod arrays. Nano Res. 4 (2011) 658–665. https://doi.org/10.1007/s12274-011-0121-1

[37] H. Du, L. Jiao, Q. Wang, J. Yang, L. Guo, Y. Si, Y. Wang, H. Yuan, Facile carbonaceous microsphere templated synthesis of Co3O4 hollow spheres and their electrochemical performance in supercapacitors. Nano Res. 6 (2013) 87–98. https://doi.org/10.1007/s12274-012-0283-5

[38] Z. Qi, A. Younis, D. Chu, S. Li, A facile and template-free on e-pot synthesis of Mn3O4 nanostructures as electrochemical supercapacitors. Nano-Micro Lett. 8 (2016) 165–173. https://doi.org/10.1007/s40820-015-0074-0

[39] J. Yang, L. Lian, H. Ruan, F. Xie, M. Wei, Nanostructured porous MnO2 on Ni foam substrate with a high mass loading via a CV electrodeposition route for supercapacitor application. Electrochimica Acta 136 (2014) 189–194. https://doi.org/10.1016/j.electacta.2014.05.074

[40] I. H. Kim, K. B. Kim, Electrochemical characterization of hydrous ruthenium oxide thin-film electrodes for electrochemical capacitor applications, J. Electrochem. Soc. 153 (2006) A383–A389. https://doi.org/10.1149/1.2147406

[41] P. Simon, Y. Gogotsi, Materials for electrochemical capacitors, Nat. Mater. 7 (2008) 845–854. https://doi.org/10.1038/nmat2297

[42] N. Wu, S. Kuo, M. Lee, Preparation and optimization of RuO2-impregnated SnO2 xerogel supercapacitor, J. Power Sources 104 (2002) 62-65. https://doi.org/10.1016/S0378-7753(01)00873-4

[43] J. K. Lee, H. M. Pathan, K. D. Jung and O. S. Joo, Electrochemical capacitance of nanocomposite films formed by loading carbon nanotubes with ruthenium oxide, J. Power Sources 159 (2006) 1527-1531. https://doi.org/10.1016/j.jpowsour.2005.11.063

[44] K. E. Swider, C. I. Merzbacher, P. L. Hagans, D. R. Rolison, Synthesis of ruthenium dioxide−titanium dioxide aerogels: redistribution of electrical properties on the nanoscale, Chem. Mater.9 (1997) 1248–1255. https://doi.org/10.1021/cm960622c

[45] T. C. Liu, W. G. Pell, B. E. Conway, Self-discharge and potential recovery phenomena at thermally and electrochemically prepared RuO2 supercapacitor electrodes, Electrochimica Acta 42 (1997) 3541-3552. https://doi.org/10.1016/S0013-4686(97)81190-5

[46]   F. Shi, L. Li, X.L. Wang, C. Gu, J.P. Tu, Metal oxide/hydroxide-based materials for supercapacitors, RSC Adv. 4 (2014) 41910–41921. https://doi.org/10.1039/C4RA06136E

[47]   J.P. Zheng, P.J. Cygan, T.R. Jow, Hydrous ruthenium oxide as an electrode material for electrochemical capacitors, J. Electrochem. Soc. 142 (1995) 2699-2703. https://doi.org/10.1149/1.2050077

[48]   X. M. Liu, X. G. Zhang, NiO-based composite electrode with RuO2 for electrochemical capacitors, Electrochimica Acta 49 (2004) 229–232. https://doi.org/10.1016/j.electacta.2003.08.005

[49]   F. Pico, J. Ibanez, T. A. Centeno, C. Pecharroman, R. M. Rojas, J. M. Amarilla, J. M. Rojo, RuO2·xH2O/NiO composites as electrodes for electrochemical capacitors: Effect of the RuO2 content and the thermal treatment on the specific capacitance, Electrochimica Acta 51 (2006) 4693–4700. https://doi.org/10.1016/j.electacta.2005.12.040

[50]   C. J. Zhang, H. H. Zhou, X. Q. Yu, D. Shan, T. T. Ye, Z. Y. Huang, Y. F. Kuang, Synthesis of RuO2 decorated quasi graphene nanosheets and their application in supercapacitors, RSC Adv. 4 (2014) 11197–11205. https://doi.org/10.1039/c3ra47641c

[51]   W. Sugimoto, T. Shibutani, Y. Murakami, Y. Takasu, Charge storage capabilities of rutile-type ruo2-vo2 solid solution for electrochemical supercapacitors, Electrochem. Solid-State Lett. 5 (2002) A170-A172. https://doi.org/10.1149/1.1483155

[52]   C. Yuan, B. Gao, X. Zhang, Electrochemical capacitance of NiO/Ru0.35V0.65O2 asymmetric electrochemical capacitor, J. Power Sources 173 (2007) 606-612. https://doi.org/10.1016/j.jpowsour.2007.04.034

[53]   Y. Wang, X. Zhang, Preparation and electrochemical capacitance of RuO2/TiO2 nanotubes composites, Electrochimica Acta 49 (2004) 1957-1962. https://doi.org/10.1016/j.electacta.2003.12.023

[54]   G. H. Deng, X. Xiao, J. H. Chen, X. B. Zeng, D. L. He, Y. F. Kuang, A new method to prepare RuO2·xH2O/carbon nanotube composite for electrochemical capacitors, Carbon 43 (2005) 1557-1563. https://doi.org/10.1016/j.carbon.2004.12.031

[55] Y. Yang, Y. Liang, Y. Zhang, Z. Zhang, Z. Li, Z. Hu, Three-dimensional graphene hydrogel supported ultrafine RuO2 nanoparticles for supercapacitor electrodes, New. J. Chem. 39 (2015) 4035-4040. https://doi.org/10.1039/C5NJ00062A

[56] V.K.A. Muniraj, C.K. Kamaja, M.V. Shelke, RuO2.nH2O nanoparticles anchored on carbon nano-onions: An efficient electrode for solid state flexible electrochemical supercapacitors, ACS Sustainable Chem. Eng. 4 (2016) 2528-2534. https://doi.org/10.1021/acssuschemeng.5b01627

[57] C. Lin, J. A. Ritter, B. N. Popov, Development of carbon-metal oxide supercapacitors from sol-gel derived carbon-ruthenium xerogels, J. Electrochem. Soc. 46 (1999) 3155-3160. https://doi.org/10.1149/1.1392448

[58] M. Min, K. Machida, J. H. Jang, K. Naoi, Hydrous RuO2/carbon black nanocomposites with 3d porous structure by novel incipient wetness methodfor supercapacitors, J. Electrochem. Soc. 153 (2006) A334-A338. https://doi.org/10.1149/1.2140677

[59] M. Ramani, B. S. Haran, R. E. White, B. N. Popov, Synthesis and characterization of hydrous ruthenium oxide-carbon supercapacitors, J. Electrochem. Soc. 148 (2001) A374-A380. https://doi.org/10.1149/1.1357172

[60] V. Barranco, F. Pico, J. Iban ez, M. A. Lillo-Rodenas, A. Linares-Solano, M. Kimura, A. Oya, R. M. Rojas, J. M. Amarilla, J. M. Rojo, Amorphous carbon nanofibres inducing high specific capacitance of deposited hydrous ruthenium oxide, Electrochimica Acta 54 (2009) 7452-7457. https://doi.org/10.1016/j.electacta.2009.07.080

[61] J. F. Zang, S. J. Bao, C. M. Li, H. J. Bian, X. Q. Cui, Q. L. Bao, C. Q. Sun, J. Guo, K. R. Lian, Well-aligned cone-shaped nanostructure of polypyrrole/RuO2 and its electrochemical supercapacitor, J. Phys. Chem. C 112 (2008) 14843-14847. https://doi.org/10.1021/jp8049558

[62] R. Liu, J. Duay, T. Lane, S. B. Lee, Synthesis and characterization of RuO2/poly(3,4-ethylenedioxythiophene) composite nanotubes for supercapacitors, Phys. Chem. Chem. Phys. 12 (2010) 4309-4316. https://doi.org/10.1039/b918589p

[63] Y. Zhu, X. Ji, C. Pan, Q. Sun, W. Song, L. Fang, Q. Chena, C. E. Banks, A carbon quantum dot decorated RuO2 network: outstanding supercapacitances under ultrafast charge and discharge, Energy Environ. Sci. 6 (2013) 3665–3675. https://doi.org/10.1039/c3ee41776j

[64]   F. Z. Amir, V. H. Pham, J. H. Dickerson, Facile synthesis of ultra-small ruthenium oxide nanoparticles anchored on reduced graphene oxide nanosheets for high-performance supercapacitors, RSC Adv. 5 (2015) 67638–67645. https://doi.org/10.1039/C5RA11772K

[65]   B. Shen, X. Zhang, R. Guo, J. Lang, J. Chen, X. Yan, Carbon encapsulated RuO2 nano-dots anchoring on graphene as an electrode for asymmetric supercapacitors with ultralong cycle life in an ionic liquid electrolyte, J. Mater. Chem. A 4 (2016) 8180–8189. https://doi.org/10.1039/C6TA02473D

[66]   J. C. Chou, Y. L.Chen, M. H. Yang, Y. Z. Chen, C. C. Lai, H. T. Chiu, C. Y. Lee, Y. L. Chueh, J. Y. Gan, RuO2/MnO2 core–shell nanorods for supercapacitors, J. Mater. Chem. A 1 (2013) 8753–8758. https://doi.org/10.1039/c3ta11027c

[67]   D. Gong, J. Zhu, B. Lu, RuO2@Co3O4 heterogeneous nanofibers: A highperformance electrode material for supercapacitors, RSC Adv. 6 (2016) 49173–49178. https://doi.org/10.1039/C6RA04884F

[68]   C. Yuan, L.Chen, B. Gao, L. Su, X. Zhang, Synthesis and utilization of RuO2.xH2O nanodots well dispersed on poly(sodium 4-styrene sulfonate) functionalized multi-walled carbon nanotubes for supercapacitors, J. Mater. Chem. 19 (2009) 246–252. https://doi.org/10.1039/B811548F

[69]   P. R. Deshmukh, R. N. Bulakhe, S. N. Pusawale, S. D. Sartale, C. D. Lokhand, Polyaniline–RuO2 composite for high performance supercapacitors: chemical synthesis and properties, RSC Adv. 5 (2015) 28687–28695. https://doi.org/10.1039/C4RA16969G

[70]   J. K. Chang, M. T. Lee, W. T. Tsai, Amorphous MnO2 supported on 3D-Ni nanodendrites for large areal capacitance supercapacitors, J. Power Sources 160 (2007) 590. https://doi.org/10.1016/j.jpowsour.2007.01.036

[71]   C. Wei, H. Pang, B. Zhang, Q. Lu, S. Liang, F. Gao, Two-Dimensional β-MnO2 nanowire network with enhanced electrochemical capacitance, Sci. Rep. 3 (2013) 2193-2197. https://doi.org/10.1038/srep02193

[72]   A. Sarkar, A. K. Satpati, V. Kumar, S. Kumar, Sol-gel synthesis of manganese oxide films and their predominant electrochemical properties, Electrochim. Acta 167 (2015) 126–131. [73] X. Zhang, X. Sun, H. Zhang, D. Zhang, Y. Ma, Microwave-assisted reflux rapid synthesis of MnO2 nanostructures and their application in supercapacitors, Electrochimica Acta 87 (2013) 637–644. https://doi.org/10.1016/j.electacta.2012.10.022

[74]   A. Bello, O. Fashedemi, M. Fabiane, J. Lekitima, K. Ozoemena, N. Manyala, Microwave assisted synthesis of MnO2 on nickel foam-graphene for electrochemical capacitor, Electrochimica Acta 114 (2013) 48–53. https://doi.org/10.1016/j.electacta.2013.09.134

[75]   B. Ming, J. Li, F. Kang, G. Pang, Y. Zhang, L. Chen, J. Xu, X. Wang, Microwave–hydrothermal synthesis of birnessite-type MnO2 nanospheres as supercapacitor electrode materials, J. Power Sources 198 (2012) 428–431. https://doi.org/10.1016/j.jpowsour.2011.10.003

[76]   S. C. Pang, M. A. Anderson, T. W. Chapman, Preparation of Manganese dioxide for electrochemical supercapacitors, J. Electrochem. Soc. 147 (2000) 444-450. https://doi.org/10.1149/1.1393216

[77]   B. Babakhani, D. G. Ivey, Improved capacitive behavior of electrochemically synthesized Mn oxide/PEDOT electrodes utilized as electrochemical capacitors, Electrochimica Acta 55 (2010) 4014–4024. https://doi.org/10.1016/j.electacta.2010.02.030

[78]   H. Kim, B. N. Popov, Synthesis and characterization of mno2-based mixed oxides as supercapacitors, J. Electrochem. Soc. 150 (2003) D56-D62. https://doi.org/10.1149/1.1541675

[79]   L. Li, Z. Y. Qin, L. F. Wang, H. J. Liu and M. F. Zhu, Anchoring alpha-manganese oxide nanocrystallites on multi-walled carbon nanotubes as electrode materials for supercapacitor, J. Nanopart. Res. 12 (2010) 2349-2353. https://doi.org/10.1007/s11051-010-9980-8

[80]   J. Yan, Z. Fan, T. Wei, W. Qian, M. Zhang, F. Wei, Fast and reversible surface redox reaction of graphene–MnO2 composites as supercapacitor electrodes, Carbon 48 (2010) 3825-3833. https://doi.org/10.1016/j.carbon.2010.06.047

[81]   M. Liu, W. W. Tjiu, J. Pan, C. Zhang, W. Gao, T. Liu, One-step synthesis of graphene nanoribbon–MnO2 hybrids and their all-solid-state asymmetric supercapacitors, Nanoscale 6 (2014) 4233–4242. https://doi.org/10.1039/c3nr06650a

[82]   M. Liu, L. Gan, W. Xiong, Z. Xu, D. Zhu and L. Chen, Development of MnO2/porous carbon microspheres with a partially graphitic structure for high performance supercapacitor electrodes, J. Mater. Chem. A 2 (2014) 2555–2562. https://doi.org/10.1039/C3TA14445C

[83] J. G. Wang, Y. Yang, Z.-H. Huang, F. Kang, A high-performance asymmetric supercapacitor based on carbon and carbon–MnO2 nanofiber electrodes, Carbon 61 (2013) 190–199 https://doi.org/10.1016/j.carbon.2013.04.084

[84] R. I. Jafri, A. K. Mishra, S. Ramaprabhu, Polyaniline–MnO2 nanotube hybrid nanocomposite as supercapacitor electrode material in acidic electrolyte, J. Mater. Chem., 21(2011) 17601–17605. https://doi.org/10.1039/c1jm13191e

[85] J. Han, L. Li, P. Fang, R. Guo, Ultrathin MnO2 nanorods on conducting polymer nanofibers as a new class of hierarchical nanostructures for high-performance supercapacitors, J. Phys. Chem. C 116 (2012) 15900–15907. https://doi.org/10.1021/jp303324x

[86] R. Liu, J. Duay, S. B. Lee, Electrochemical formation mechanism for the controlled synthesis of heterogeneous MnO2/poly(3,4-ethylenedioxythiophene) nanowires, ACS Nano 5 (2011) 5608–5619. https://doi.org/10.1021/nn201106j

[87] A. Bahloul, B. Nessark, E. Briot, H. Groult, A. Mauger, K. Zaghi, C. M. Julien, Polypyrrole-covered MnO2 as electrode material for hybrid supercapacitor, J. Power Sources 240 (2013) 267–272. https://doi.org/10.1016/j.jpowsour.2013.04.013

[88] C. Wang, Y. Zhan, L. Wu, Y. Li, J. Liu, High-voltage and high-rate symmetric supercapacitor based on MnO2-polypyrrole hybrid nanofilm, Nanotechnol. 25 (2014) 305401. https://doi.org/10.1088/0957-4484/25/30/305401

[89] J. X. Zhu, W. H. Shi, N. Xiao, X. H. Rui, H. T. Tan, X. H. Lu, H. H. Hng, J. Ma, Q. Y. Yan, Oxidation-etching preparation of MnO2 tubular nanostructures for high-performance supercapacitors, ACS Appl. Mater. Interfaces 4 (2012) 2769–2774. https://doi.org/10.1021/am300388u

[90] S. Maiti, A. Pramanik, S. Mahanty, Influence of imidazolium-based ionic liquidelectrolytes on the performance of nanostructured MnO2 hollow spheres as electrochemical supercapacitor, RSC Adv. 5 (2015) 41617–41626. https://doi.org/10.1039/C5RA05514H

[91] N. Phattharasupakun, J. Wutthiprom, P. Chiochan, P. Suktha, M.Suksomboon, S. Kalasina, M. Sawangphruk, Turning conductive carbon nanospheres into nanosheets for high-performance supercapacitors of MnO2 nanorods, Chem. Commun. 52 (2016) 2585-2588. https://doi.org/10.1039/C5CC09648K

[92] H. Zhang, G. Cao, Z. Wang, Y. Yang, Z. Shi, Z. Gu, Growth of manganese oxide nanoflowers on vertically-aligned carbon nanotube arrays for high-

rateelectrochemical capacitive energy storage, Nano Lett. 8 (2008) 2664–2668. https://doi.org/10.1021/nl800925j

[93] K. Dai, L. Lu, C. Liang, J. Dai, Q. Liu, Y. Zhang, G. Zhu, Z. Liu, In situ assembly of MnO2 nanowires/graphene oxide nanosheets composite with high specific capacitance, Electrochimica Acta 116 (2014) 111–117. https://doi.org/10.1016/j.electacta.2013.11.036

[94] S. Zhu, H. Zhang, P. Chen, L.H. Nie, C.H. Li, S.K. Li, Self-assembled three-dimensional hierarchical graphene hybrid hydrogels with ultrathin ƐMnO2 nanobelts for high performance supercapacitors, J. Mater. Chem. A 3 (2015) 1540–1548. https://doi.org/10.1039/C4TA04921G

[95] L. Wang, Y. Zheng, S. Chen, Y. Ye, F. Xu, H. Tan, Z. Li, H. Hou, Y. Song, Three-dimensional kenaf stem-derived porous carbon/mno2 for high-performance supercapacitors, Electrochimica Acta 135 (2014) 380–387. https://doi.org/10.1016/j.electacta.2014.05.044

[96] H. Jiang, J. Ma, C. Li, Polyaniline–MnO2 coaxial nanofiber with hierarchical structure for high-performance supercapacitors, J. Mater. Chem. 22 (2012) 16939–16942. https://doi.org/10.1039/c2jm33249c

[97] P. Tang, L. Han, L. Zhang, S. Wang, W. Feng, G. Xu, L. Zhang, Controlled construction of hierarchical nanocomposites consisting of MnO2 and PEDOT for high-performance supercapacitor applications, Chem. Electro. Chem. 2 (2015) 949–957.

[98] A. Q. Zhang, Y. H. Xiao, L. Z. Lu, L. Z. Wang, F. Li, Polypyrrole/MnO2 composites and their enhanced electrochemical capacitance, J. Appl. Polym. Sci. 128 (2013) 1327–1331.

[99] H. Chen, S. Zhou, L. Wu, Porous nickel hydroxide−manganese dioxide-reduced graphene oxide ternary hybrid spheres as excellent supercapacitor electrode materials, ACS Appl. Mater. Interfaces 6 (2014) 8621–8630. https://doi.org/10.1021/am5014375

[100] M. S. Wu, Y. A. Huang, J. J. Jow, W. D. Yang, C. Y. Hsieh, H. M. Tsai, Int. J. Hydrogen Energy 33 (2008) 2921–2926. https://doi.org/10.1016/j.ijhydene.2008.04.012

[101] V. Srinivasan, J. W. Weidner, An electrochemical route for making porous nickel oxide electrochemical capacitors, J. Electrochem. Soc. 144 (1997) L210–L213. https://doi.org/10.1149/1.1837859

[102] J. Cheng, G. P. Cao, Y. S. Yang, Characterization of sol–gel-derived NiOx xerogels as supercapacitors, J. Power Sources 159 (2006) 734-741. https://doi.org/10.1016/j.jpowsour.2005.07.095

[103] M. S. Wu, M. J. Wang, J. J. Jow, Fabrication of porous nickel oxide film with open macropores by electrophoresis and electrodeposition for electrochemical capacitors, J. Power Sources 195 (2010) 3950–3955. https://doi.org/10.1016/j.jpowsour.2009.12.136

[104] Y. Y. Xi, D. Li, A. B. Djurišića, M. H. Xie, K. Y. K. Man, W. K. Chan, Hydrothermal synthesis vs electrodeposition for high specific capacitance nanostructured NiO films, Electrochem. Solid-State Lett. 11 (2008) D56-D59. https://doi.org/10.1149/1.2903345

[105] K. C. Liu, M. A. Anderson, Porous nickel oxide/nickel films for electrochemical capacitors, J. Electrochem. Soc. 143 (1996) 124-130. https://doi.org/10.1149/1.1836396

[106] G. W. Yang, C. L. Xu, H. L. Li, Electrodeposited nickel hydroxide on nickel foam with ultrahigh capacitance, Chem. Commun. (2008) 6537–6539. https://doi.org/10.1039/b815647f

[107] V. Gupta, T. Kawaguchi, N. Miura, Synthesis and electrochemical behavior of nanostructured cauliflower-shape Co–Ni/Co–Ni oxides composites, Mater. Res. Bull. 44 (2009) 202-206. https://doi.org/10.1016/j.materresbull.2008.04.020

[108] Y. Shan, L. Gao, Formation and characterization of multi-walled carbon nanotubes/Co3O4 nanocomposites for supercapacitors, Mater. Chem. Phys. 103 (2007) 206–210. https://doi.org/10.1016/j.matchemphys.2007.02.038

[109] L. Wang, X. Liu, X. Wang, X. Yang, L. Lu, Preparation and electrochemical properties of mesoporous Co3O4 crater-like microspheres as supercapacitor electrode materials, Curr. Appl. Phys. 10 (2010) 1422-1426. https://doi.org/10.1016/j.cap.2010.05.007

[110] X. H. Xia, J. P. Tu, Y. Q. Zhang, Y. J. Mai, X. L. Wang, C. D. Gu, X. B. Zhao, Self-supported hydrothermal synthesized hollow Co3O4 nanowire arrays with high supercapacitor capacitance, RSC Adv. (2012) 1835–1841.

[111] X. C. Dong, H. Xu, X. W. Wang, Y. X. Huang, M. B. Chan Park, H. Zhang, L.-H. Wang, W. Huang, P. Chen, 3D graphene–cobalt oxide electrode for high-performance supercapacitor and enzymeless glucose detection, ACS Nano 6 (2012) 3206–3213. https://doi.org/10.1021/nn300097q

[112] V. Gupta, T. Kusahara, H. Toyama, S. Gupta, N. Miura, Potentiostatically deposited nanostructured α-Co(OH)2: A high performance electrode material for redox-capacitors, Electrochem. Commun. 9 (2007) 2315–2319. https://doi.org/10.1016/j.elecom.2007.06.041

[113] X. H. Xia, Y. Q. Zhang, D. L. Chao, G. Cao, Y. J. Zhang, L. Li, X. Ge, I. M. Bacho, J. P. Tu, H. J. Fan, Solution synthesis of metal oxides for electrochemical energy storage applications, Nanoscale 6 (2014)5008–5048. https://doi.org/10.1039/C4NR00024B

[114] H. Y. Lee and J. B. Goodenough, Ideal supercapacitor behavior of amorphous v2o5·nh2o in potassium chloride (kcl) aqueous solution, Solid State Chem. 148 (1999) 81-84. https://doi.org/10.1006/jssc.1999.8367

[115] M. Jayalakshmi, M. Mohan Rao, N. Venugopal, K. B. Kim, Hydrothermal synthesis of SnO2–V2O5 mixed oxide and electrochemical screening of carbon nano-tubes (CNT), V2O5, V2O5–CNT, SnO2–V2O5–CNT electrodes for supercapacitor applications, J. Power Sources 166 (2007) 578-583. https://doi.org/10.1016/j.jpowsour.2006.11.025

[116] K. R. Prasad, N. Miura, Electrochemical synthesis and characterization of nanostructured tin oxide for electrochemical redox supercapacitors, Electrochem. Commun. 6 (2004) 849-852. https://doi.org/10.1016/j.elecom.2004.06.009

[117] X. Zhao, C. Johnston, P. S. Grant, A novel hybrid supercapacitor with a carbon nanotube cathode and an iron oxide/carbon nanotube composite anode, J. Mater. Chem. 19 (2009) 8755-8760. https://doi.org/10.1039/b909779a

[118] H. Zhu, D. Yang, L. Zhu, Hydrothermal growth and characterization of magnetite (Fe3O4) thin films, Surf. Coat. Technol. 201 (2007) 5870. https://doi.org/10.1016/j.surfcoat.2006.10.037

[119] Y. Q. Zhang, X. H. Xia, J. P. Tu, Y. J. Mai, S. J. Shi, X. L. Wang, C. D. Gu, Self-assembled synthesis of hierarchically porous NiO film and its application for electrochemical capacitors, J. Power Sources 199 (2012) 413–417. https://doi.org/10.1016/j.jpowsour.2011.10.065

[120] Y. Han, S. Zhang, N. Shen, D. Li, X. Li, MOF-derived porous NiO nanoparticle architecture for high-performance supercapacitors, Mater. Lett. 188 (2017) 1–4. https://doi.org/10.1016/j.matlet.2016.09.051

[121] H. L. Wang, H. S. Casalongue, Y. Y. Liang, H. J. Dai, Ni(OH)2 nanoplates grown on graphene as advanced electrochemical pseudocapacitor materials, J. Am. Chem. Soc. 132 (2010) 7472–7477. https://doi.org/10.1021/ja102267j

[122] H. Pang, X. Li, Q. Zhao, H. Xue, W.Y. Lai, Z. Hu, W. Huang, One-pot synthesis of heterogeneous Co3O4-Nanocube/Co(OH)2-nanosheet hybrids for high-performance flexible asymmetric all-solid-state supercapacitors, Nano Energy 35 (2017) 138-145. https://doi.org/10.1016/j.nanoen.2017.02.044

[123] M. Gopalakrishnan, G. Srikesh, A. Mohan, V. Arivazhagan, In-situ synthesis of Co3O4/graphite nanocomposite for high-performance supercapacitor electrode applications, Appl. Surf. Sci. 403 (2017) 578–583. https://doi.org/10.1016/j.apsusc.2017.01.092

[124] L. Y. Liu, X. Zhang, H. X. Li, B. Liu, J. W. Lang, L. B. Kong, X. B. Yan, Synthesis of Co–Ni oxide microflowers as a superior anode for hybrid supercapacitors with ultralong cycle life, Chinese Chem. Lett. 28 (2017) 206–212. https://doi.org/10.1016/j.cclet.2016.07.027

[125] N. V. Hoa, T. T. H. Quyen, N. H. Nghia, N.V. Hieu, J. J. Shim, In situ growth of flower-like V2O5 arrays on graphene@nickel foam as high-performance electrode for supercapacitors, J. Alloy. Compds. 702 (2017) 693-699. [126] N. G. Prakash, M. Dhananjaya, B. P. Reddy, K. S. Ganesh, A. L. Narayana, O. M. Hussain, Molybdenum doped V2O5 thin films electrodes for supercapacitors, Materials Today: Proceedings 3 (2016) 4076–4081.

[127] X. Wang, C. Zuo, L. Jia, Q. Liu, X. Guo, X. Jing, Synthesis of sandwich-like vanadium pentoxide/carbon nanotubes composites for high performance supercapacitor electrodes, J. Alloy. Compds. 708 (2017) 134-140. https://doi.org/10.1016/j.jallcom.2017.02.306

Chapter 3

Activated Carbon/Transition Metal Oxides Thin Films for Supercapacitors

F. F.M. Shaikh[1, 2,*], S. R. Jadakar[1], R. K. Kamat[2], H. M. Pathan[1]

[1]Advanced Physics Laboratory, Department of Physics, Savitribai Phule Pune University, Pune-411007, Maharashtra, India

[2]Department of Electronics, Shivaji University, Kolhapur-416004, Maharashtra, India

fouziashaikh18@gmail.com

Abstract

A supercapacitor is considered as the device capable of alleviating the current energy crisis. These devices are classified as Electric Double Layer Capacitor (EDLCs), pseudo-capacitor or hybrid supercapacitor. Hybrid devices cover larger surface area, higher specific capacitance, stability, better electrical conductivity, etc. This chapter presents a brief analysis of different metal oxide/hydroxide along with activated carbon as a composite material or as an anode or cathode material for supercapacitors.

Keywords

Energy Storage Device, Supercapacitor, Electric Double Layer Capacitor (EDLCs), Pseudo-capacitor, Hybrid Supercapacitor, Activated Carbon, Transition Metal Oxide

Contents

1. Introduction

The drastic change in weather has made the whole world to focus on the issue of global warming that concludes in utilizing the renewable and sustainable resources. This has led to tremendous pressure in storing electric energy devices such as supercapacitor [1]. Electrochemical capacitors stand out when compared to batteries and capacitors as they are highly reversible and show higher power densities [2-3]. With power density more than 1 k W kg^{-1} and better life cycle, the supercapacitor can be charged and discharged quickly [4]. Harvesting and storing energy and the minimal utility of fuels obtained from hydrocarbon are the key factors of supercapacitors. Recently, consumer electronics, industrial energy and power management systems, memory backup systems, etc. are demanding supercapacitors [5]. A supercapacitor is safe and can be trusted. It is applied in emergency doors of Airbus A380. It is also used in hybrid electric vehicle along with batteries as it can store energy quickly on the application of brakes [6]. However, the supercapacitor is governed by the capacitance of the material and cell voltage. These two properties limit the performance of the device [7].

There are different classes of supercapacitors. The first one is Electric Double Layer Capacitor (EDLCs), which was first introduced by Von Helmholtz [8] and further modified by Gouy and Chapman [9-10]. However, this model was an extravagance. Finally, Stern combines Helmholtz as well as Gouy and Chapman models with the introduction of a compact layer also known as Stern layer [11]. The behavior of EDLCs is determined by the type of electrolyte, the solvent of electrolyte, the electric field across the electrode and the chemical relation between the ions and the surface of electrode [6]. EDLC has seen great advancement due to the development in the pore size i.e. less than 1 nm of carbon [12-13]. Moreover, the microporous structure has been studied theoretically for describing the surface area for better capacitive effect and model of electrolyte absorbed into the porous structure [14-15]. This type of supercapacitor shows non-Faradaic behavior, possesses stability but offers low specific capacitance as well as energy density [7]. The traditional equation used to express the relationship between specific area of electrode and distance between the two electrodes is given by

$$C = \frac{\varepsilon_o \varepsilon_r}{d} A \qquad\qquad (1)$$

Where ε_r is the electrolyte dielectric constant, ε_o is the permittivity of a vacuum, A is the specific surface area of the electrode accessible to the electrolyte ions, and d is the effective thickness of the EDL (the Debye length) [6, 16]. Many reports suggest that the pore size determines or can amend the specific capacitance [17-18]. EDLCs applications are expanding in batteries and another storage cell. Some of these applications are electric bus (China) charges in 1 min, hybrid energy-efficient forklift (Still, Germany), wind turbines, cranes (Gottwald, Germany) and electric ferry (STX Europe, South Korea), etc. It can be also used as one of the electrodes in organic asymmetric capacitors, aqueous asymmetric capacitor, organic EDLCs, etc. There has been an increase in sales of EDLCs and decrease in manufacturing cost [19]. Three different types of electrolytes are used namely aqueous, organic and ionic. Ionic electrolyte makes EDLCs less inflammable as compared to batteries and can be used in many applications [20-22].

The pseudo-capacitor which is based on metal oxide and hydroxide shows redox reaction and Faradaic process possesses low stability but offers high specific capacitance and energy density [23].

$$C = \frac{dq}{dv} \qquad\qquad (2)$$

Where C is pseudocapacitance calculated by a change in charge acceptance due to change in potential [6, 24]. Metal oxides are well-known candidates in terms of cost-effectiveness, environment-friendly and large specific capacitance [25-26]. Nevertheless, pseudocapacitive materials have shown remarkable progress when compared to carbon-based materials [27]. Pseudosupercapacitor materials include metal oxides such as ruthenium oxide, nickel oxide, magnesium oxide, tin oxide, cobalt oxide, titanium oxide, zinc oxide, copper oxide, vanadium oxide, tungsten oxide, ferrite oxide, etc [28-40]. However, the poor electronic conductivity of metal oxides limits their use in high-performance supercapacitors. One of the other challenging issues is to tackle their capacity decay with cycling, arising from large volume expansion during the charge uptake/ release process.

Covering it with carbon has been reported as the solution to make the best use of transition metal oxides for increasing the conductivity [41-42]. In recent times, researchers are more focused on transition metal oxide and activated carbon [43-45] as the combo provides supreme properties which are not only limited to distinct specific capacitance but also many other chemical and electrical properties [46-47]. Change in

Faradic reactions results in large power density and stability of the material on combining carbon with metal oxide [48-49]. Their surface area can be controlled, specific capacitance is higher than carbon electrode alone, its conductivity can be adjusted, its stability is higher than metal oxide electrode and it's cost effective.

2.   Activated carbon

The use of carbon as electrode material for supercapacitors is restricted. The surface area in case of carbons is not that high so as to make them porous and because of this, they offer low specific capacitance. The disorganized residue blocks the pores. The process of improving the carbon electrode by increasing surface area and also porosity is called activation [53]. Activated carbons (ACs) have been used long back in Egypt in 155- BC [54] and are also being used in the US since 1913 in industries [56]. Activation carbonizes precursor in two ways either by chemical activation or by thermal activation. Thermal activation (the process of keeping only carbon material and developing porosity) is carried out at a temperature above 700 °C in presence of an oxidizing agent such as steam, air, etc. Due to this, the porosity and the surface area increases and can be controlled by carbon burn-off which is decided by the duration of activation and temperature. This process, however, decreases the strength of carbon by lowering the pore widening and low density [55, 57]. In the case of chemical activation, dehydration process takes place by using chemical agents such as zinc chloride, potassium hydroxide, phosphoric acid, etc. at relatively lower temperature i.e. below 700 °C [58, 5, 8-11, 16-18, 24, 59-66]. A.T. Mohd Din et al., have reported a combination of these two processes i.e. physicochemical method. In this method, the temperature was selected to be above 300 °C and oxidizing and chemical agents both were introduced [67]. Fig. 1 encapsulates the different synthesis methods of activated carbon.

Better supercapacitors in terms of power and energy density can be achieved by using activated carbons. In 1981, a supercapacitor with 36.2 W h kg^{-1}, 11.1 k W kg^{-1} and 36.5 F g^{-1} specific energy, specific power and specific capacitance, respectively was made of commercially available AC electrodes [68]. Further research revealed the connection between surface area and specific capacitance and many reports mention this relationship until the specific capacitance reached 270 F g^{-1} [69-73]. ACs production through agricultural materials has become a new field of research [74]. ACs were also synthesized by different natural materials like firewood, coconut shell, coal, phenolic resin, etc. [70, 75]. Other precursors are petroleum coke, coal, pitch, nut shells, wood, peat, starch, lignite, sucrose, leaves, corn grain, coffee grounds and straw [58, 76-86].

The literature review shows an easy and cost-effective method to synthesize ACs by using biomass [61-66]. These raw materials having an abundant amount of carbon are easily available in nature. Thus, activated carbons (ACs) offering porous structure, higher surface area, economic and commercial can be produced on a large scale [87-89]. The main drawback of ACs is that the pore size varies from micro to macro size [87]. But this can be overcome by utilizing nitrogen so as to tune the pore size for the effective functionality of the surface area [88-91]. The biggest advantage of ACs over CNT is that they are cost-effective and can be easily used commercially [92]. ACs show a surface area of around 3500 m^2 g^{-1}. ACs can be classified as powder, films and monoliths and fibers [71,93-98].

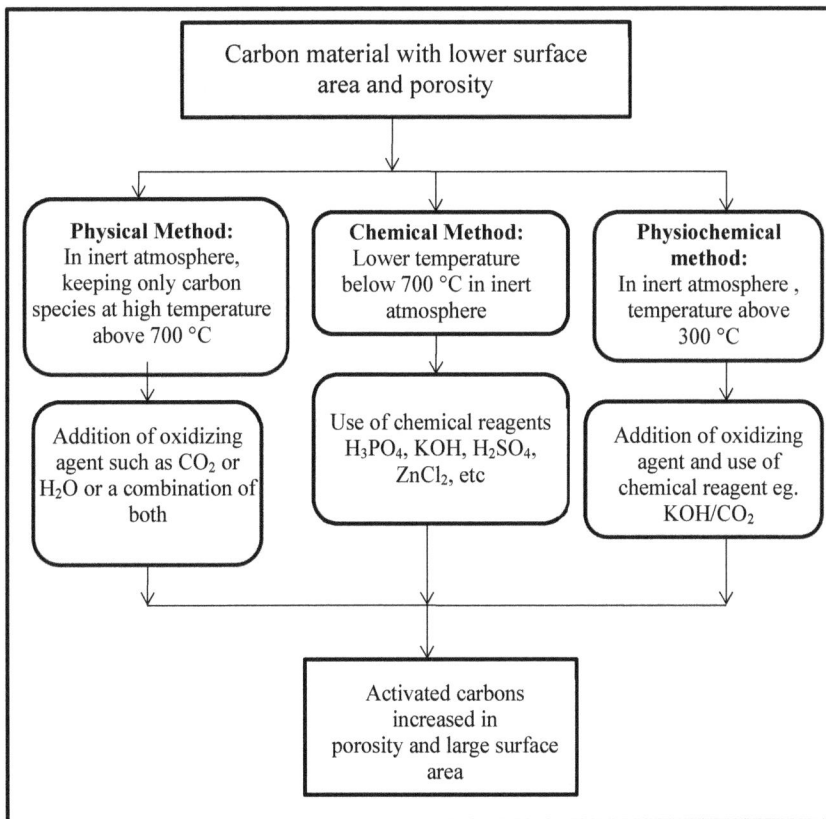

Fig. 1. Summary of the different synthesis methods of activated carbon.

Table 1 Brief description of the supercapacitors made from ACs.

Source of Carbon	Activation technique	Specific Capacitance achieved [F g⁻¹]	Ref.
Naturally obtained carbon from burned plant stocks	Chemically activated using KOH	160 to 180	[112]
Utilizing corncobs	KOH followed by heat treatment at different temperatures under N_2 flow	185	[113]
Commercially available activated carbon	-	185	[114]
Waste tea	H_3PO_4 AC-H K_2CO_3 AC-K	AC-H 123 AC-K 203	[115]
Spanish anthracite	Three different electrodes chemically activated by KOH, KOH with heat treatment and Commercially available	16 to 40	[50]
Commercially available activated carbon	-	150	[99]
Cotton stalks	Chemically activated by phosphoric acid	114	[116]
Coconut shell	Addition of functional group	154	[117]
Rice straw	Precarbonization and chemical activated by H_3PO_4	56–112	[118]
Commercially available activated carbon	Sulfur	44 for AC and 64 for SAC	[119]

However, it has been noted that increase in the surface area can decrease the overall performance of the device. It is clear from this that other parameters such as pore structure and shape, pore size distribution, electrical conductivity, surface functionality, etc., play an important role in the capacitance. If activation is further increased, the material density decreases due to the increase of pore size which may also lead to degradation of electrolyte due to dangling band positions [6, 18]. As in all other electrode material based supercapacitors, electrolyte plays a crucial role even in ACs based supercapacitor devices. Studies have shown that organic electrolytes result in better performance as compared to aqueous electrolytes. This may be due to the large effectual size of ions of electrolyte and wettability of the carbon surface by the organic electrolyte [99]. However, not only the conductivity but also the process of drying and purification of the aqueous electrolyte is better than the organic electrolyte. ACs with pore size 0.4 to 0.7 nm is suitable for aqueous electrolyte and 0.8 nm for organic electrolyte [18]. The optimizing functionality of surface of the ACs electrode and porosity can enhance capacitance as well as the aging of the electrode [60].

B. Li, et al. reported that nitrogen-doped ACs showed after 8000 cycles, 76.3% capacity retention tested at 1.6 A g^1[100]. V. B. Kumar et al. have discussed the composite of AC and carbon dots with 100% Coulombic efficiency over thousand cycles [101]. ACs are believed to be of three types depending upon their pore size i.e. micro, meso and macro pores. In case of EDLC mesopores strongly contributes [75, 111].

Table 1 showed the brief description of supercapacitors made from ACs.

3.   Hybrid supercapacitor

Hybrid supercapacitors are the latest problem-solving strategy. They combine pseudo capacitor (high capacitance material) with carbon (high stability material) for excellent performance. This configuration results in high capacitance along with wider operating voltage range [50-52]. There have been many reports on composites of carbon and metal oxide. Mingjia Zhi, et al. suggested that the structure, properties, and constituents of carbon-metal oxide composites play an important role on the synergistic effects in terms of supercapacitor performance [23].

Fig. 2 shows the schematic diagram of activated carbon and transition metal oxide supercapacitor. In the case of hybrid supercapacitors, ACs are either the positive electrode or negative electrode. This depends on the nature of the oxide material.

Table 2 to 4 encapsulates the hybrid supercapacitors with a composite material (Ac/metal oxide), AC as an anode and metal oxides as cathode and AC as cathode and metal oxides as an anode, respectively.

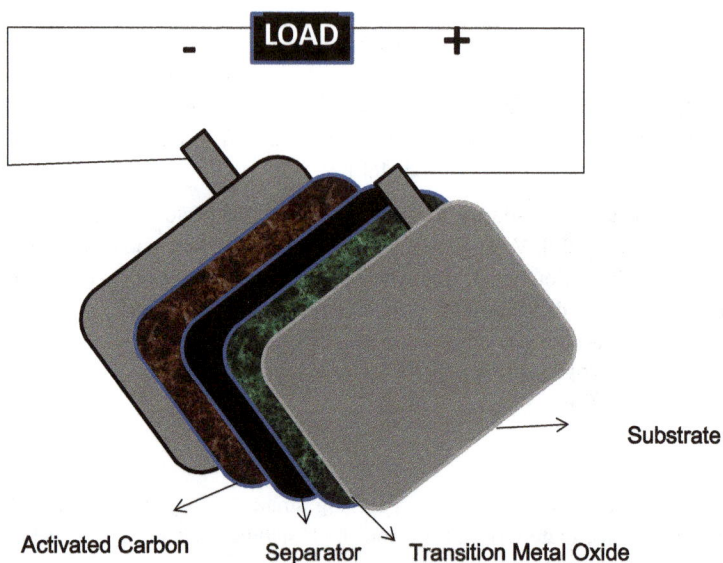

Fig. 2. Schematic diagram of activated carbon and transition metal oxide supercapacitor.

Table 2 Hybrid supercapacitor made up of composite material (Ac/metal oxide) and their electrochemical properties.

Metal Oxide	Electrochemical property	Ref.
Ruthenium oxide	350 [F g^{-1}]	[120]
	324 [F g^{-1}]	[121]
	250 [F g^{-1}]	[122]
	980 [F g^{-1}]	[123]
	180 [F g^{-1}]	[124]
	180 [F g^{-1}]	[125]
	260 [F g^{-1}]	[126]
	116 [F g^{-1}]	[127]
	1450 [F g^{-1}]	[128]
Manganese oxide	242 [F g^{-1}]	[129]
	375 [F g^{-1}]	[130]
	290 [F g^{-1}]	[131]
	250 [F g^{-1}]	[132]
	332.6 [F g^{-1}]	[133]
	485.4 [F g^{-1}]	[134]

Titanium dioxide	122 [F g^{-1}]	[135]
Nickel hydroxide	530 [F g^{-1}]	[136]
Ni-decorated with ACs composite	95 [F g^{-1}]	[137]
Cobalt hydroxide carbonate	301 [F g^{-1}]	[138]
	178.2 [F g^{-1}]	[139]
Zinc oxide	160 [F g^{-1}]	[140]
Vanadium Oxide (V$_2$O$_5$)	73 [F g^{-1}]	[141]
V$_2$O$_5$/polyindole (V$_2$O$_5$/PIn)	535.5 [F g^{-1}]	[142]
Ternary composite of V$_2$O$_5$/carbon nanotubes/super activated carbon (V$_2$O$_5$/CNTs–SAC)	357.5 [F g^{-1}]	[143]
Molybdenum oxide	177 [F g^{-1}]	[144]
Reduced graphene oxide (RGO)	80 % specific capacitance is preserved after 1000 galvanostatic charge-discharge cycles	[145]
Graphene oxide	602.36 [F g^{-1}]	[146]
	225 [F g^{-1}]	[147]
Fe$_3$O$_4$	4.36 Ω	[148]
	154 [F g^{-1}]	[149]
	168 [F g^{-1}]	[150]
Polyaniline (PANI), activated carbon and TiO$_2$	286 [F g^{-1}]	[151]
Mesopore nickel-based mixed rare-earth oxide (NMRO) and activated carbon (AC)	Power density 458 [W kg^{-1}], energy density 50 [W h kg^{-1}]	[152]
AC-coated Li$_4$Ti$_5$O$_{12}$ electrode	Maximum volumetric energy and power density of 57 [W h L^{-1}] and 2600 [W L^{-1}]	[153]
Indium tin oxide/AC composite	75% increase in capacitance	[154]

Table 3 Hybrid supercapacitor made up of AC as anode and metal oxides as cathode and their specific capacitance.

Metal Oxide	Specific Capacitance [F g^{-1}]	Ref.
Manganese oxide	57	[155]
	33	[156]
	244.7	[157]
Nickel oxide	105.8	[158]
Cobalt oxide	61	[159]
Graphene oxide- cobalt oxide	114.1	[160]
NaMnO$_2$ and an aqueous Na$_2$SO$_4$ solution as electrolyte	38	[161]
Li$_2$FeSiO$_4$ (LFSO)	49	[162]

Table 4 Hybrid supercapacitor with AC as cathode and metal oxides as anode and their electrochemical properties.

Metal Oxide	Supercapacitive property	Ref.
Manganese oxide	140 [F g^{-1}]	[44]
	21 ± 2 [F g^{-1}]	[45]
	20 [F g^{-1}]	[163]
	Energy density 21 [W h Kg^{-1}]	[164]
	62.4 [F g^{-1}]	[165]
	22 [F g^{-1}]	[166]
	228 [F g^{-1}]	[167]
Nickel hydroxide	153 [F g^{-1}]	[168]
	The discharge capacitance at 20,000th cycle of the HC cell was over 90% of initial value, which is much higher than that of the ECLC cell.	[169]
NiO$_2$	194 [F g^{-1}]	[170]
Nickel and nickel oxide materials	250 [F g^{-1}]	[171]
α-Ni(OH)$_2$ materials	127 [F g^{-1}]	[172]
Co(OH)$_2$ nanoflakes	735 [F g^{-1}]	[173]
	72.4 [F g^{-1}]	[174]
	416 [F g^{-1}]	[175]
Vanadium oxide (V$_2$O$_5$)	67 [F g^{-1}]	[176]
	32 [F g^{-1}]	[177]
Graphene oxide/polypyrrole (GO/PPy)	57.3 [F g^{-1}]	[178]
Graphene/MnO$_2$	113.5 [F g^{-1}]	[179]
RuO$_2$/TiO$_2$ /AC	46 [F g^{-1}]	[180]
Silicon carbide–MnO$_2$	59 [F g^{-1}]	[181]
Ni$_x$Co$_{3x}$O$_4$	105 [F g^{-1}]	[182]
A lithium-ion intercalated compound LiMn$_2$O$_4$	Specific energy 35 [W h kg^{-1}]	[183]
Co(OH)$_2$/ultra-stable Y zeolite	110 [F g^{-1}]	[184]
Graphene oxide/polypyrrole composite	64 [F g^{-1}]	[185]
Fe$_3$O$_4$ positive electrode	37 [F g^{-1}]	[186]

3.1   The behavior of various metal oxides with AC

3.1.1   Ruthenium oxide

Ruthenium oxide offers the best result as a composite material with ACs. C-C. Hu, et al. reports highest capacitance (1450 F g^{-1}) for this type of composite material [128]. Ruthenium will always be the best candidate for pseudocapacitor but due to its cost, it is not being commercially used on a large scale. However, ACS can be synthesized at a very low cost and a small amount of ruthenium can be added. This composition improves the overall performance of the supercapacitor device and the manufacturing price is kept low.

3.1.2   Manganese oxide

This electrode material shows all-around good performance. It can be used as a composite material or a cathode or anode for the device along with AC. Manganese oxide is an easily available material and is considered environment friendly. Many reports have suggested that it is a reliable material for supercapacitor devices. Graphene, sodium, silicon, lithium etc. are being combined with manganese oxide for good performance.

3.1.3   Nickel oxide

Nickel has proved to be dynamic as supercapacitor material. Only Ni, nickel oxide or hydroxide in alpha or beta phases can be used for best performance. It can also be used as a composite material and cathode or anode of the device. Nickel and its derivatives are well known for stability and high specific capacitance [135,158,168].

3.1.4   Cobalt hydroxide

Cobalt hydroxide shows specific capacitance approximately 700 F g^{-1}. In the case of the hybrid device, cobalt hydroxide shows appreciable performance when used as the anodic material. However, it can be used as a cathode or can be combined with AC.

3.1.5   Vanadium oxide

Vanadium oxide offers a wider potential range. Thus, when combined with ACs, it shows better performance as anodic material. The study reveals that a device made up of AC and vanadium oxide along with polymer shows better electrochemical properties [142].

3.1.6 Graphene oxide

Graphene oxide is being a new sensation in the field of supercapacitors. In hybrid supercapacitor devices, it proves itself to be the better choice as combined electrode material along with ACs.

Another metal oxide such as titanium dioxide, zinc oxide, molybdenum oxide, $Li_4Ti_5O_{12}$, Li_2FeSiO_4, iron oxide, etc. have been reported as useful materials for hybrid supercapacitors.

Conclusion

Supercapacitors device can be easily fabricated using AC and various metal oxides. Among these ruthenium offers the best result as a composite material. Manganese, nickel and graphene and their derivatives can be used either as a composite material or an anode or cathode. ACs are mostly made of agricultural raw material. AC are flexible in its usage which makes it possible to enhance the performance of the device. AC avoids the use of binders and conductive promoters and felicitates electrochemical performance due to their porous surface, Whereas metal oxide offers high specific capacitance. A hybrid supercapacitor is a promising candidate for energy storage devices.

References

[1]   P. Simon, Y. Gogotsi, Materials for electrochemical capacitors, Nature Materials 7 (2008) 845-854. https://doi.org/10.1038/nmat2297

[2]   M. Salanne, B. Rotenberg, K. Naoi, K. Kaneko, P-L. Taberna, C. P. Grey, B. Dunn, P. Simon, Efficient storage mechanisms for building better supercapacitors, Nature Energy 1 (2016) 16070. https://doi.org/10.1038/nenergy.2016.70

[3]   S.Chu, A.Majumdar, Opportunities and challenges for a sustainable energy future, Nature, 488 (2012) 294−303. https://doi.org/10.1038/nature11475

[4]   A. Shukla, Electrochemical Power Sources - Fuel Cells and Supercapacitors, Resonance 6 (8) (2001) 72–81. https://doi.org/10.1007/BF02902517

[5]   J. R. Miller, A. F. Burke, Electrochemical capacitors: challenges and opportunities for real-world applications, Electrochem. Soc. Interface Spring 17 (2008) 53-57.

[6]   L. L. Zhang, X. S. Zhao, Carbon-based materials as supercapacitor electrodes, Chem. Soc. Rev. 38 (2009) 2520–2531. https://doi.org/10.1039/b813846j

[7]   M. Inagaki, H. Konno, O. Tanaike, Carbon materials for electrochemical capacitors, Journal of Power Sources 195 (2010) 7880–7903. https://doi.org/10.1016/j.jpowsour.2010.06.036

[8]   H. V. Helmholtz, Ann. Phys. (Leipzig), 1853, 89, 211. https://doi.org/10.1002/andp.18531650603

[9]   D. L. Chapman, Philos. Mag., 1913, 6, 475. https://doi.org/10.1080/14786440408634187

[10]   G. Gouy, J. Phys., 1910, 4, 457.

[11]   O. Stern, Z. Electrochem., 1924, 30, 508.

[12]   J. Chmiola, G. Yushin, Y. Gogotsi, C. Portet, P. Simon, P. L. Taberna, Anomalous increase in carbon capacitance at pore sizes less than 1 nanometer, Science 313 (2006)1760–1763. https://doi.org/10.1126/science.1132195

[13]   E. Frackowiak, F. Béguin, Carbon materials for the electrochemical storage of energy in capacitors, Carbon 39 (2001) 937–950. https://doi.org/10.1016/S0008-6223(00)00183-4

[14]   A. V. Neimark, Y. Lin, P. I. Ravikovitch, M. Thommes, Quenched solid density functional theory and pore size analysis of micro-mesoporous carbons, Carbon 47 (2009) 1617–1628. https://doi.org/10.1016/j.carbon.2009.01.050

[15]   C. Merlet, B. Rotenberg, P. A. Madden, P-L. Taberna, P. Simon, Y. Gogotsi, M. Salanne, On the molecular origin of supercapacitance in nanoporous carbon electrode, Nature Mater. 11 (2012)306–310. https://doi.org/10.1038/nmat3260

[16]   E. Frackowiak, Carbon materials for supercapacitor application, Phys. Chem. Chem. Phys. 9 (2007) 1774-1785. https://doi.org/10.1039/b618139m

[17]   J. Huang, B. G. Sumpter, V. Meunier, A universal model for nanoporous carbon supercapacitors applicable to diverse pore regimes, carbon materials, and electrolytes, Chem. Eur. J. 14 (2008) 6614– 6626. https://doi.org/10.1002/chem.200800639

[18]   E. Raymundo-Pinero, K. Kierzek, J. Machnikowski, F. Beguin, Relationship between the nanoporous texture of activated carbons and their capacitance properties in different electrolytes, Carbon 44 (2006) 2498 –2507. https://doi.org/10.1016/j.carbon.2006.05.022

[19]   W. Gu, G. Yushin, Review of nanostructured carbon materials for electrochemical capacitor applications: advantages and limitations of activated carbon, carbide-

derived carbon, zeolite-templated carbon, carbon aerogels, carbon nanotubes, onion-like carbon, and graphene, WIREs Energy Environ. 3 (2014) 424–473. https://doi.org/10.1002/wene.102

[20]   S.G. Lee, Functionalized imidazolium salts for task specific ionic liquids and their applications, Chem Commun, 10 (2006) 1049–1063. https://doi.org/10.1039/b514140k

[21]   M Galinski, A Lewandowski, I Stepniak, Ionic liquids as electrolytes, Electrochim. Acta 51 (2006) 5567–5580. https://doi.org/10.1016/j.electacta.2006.03.016

[22]   S. Pandey, Analytical applications of room-temperature ionic liquids: A review of recent efforts, Anal. Chim. Acta 556 (2006) 38–45. https://doi.org/10.1016/j.aca.2005.06.038

[23]   M. Zhi, C. Xiang, J.Li, M. Li, N. Wu, Nanostructured carbon–metal oxide composite electrodes for supercapacitors: A review, Nanoscale 5 (2013) 72–88. https://doi.org/10.1039/C2NR32040A

[24]   B. E. Conway, Electrochemical Supercapacitors: Scientific Fundamentals and Technological Applications, Kluwer Academic/Plenum Publisher, New York, 1999. https://doi.org/10.1007/978-1-4757-3058-6

[25]   K. T. Nam, D. W. Kim, P. J. Yoo, C. Y. Chiang, N. Meethong, P. T. Hammond, Y. M. Chiang, A. M. Belcher, Virus-enabled synthesis and assembly of nanowires for lithium ion battery electrodes, Science 312 (2006) 885-888. https://doi.org/10.1126/science.1122716

[26]   M. Toupin, T. Brousse, D. Belanger, Charge storage mechanism of MnO_2 electrode used in aqueous electrochemical capacitor, Chem. Mater. 16 (2004) 3184–3190. https://doi.org/10.1021/cm049649j

[27]   S. H. Ng, J. Wang, D. Wexler, K. Konstantinov, Z. P. Guo, H. K. Liu, Highly reversible lithium storage in spheroidal carbon-coated silicon nanocomposites as anodes for lithium-ion batteries, Angew. Chem. Int. Ed. 45 (2006) 6896 –6899. https://doi.org/10.1002/anie.200601676

[28]   U. M. Patil, R. R. Salunkhe, K. V. Gurav, C. D. Lokhande, Chemically deposited nanocrystalline NiO thin films for supercapacitor application, Appl. Surf. Sci. 255 (2008) 2603–2607. https://doi.org/10.1016/j.apsusc.2008.07.192

[29]   D. L. Yan, Z. L. Guo, G. S. Zhu, Z. Z. Yu, H. R. Xu, A. B. Yu, MnO_2 film with three-dimensional structure prepared by hydrothermal process for supercapacitor,

J. Power Sources 199 (2012) 409–412.
https://doi.org/10.1016/j.jpowsour.2011.10.051

[30] T. Lu, Y. P. Zhang, H. B. Li, L. K. Pan, Y. L. Li, Z. Sun, Electrochemical behaviors of graphene–ZnO and graphene–SnO2 composite films for supercapacitors, Electrochim. Acta 55 (2010) 4170–4173. https://doi.org/10.1016/j.electacta.2010.02.095

[31] R. Z. Li, X. Ren, F. Zhang, C. Du, J. P. Liu, Synthesis of Fe3O4@SnO2 core–shell nanorod film and its application as a thin-film supercapacitor electrode, Chem. Commun. 48 (2012) 5010–5012. https://doi.org/10.1039/c2cc31786a

[32] B. R. Duan, Q. Cao, Hierarchically porous Co3O4 film prepared by hydrothermal synthesis method based on colloidal crystal template for supercapacitor application, Electrochim. Acta 64 (2012) 154–161. https://doi.org/10.1016/j.electacta.2012.01.004

[33] X. Y. Chen, E. Pomerantseva, P. Banerjee, K. Gregorczyk, R. Ghodssi, G. Rubloff, Ozone-based atomic layer deposition of crystalline V2O5 films for high performance electrochemical energy storage, Chem. Mater. 24 (2012) 1255– 1261. https://doi.org/10.1021/cm202901z

[34] X. Sun, M. Xie, G. K. Wang, H. T. Sun, A. S. Cavanagh, J. J. Travis, S. M. George, J. Lian, Atomic Layer Deposition of TiO2 on Graphene for Supercapacitors, J. Electrochem. Soc. 159 (2012) A364–A369. https://doi.org/10.1149/2.025204jes

[35] J. S. Shaikh, R. C. Pawar, R. S. Devan, Y. R. Ma, P. P. Salvi, S. S. Kolekar, P. S. Patil, Synthesis and characterization of Ru doped CuO thin films for supercapacitor based on Bronsted acidic ionic liquid, Electrochim. Acta 56 (2011) 2127–2134. https://doi.org/10.1016/j.electacta.2010.11.046

[36] X. B. Ren, H. Y. Lu, H. B. Lin, Y. N. Liu, Y. Xing, Preparation and characterization of the Ti/IrO2/WO3 as supercapacitor electrode materials, Russ. J. Electrochem. 46 (2010) 77–80. https://doi.org/10.1134/S102319351001009X

[37] P. M. Kulal, D. P. Dubal, C. D. Lokhande, V. J. Fulari, Chemical synthesis of Fe2O3 thin films for supercapacitor application, J. Alloys Compd. 509 (2011) 2567–2571. https://doi.org/10.1016/j.jallcom.2010.11.091

[38] P. Lu, D. Xue, H. Yang, Y. Liu, Supercapacitor and nanoscale research towards electrochemical energy storage Int. J. Smart Nano Mater. 1 (2012) 1–25.

[39] X. Y. Lang, A. Hirata, T. Fujita, M. W. Chen, Nanoporous metal/oxide hybrid electrodes for electrochemical supercapacitors, Nat. Nanotechnol. 6 (2011) 232–236. https://doi.org/10.1038/nnano.2011.13

[40] P. Poizot, S. Laruelle, S. Grugeon, L. Dupont, J. M. Tarascon, Nano-sized transition-metal oxides as negative-electrode materials for lithium-ion batteries, Nature 407 (2000) 496-499. https://doi.org/10.1038/35035045

[41] G. Derrien, J. Hassoun, S. Panero, B. Scrosati, Nanostructured Sn–C composite as an advanced anode material in high-performance lithium-ion batteries, Adv. Mater. 19 (2007) 2336–2340. https://doi.org/10.1002/adma.200700748

[42] C. C. Hu, W. C. Chen, K. H. Chang, How to achieve maximum utilization of hydrous ruthenium oxide for supercapacitors, J. Electrochem. Soc. 151 (2004) A281−A290. https://doi.org/10.1149/1.1639020

[43] P.-C. Chen, G. Shen, Y. Shi, H. Chen, C. Zhou, Preparation and characterization of flexible asymmetric supercapacitors based on transition-metal-oxide nanowire/single-walled carbon nanotube hybrid thin-film electrodes, ACS Nano 4 (2010) 4403-4411. https://doi.org/10.1021/nn100856y

[44] V. Khomenko, E. Raymundo-Pi-ero, F. Béguin, Optimisation of an asymmetric manganese oxide/activated carbon capacitor working at 2V in aqueous medium, Journal of Power Sources 153 (2006) 183-190. https://doi.org/10.1016/j.jpowsour.2005.03.210

[45] T. Brousse, P.-L. Taberna, O. Crosnier, R. Dugas, P. Guillemet, Y. Scudeller, Y. Zhou, F. Favier, D. Bélanger, P. Simon, Long-term cycling behavior of asymmetric activated carbon/MnO2 aqueous electrochemical supercapacitor, Journal of Power Sources, 173 (2007) 633-641. https://doi.org/10.1016/j.jpowsour.2007.04.074

[46] S.W. Lee, J. Kim, S. Chen, P.T. Hammond, Y.S. Horn, carbon nanotube/manganese oxide ultrathin film electrodes for electrochemical capacitors, ACS Nano 4 (2010) 3889 –3896. https://doi.org/10.1021/nn100681d

[47] Z.S. Wu, D.W. Wang, W. Ren, J. Zhao, G. Zhou, F. Li, H.M. Cheng, Anchoring Hydrous RuO2 on graphene sheets for high-performance electrochemical capacitors, Adv. Funct. Mater. 20 (2010) 3595-3602. https://doi.org/10.1002/adfm.201001054

[48] Y. Chen, X. Zhang, D. Zhang, P. Yu, Y. Ma, High performance supercapacitors based on reduced graphene oxide in aqueous and ionic liquid electrolytes, Carbon 49 (2011) 573–580. https://doi.org/10.1016/j.carbon.2010.09.060

[49] C. D. Lokhande, D. P. Dubal, O. S. Joo, Metal oxide thin film based supercapacitors, Curr. Appl. Phys. 11 (2011) 255-270. https://doi.org/10.1016/j.cap.2010.12.001

[50] I. Pi-eiro-Prado, D. Salinas-Torres, R. Ruiz-Rosas, E. Morallón, D. Cazorla-Amorós, Design of activated carbon/activated carbon asymmetric capacitors, Frontiers in Materials (2016) doi: 10.3389/fmats.2016.00016. https://doi.org/10.3389/fmats.2016.00016

[51] V. Khomenko, E. Raymundo-Pinero, F. Beguin, High-energy density graphite/AC capacitor in organic electrolyte, Journal of Power Sources 177 (2008) 643–651. https://doi.org/10.1016/j.jpowsour.2007.11.101

[52] G.G. Amatucci, F. Badway, A. Du Pasquier, T. Zheng, An asymmetric hybrid nonaqueous energy storage cell. J. Electrochem. Soc. 148 (2001) A930–A93. https://doi.org/10.1149/1.1383553

[53] H.O. Pierson, Handbook of Carbon, Graphite, Diamond and Fullerenes, Noyes Publications, NJ, USA, 1993.

[54] D.O. Cooney, Activated Charcoal: Antidotal and other Medical Uses, New York: Dekker; 1980.

[55] A. G. Pandolfo, A. F. Hollenkamp, Carbon properties and their role in supercapacitors, J. Power Sources 157 (2006) 11–27. https://doi.org/10.1016/j.jpowsour.2006.02.065

[56] F.S. Baker, C.E. Miller, E.D. Repik. Kirk-Othme, Encyclopedia of Chemical Technology, New York: John Wiley, 4. 4 (1992) 1015–1037.

[57] R.C. Bansal, J.B. Donnet, F. Stoeckli, Active Carbon, Marcel Dekker, New York, 1988 (Chapter 2).

[58] D Lozano-Castello, D Cazorla-Amoros, A Linares-Solano, S Shiraishi, H Kurihara, A Oya, Influence of pore structure and surface chemistry on electric double layer capacitance in non-aqueous electrolyte, Carbon 41(2003) 1765–1775. https://doi.org/10.1016/S0008-6223(03)00141-6

[59] G. Salitra, A. Soffer, L. Eliad, Y. Cohen, D. Aurbach, Carbon electrodes for double-layer capacitors i. relations between ion and pore dimensions, J. Electrochem. Soc. 147 (2000) 2486-2493. https://doi.org/10.1149/1.1393557

[60] M. Zhu, C. J. Weber, Y. Yang, M. Konuma, U. Starke, K. Kern, A. M. Bittner, Factors affecting the size and deposition rate of the cathode deposit in an anodic arc used to produce carbon nanotubes, Carbon 46 (2008) 1826-1828. https://doi.org/10.1016/j.carbon.2008.07.025

[61] L. Weinstein, R. Dash, Supercapacitor carbons, Mater. Today 16 (2013) 356–357. https://doi.org/10.1016/j.mattod.2013.09.005

[62] X. He, P. Ling, M. Yu, X. Wang, X. Zhang, M. Zheng, Rice husk-derived porouscarbons with high capacitance by ZnCl2 activation for supercapacitors, Electrochim. Acta 105 (2013) 635–641. https://doi.org/10.1016/j.electacta.2013.05.050

[63] M. Wu, P. Ai, M. Tan, B. Jiang, Y. Li, J. Zheng, W. Wu, Z. Li, Q. Zhang, X. He, Synthesis of starch-derived mesoporous carbon for electric double layer capacitor, Chem. Eng. J. 245 (2014) 166–172. https://doi.org/10.1016/j.cej.2014.02.023

[64] J. Xua, L. Chena, H. Qua, Y. Jiaoa, J. Xiea, G. Xing, Preparation and characterization of activated carbon from reedy grass leaves by chemical activation with H3PO4, Appl. Surf. Sci. 320 (2014) 674–680. https://doi.org/10.1016/j.apsusc.2014.08.178

[65] E. Frackowiak, Q. Abbas, F. Béguin, Carbon/carbon supercapacitors, J. Energy Chem. 22 (2013) 226–240. https://doi.org/10.1016/S2095-4956(13)60028-5

[66] A. Mestre, E. Tyszko, M. Andrade, M. Galhetas, C. Freire, A. Carvalho, Sustainable activated carbons prepared from a sucrose-derived hydrochar: Remarkable adsorbents for pharmaceutical compounds, RSC Adv. 5 (2015) 19696–19707. https://doi.org/10.1039/C4RA14495C

[67] A.T. Mohd Din, B. Hameed, A.L. Ahmad, Batch adsorption of phenol onto physiochemical-activated coconut shell, J. Hazard. Mater. 161 (2009) 1522–1529. https://doi.org/10.1016/j.jhazmat.2008.05.009

[68] M. Nawa, T. Nogami, H. Mikawa, Application of activated carbon fiber fabrics to electrodes of rechargeable battery and organic electrolyte capacitor, j. Electrochem. Soc. 131 (1984) 1457-1459. https://doi.org/10.1149/1.2115872

[69] A. Yoshida,I. Tanahashi, Y. Takeuchi, A. Nishino, An electric double-layer capacitor with activated carbon fiber electrodes, IEEE Trans. Compon. Hybrids Manuf. Technol. 10 (1987) 100-102. https://doi.org/10.1109/TCHMT.1987.1134717

[70] M. Nakamura, M. Nakanishi, K. Yamamoto, Influence of physical properties of activated carbons on characteristics of electric double-layer capacitors, J. Power Sources 60 (1996) 225-231. https://doi.org/10.1016/S0378-7753(96)80015-2

[71] M. Endo, T. Maeda, T. Takeda, Y. J. Kim, K. Koshiba, H. Hara, M. S. Dresselhaus, capacitance and pore-size distribution in aqueous and nonaqueous electrolytes using various activated carbon electrodes, J. Electrochem. Soc. 148 (2001) A910-A914. https://doi.org/10.1149/1.1382589

[72] A. Yoshida, S. Nonaka, I. Aoki, A. Nishino, Electric double-layer capacitors with sheet-type polarizable electrodes and application of the capacitors, J. Power Sources 60 (1996) 213-218. https://doi.org/10.1016/S0378-7753(96)80013-9

[73] H. Yang, M. Yoshio, K. Isono, R. Kuramoto, Improvement of commercial activated carbon and its application in electric double layer capacitors, Electrochem. Solid State Lett. 5 (2002) A141-A144. https://doi.org/10.1149/1.1477297

[74] T. M. Alslaibi, I. Abustan, M.A. Ahmad, A. A. Foul, A review: production of activated carbon from agricultural byproducts via conventional and microwave heating, J. Chem. Technol. Biotechnol. 88 (2013) 1183–1190. https://doi.org/10.1002/jctb.4028

[75] G. Gryglewicz, J. Machnikowski, E. Lorenc-Grabowska, G. Lota, E. Frackowiak, Effect of pore size distribution of coal-based activated carbons on double layer capacitance, Electrochim. Acta 50 (2005) 1197–1206. https://doi.org/10.1016/j.electacta.2004.07.045

[76] H. Marsh, Activated Carbon Compendium: A Collection of Papers from the Journal Carbon 1996–2000, Elsevier; 2001, Gulf Publishing, Texas, USA.

[77] V. Subramanian, C. Luo, A.M. Stephan, K.S. Nahm, S. Thomas, B.Q. Wei, Supercapacitors from activated carbon derived from banana fibers, J. Phys. Chem. C 111 (2007) 7527–7531. https://doi.org/10.1021/jp067009t

[78] Q.Y. Li, H.Q. Wang, Q.F. Dai, J.H. Yang, Y.L. Zhong, Novel activated carbons as electrode materials for electrochemical capacitors from a series of starch, Solid State Ionics 179 (2008) 269–273. https://doi.org/10.1016/j.ssi.2008.01.085

[79] L. Wei, G. Yushin, Electrical double layer capacitors with activated sucrose-derived carbon electrodes, Carbon 49 (2011) 4830–4838. https://doi.org/10.1016/j.carbon.2011.07.003

[80]   L. Wei, M. Sevilla, A.B. Fuertes, R. Mokaya, G. Yushin, Hydrothermal carbonization of abundant renewable natural organic chemicals for high-performance supercapacitor electrodes, Adv. Energy Mater. 1 (2011) 356–361. https://doi.org/10.1002/aenm.201100019

[81]   T. E. Rufford, D. Hulicova-Jurcakova, K. Khosla, Z.H. Zhu, G.Q. Lu, Microstructure and electrochemical double-layer capacitance of carbon electrodes prepared by zinc chloride activation of sugar cane bagasse, J. Power Sources 195 (2010) 912–918. https://doi.org/10.1016/j.jpowsour.2009.08.048

[82]   B Xu, Y.F. Chen, G. Wei, G.P. Cao, H. Zhang, Y.S. Yang, Activated carbon with high capacitance prepared by NaOH activation for supercapacitors, Mater. Chem. Phys. 124 (2010) 504–509. https://doi.org/10.1016/j.matchemphys.2010.07.002

[83]   D. Lozano-Castello, D. Cazorla-Amoros, A. Linares-Solano, S. Shiraishi, H. Kurihara, A. Oya, Influence of pore structure and surface chemistry on electric double layer capacitance in non-aqueous electrolyte, Carbon 41 (2003) 1765–1775. https://doi.org/10.1016/S0008-6223(03)00141-6

[84]   R Wang, P.Y. Wang, X.B. Yan, J.W. Lang, C. Peng, Q.J. Xue, Promising porous carbon derived from celtuce leaves with outstanding supercapacitance and CO2 capture performance, ACS Appl. Mater. Interf. 4 (2012) 5800–5806. https://doi.org/10.1021/am302077c

[85]   L.G. Juntao Zhang, S. Kang, J. Jianchun, Z. Xiaogang, Preparation of activated carbon from waste Camellia oleifera shell for supercapacitor application, J. Solid State Electrochem. 16 (2012) 2179. https://doi.org/10.1007/s10008-012-1639-1

[86]   Z. Li, L. Zhang, B.S. Amirkhiz, X.H. Tan, Z.W. Xu, H.L. Wang, B.C. Olsen, C.M. Holt, D. Mitlin, Carbonized chicken eggshell membranes with 3d architectures as high-performance electrode materials for supercapacitors, Adv Energy Mater. 2 (2012) 431–437. https://doi.org/10.1002/aenm.201100548

[87]   G. Lota, B. Grzyb, H. Machnikowska, J. Machnikowski, E. Frackowiak, Effect of nitrogen in carbon electrode on the supercapacitor performance, Chem. Phys. Lett. 404 (2005) 53–58. https://doi.org/10.1016/j.cplett.2005.01.074

[88]   J. Lee, J. Kim, T. Hyeon, Recent progress in the synthesis of porous carbon materials, Adv. Mater. 18 (2006) 2073–2094. https://doi.org/10.1002/adma.200501576

[89]   M. Seredych, D. Hulicova-Jurcakova, G. Q. Lu, T. J. Bandosz, Surface functional groups of carbons and the effects of their chemical character, density and

accessibility to ions on electrochemical performance, Carbon 46 (2008) 1475–1488. https://doi.org/10.1016/j.carbon.2008.06.027

[90] K. Jurewicz, K. Babeł, A. Z´io´łkowski, H. Wachowska, Ammoxidation of active carbons for improvement of supercapacitor characteristics, Electrochim. Acta 48 (2003) 1491–1498. https://doi.org/10.1016/S0013-4686(03)00035-5

[91] H. Guo, Q. Gao, Boron and nitrogen co-doped porous carbon and its enhanced properties as supercapacitor, J. Power Sources 186 (2009) 551–556. https://doi.org/10.1016/j.jpowsour.2008.10.024

[92] V.V.N. Obreja, On the performance of supercapacitors with electrodes based on carbon nanotubes and carbon activated material—A review, Physica E 40 (2008) 2596–2605. https://doi.org/10.1016/j.physe.2007.09.044

[93] F. Beguin, E. Frackowiak, Supercapacitors: Materials, Systems, and Applications, Berlin, Germany: Wiley- VCH; 2013. https://doi.org/10.1002/9783527646661

[94] G. Hasegawa, Monolithic electrode for electric double layer capacitors based on macro/meso/microporous S-Containing activated carbon with high surface area, J. Mater. Chem. 21 (2011) 2060–2063. https://doi.org/10.1039/c0jm03793a

[95] D. Carriazo, F. Pico, M.C. Gutierrez, F. Rubio, J.M. Rojo, F. Monte, Block-Copolymer assisted synthesis of hierarchical carbon monoliths suitable as supercapacitor electrodes, J. Mater. Chem. 20 (2010) 773-780. https://doi.org/10.1039/B915903G

[96] C. Kim, Electrochemical characterization of electrospun activated carbon nano fibres as an electrode in supercapacitors, J Power Sources, 142 (2005) 382–388. https://doi.org/10.1016/j.jpowsour.2004.11.013

[97] K.S. Hung, C. Masarapu, T.H. Ko, B.Q. Wei, Wide temperature range operation supercapacitors from nanostructured activated carbon fabric, J. Power Sources 193 (2009) 944–949. https://doi.org/10.1016/j.jpowsour.2009.01.083

[98] Q. Zhang, J.P. Rong, D.S. Ma, B.Q. Wei, The governing self-discharge processes in activated carbon fabric-based supercapacitors with different organic electrolytes, Energy Environ. Sci. 4 (2011) 2152–2159. https://doi.org/10.1039/c0ee00773k

[99] P. Ratajczak, K. Jurewicz, F. Be´guin, Factors contributing to ageing of high voltage carbon/carbon supercapacitors in salt aqueous electrolyte, J. Appl. Electrochem. 44 (2014) 475–480. https://doi.org/10.1007/s10800-013-0644-0

[100] V. D. Patake, C. D. Lokhande, O. S. Joo, Electrodeposited ruthenium oxide thin films for supercapacitor: Effect of surface treatments, Appl. Surf. Sci. 255 (2009) 4192–4196. https://doi.org/10.1016/j.apsusc.2008.11.005

[101] V. B. Kumar, A. Borenstein, B. Markovsky, D. Aurbach, A. Gedanken, M. Talianker, Z. Porat, Activated carbon modified with carbon nanodots as novel electrode material for supercapacitors, J. Phys. Chem. C 120 (2016) 13406−13413. https://doi.org/10.1021/acs.jpcc.6b04045

[111] A. Davies, A. Yu, Material Advancements in Supercapacitors: From activated carbon to carbon nanotube and graphene, The Canadian Journal Of Chemical Engineering 89 (2011) 1342-1357. https://doi.org/10.1002/cjce.20586

[112] B. K Ostafiychuk, I. M Budzulyak, B. I Rachiy, V.M Vashchynsky, V. I Mandzyuk, R. P Lisovsky, L. O Shyyko, Thermochemically activated carbon as an electrode material for supercapacitors, Nanoscale Research Letters 10 (2015) 65. https://doi.org/10.1186/s11671-015-0762-1

[113] B. Li, F. Dai, Q. Xiao, L. Yang, J. Shen, C. Zhang, M. Cai, Nitrogen-doped activated carbon for a high energy hybrid supercapacitor, Energy Environ. Sci. 9 (2016) 102-106. https://doi.org/10.1039/C5EE03149D

[114] J. Su'arez-Guevara, V. Ruiz, P. Gomez-Romero, Hybrid energy storage: high voltage aqueous supercapacitors based on activated carbon–phosphotungstate hybrid materials, J. Mater. Chem. A 2 (2014) 1014–1021. https://doi.org/10.1039/C3TA14455K

[115] I. I. Gurten Ina, S. M. Holmes, A. Banford, Z. Aktasa, The performance of supercapacitor electrodes developed from chemically activated carbon produced from waste tea, Appl. Surf. Sci. 357 (2015) 696–703. https://doi.org/10.1016/j.apsusc.2015.09.067

[116] M. Chen, X. Kang, T. Wumaier, J. Dou, B. Gao, Y.Han, G. Xu, Z. Liu, L.Zhang, Preparation of activated carbon from cotton stalk and its application in supercapacitor, J. Solid State Electrochem. 17 (2013) 1005–1012. https://doi.org/10.1007/s10008-012-1946-6

[117] Y. Jang, J. Jo, Y-M Choi, I. Kim, S-H. Lee, D. Kim, S. M. Yoon, Activated carbon nanocomposite electrodes for high performance supercapacitors, Electrochimica Acta 102 (2013) 240– 245. https://doi.org/10.1016/j.electacta.2013.04.020

[118] T. Adinaveen, L. J. Kennedy, J. J. Vijaya, G. Sekaran, Surface and porous characterization of activated carbon prepared from pyrolysis of biomass (rice

straw) by two-stage procedure and its applications in supercapacitor electrodes, J. Mater. Cycles Waste Manag. 17 (2015) 736–747. https://doi.org/10.1007/s10163-014-0302-6

[119] Y. Huang, S. L. Candelaria, Y. Li, Z. Li, J. Tian, L. Zhang, G. Cao, Sulfurized activated carbon for high energy density supercapacitors, J. Power Sources 252 (2014) 90-97. https://doi.org/10.1016/j.jpowsour.2013.12.004

[120] J. Zhang, D. Jiang, B. Chen, J. Zhu, L. Jiang, H. Fang, Preparation and electrochemistry of hydrous ruthenium oxide/active carbon electrode materials for supercapacitor, J. Electrochem. Soc. 148 (2001) A1362-A1367. https://doi.org/10.1149/1.1417976

[121] T. Nanaumi, Y. Ohsawa, K. Kobayakawa, Y. Sato, High energy electrochemical capacitor materials prepared by loading ruthenium oxide on activated carbon for supercapacitor, Electrochem. 70 (2002) 681-685.

[122] M. S. Dandekar, G. Arabale, K. Vijayamohanan, Preparation and characterization of composite electrodes of coconut-shell-based activated carbon and hydrous ruthenium oxide for supercapacitors, J. Power Sources 141 (2005) 198–203. https://doi.org/10.1016/j.jpowsour.2004.09.008

[123] W-C. Chen, C-C. Hu, C-C. Wang, C-K. Min, Electrochemical characterization of activated carbon–ruthenium oxide nanoparticles composites for supercapacitors, J. Power Sources 125 (2004) 292–298. https://doi.org/10.1016/j.jpowsour.2003.08.001

[124] C-C. Wang, C-C. Hu, Electrochemical catalytic modification of activated carbon fabrics by ruthenium chloride for supercapacitors, Carbon 43 (2005) 1926–1935. https://doi.org/10.1016/j.carbon.2005.02.041

[125] J.M. Sieben, E. Morallón, D. Cazorla-Amorós, Flexible ruthenium oxide-activated carbon cloth composites prepared by simple electrodeposition methods, Energy 58 (2013) 519-526. https://doi.org/10.1016/j.energy.2013.04.077

[126] M. Ramani, B. S. Haran, R. E. White, B. N. Popov, L. Arsov, Studies on activated carbon capacitor materials loaded with different amounts of ruthenium oxide, J. Power Sources 93 (2001) 209-214. https://doi.org/10.1016/S0378-7753(00)00575-9

[127] B-H. Kima, C. H. Kimb, D. G. Le, Mesopore-enriched activated carbon nanofiber web containing RuO2 as electrode material for high-performance supercapacitors,

J. Electroanal. Chem. 760 (2016) 64–70.
https://doi.org/10.1016/j.jelechem.2015.12.001

[128] C-C. Hu, W-C. Chen, Effects of substrates on the capacitive performance of RuOx·nH2O and activated carbon–RuOx electrodes for supercapacitors, Electrochimica Acta 49 (2004) 3469–3477. https://doi.org/10.1016/j.electacta.2004.03.017

[129] H. Lee, S. H. Park, S-J. Kim, Y-K. Park, B-J. Kim, K-H. An, S. J. Ki, S-C. Jung, Synthesis of manganese oxide/activated carbon composites for supercapacitor application using a liquid phase plasma reduction system, Int. J. Hyd. Ener. 40 (2015) 754–759. https://doi.org/10.1016/j.ijhydene.2014.08.085

[130] Y. Zhang, Q. Yao, H. Gao, L. Wang, X. Jia, A. Zhang, Y. Song, T. Xia, H. Dong, Facile synthesis and electrochemical performance of manganese dioxide doped by activated carbon, carbon nanofiber and carbon nanotube, Powder Tech. 262 (2014) 150–155. https://doi.org/10.1016/j.powtec.2014.04.080

[131] Y. Jang, J. Jo, H. Jang, I. Kim, D. Kang, K-Y. Kim, Activated carbon/manganese dioxide hybrid electrodes for high performance thin film supercapacitors, Appl. Phys. Lett. 104 (2014) 243901. https://doi.org/10.1063/1.4884391

[132] J. M. Koa, K.M. Kim, Electrochemical properties of MnO2/activated carbon nanotube composite as an electrode material for supercapacitor, Mater. Chem. Phys. 114 (2009) 837–841. https://doi.org/10.1016/j.matchemphys.2008.10.047

[133] J-W. Wang, Y. Chen, B-Z. Chen, A Synthesis Method of MnO2/Activated Carbon Composite for Electrochemical Supercapacitors, J. Electrochem. Soc. 8 (2015) A1654-A1661. https://doi.org/10.1149/2.0031509jes

[134] Y. Qiu, P. Xu, B. Guo, Z. Cheng, H. Fan, M. Yang, X. Yang, J. Li, Electrodeposition of manganese dioxide film on activated carbon paper and application for supercapacitor with high rate capability, RSC Adv. 4 (2014) 64187-64192. https://doi.org/10.1039/C4RA11127C

[135] M. Selvakumar, D. K. Bhat, Microwave synthesized nanostructured TiO2-activated carbon composite electrodes for supercapacitor, Appl. Surf. Sci. 263 (2012) 236–241. https://doi.org/10.1016/j.apsusc.2012.09.036

[136] J.H. Park, O.O. Park, K.H. Shin, C.S. Jin, J.H. Kim, An electrochemical capacitor based on a Ni(OH)2/activated carbon composite electrode, Electrochem. Solid-State Lett. 5 (2002) H7-H10. https://doi.org/10.1149/1.1432245

[137] M-S. Wu, K-H. Lin, Step electrophoretic deposition of ni-decorated activated-carbon film as an electrode material for supercapacitors, J. Phys. Chem. C 114 (2010) 6190–6196. https://doi.org/10.1021/jp9109145

[138] T. M. Masikhwa, J. K. Dangbegnon, A. Bello, M. J. Madito, D. Momodu, N. Manyala, Preparation and electrochemical investigation of the cobalt hydroxide carbonate/activated carbon nanocomposite for supercapacitor applications, J. Phys. Chem. Solids 88 (2016) 60–67. https://doi.org/10.1016/j.jpcs.2015.09.015

[139] C. H. Kim, B-H. Kim, Zinc oxide/activated carbon nanofiber composites for high-performance supercapacitor electrodes, J. Power Sour. 274 (2015) 512-520. https://doi.org/10.1016/j.jpowsour.2014.10.126

[140] M. Selvakumar, D.K. Bhat, A.M. Aggarwal, S.P. Iyer, G. Sravani, Nano ZnO activated carbon composite electrodes for supercapacitors, Phys. B 405 (2010) 2286-2289. https://doi.org/10.1016/j.physb.2010.02.028

[141] B. H. Kim, K. S. Yang, D. J. Yang, Electrochemical behavior of activated carbon nanofiber-vanadium pentoxide composites for double-layer capacitors, Electrochimica Acta 109 (2013) 859–865. https://doi.org/10.1016/j.electacta.2013.07.180

[142] X. Zhou, Q. Chen, A. Wang, J. Xu, S. Wu, J. Shen, The bamboo-like composites of V2O5/polyindole and activated carbon cloth as electrodes for all-solid-state flexible asymmetric supercapacitors, ACS Appl. Mater. Interfaces 6 (2016) 3776–3783. https://doi.org/10.1021/acsami.5b10196

[143] Q. Wang, Y. Zou, C. Xiang, H. Chu, H. Zhang, F. Xu, L. Sun, C. Tang, High-performance supercapacitor based on V2O5/carbon nanotubes-super activated carbon ternary composite, Ceram. Int. 42 (2016) 12129–12135. https://doi.org/10.1016/j.ceramint.2016.04.145

[144] W. Sugimoto, T. Ohnuma, Y. Murakami, Y. Takasu, Molybdenum oxide/carbon composite electrodes as electrochemical supercapacitors, Electrochem. Solid-State Lett. 9 (2001) A145-A147. https://doi.org/10.1149/1.1388995

[145] M. Zhong, Y. Song, Y. Li, C. Ma, X. Zhai, J. Shi, Q. Guo, L. Liu, Effect of reduced graphene oxide on the properties of an activated carbon cloth/polyaniline flexible electrode for supercapacitor application, J. Power Sources 217 (2012) 6-12. https://doi.org/10.1016/j.jpowsour.2012.05.086

[146] M. Ates, D. Cinar, S. Caliskan, U. Gecgel, O. Uner, Y. Bayrak, I. Candan, Active carbon/graphene hydrogel nanocomposites as a symmetric device for

supercapacitors, J. Fullerenes, Nanotubes and Carbon Nanostructures 24 (2016) 427-434. https://doi.org/10.1080/1536383X.2016.1174115

[147] M. Enterría, F.J. Martín-Jimeno, F. Suárez-García, J.I. Paredes, M.F.R. Pereira, J.I. Martins, A. Martínez-Alonso, J.M.D. Tascón, J.L. Figueiredo, Effect of nanostructure on the supercapacitor performance of activated carbon xerogels obtained from hydrothermally carbonized glucose-graphene oxide hybrids, Carbon 105 (2016) 474–483. https://doi.org/10.1016/j.carbon.2016.04.071

[148] M.Y. Ho, P.S. Khiew, Heat-treated Fe3O4—activated carbon nanocomposite for high performance electrochemicalcapacitor, Adv. Mater. Res. 894 (2014) 349–354. https://doi.org/10.4028/www.scientific.net/AMR.894.349

[149] P. He, K. Yang, W. Wang, F. Dong, L. Du, H. Liu, Nanosized Fe3O4-modified activated carbon for supercapacitor electrodes. Russ. J. Electrochem. 49 (2013) 354–358. https://doi.org/10.1134/S1023193513040095

[150] I. Oh, M. Kim, J. Kim, Controlling hydrazine reduction to deposit iron oxides on oxidized activated carbon for supercapacitor application, Energy 86 (2015) 292-299. https://doi.org/10.1016/j.energy.2015.04.040

[151] Q. Tan, Y. Xu, J. Yang, L. Qiu, Y. Chen, X. Chen, Preparation and electrochemical properties of the ternary nanocomposite of polyaniline/activated carbon/TiO2 nanowires for supercapacitors, Electrochimica Acta 88 (2013) 526–529. https://doi.org/10.1016/j.electacta.2012.10.126

[152] H. Liu, P. He, Z. Li, Y. Liu, J. Li, A novel nickel-based mixed rare-earth oxide/activated carbon supercapacitor using room temperature ionic liquid electrolyte, Electrochimica Acta 51 (2006) 1925–193. https://doi.org/10.1016/j.electacta.2005.06.034

[153] H. G. Jung, N. Venugopal, B. Scrosati, Y. K. Sun, A high energy and power density hybrid supercapacitor based on an advanced carbon-coated Li4Ti5O12 electrode, J. Power Sources 221 (2013) 266-271. https://doi.org/10.1016/j.jpowsour.2012.08.039

[154] K. Karthikeyana, V. Aravindanb, S. B. Leea, I.C. Janga, H. H. Lima, G. J. Parkc, M. Yoshioc, Y. S. Lee, A novel asymmetric hybrid supercapacitor based on Li2FeSiO4 and activated carbon electrodes, J. Alloy Comp. 504 (2010) 224–227. https://doi.org/10.1016/j.jallcom.2010.05.097

[155] Q. Qu, L. Li , S. Tian, W. Guo, Y. Wu, R. Holze, A cheap asymmetric supercapacitor with high energy at high power: Activated

carbon//K0.27MnO2·0.6H2O, J. Power Sources 195 (2010) 2789–2794. https://doi.org/10.1016/j.jpowsour.2009.10.108

[156] P. C. Gao, A. H. Lu, W. C. Li, Dual functions of activated carbon in a positive electrode for MnO2-based hybrid supercapacitor, J. Power Sources 196 (2011) 4095–4101. https://doi.org/10.1016/j.jpowsour.2010.12.056

[157] M. Kim, Y. Hwang, K. Min, J. Kim, Introduction of MnO2 nanoneedles to activated carbon to fabricate high-performance electrodes as electrochemical supercapacitors, Electrochimica Acta 113 (2013) 322–331. https://doi.org/10.1016/j.electacta.2013.09.058

[158] Z. Lin, X. Yan, J. Lang, R. Wang, L-B. Kong, Adjusting electrode initial potential to obtain high-performance asymmetric supercapacitor based on porous vanadium pentoxide nanotubes and activated carbon nanorods, J. Power Sourc. 279 (2015) 358-364. https://doi.org/10.1016/j.jpowsour.2015.01.034

[159] G. Godillot, P-L. Taberna, B. Daffos, P. Simon, C. Delmas, L. Guerlou-Demourgues, High power density aqueous hybrid supercapacitor combining activated carbon and highly conductive spinel cobalt oxide, J. Power Sourc. 331 (2016) 277-284. https://doi.org/10.1016/j.jpowsour.2016.09.035

[160] L-J. Xie, J-F. Wu, C-M. Chen , C-M Zhang,, L. Wan, J-L Wang, Q-Q. Kong, C-X. Lv, K-X. Li, G-H. Sun, A novel asymmetric supercapacitor with an activated carbon cathode and a reduced graphene oxide-cobalt oxide nanocomposite anode, J. Power Sourc. 242 (2013) 148-156 https://doi.org/10.1016/j.jpowsour.2013.05.081

[161] Q.T. Qu, Y. Shi, S. Tian, Y.H. Chen, Y.P. Wu, R. Holze, A new cheap asymmetric aqueous supercapacitor: Activated carbon//NaMnO2, J. Power Sourc. 194 (2009) 1222–1225. https://doi.org/10.1016/j.jpowsour.2009.06.068

[162] K. Karthikeyan, V. Aravindan, S.B. Lee, I.C. Jang, H.H. Lim, G.J. Park, M. Yoshio, Y.S. Lee, A novel asymmetric hybrid supercapacitor based on Li2FeSiO4 and activated carbon electrodes, J. Alloy. Comp. 504 (2010) 224–227. https://doi.org/10.1016/j.jallcom.2010.05.097

[163] Q. Qu, P. Zhang, B. Wang, Y. Chen, S. Tian, Y. Wu, R. Holze, Electrochemical Performance of MnO2 Nanorods in Neutral Aqueous Electrolytes as a Cathode for Asymmetric Supercapacitors, J. Phys. Chem. C 113 (2009) 14020-14027. https://doi.org/10.1021/jp8113094

[164] C. Xu, H. Du, B. Li, F. Kang, Y. Zeng, Asymmetric activated carbon-manganese dioxide capacitors in mild aqueous electrolytes containing alkaline-earth cations, J. Electrochem. Soc. 156 (2009) A435-A441. https://doi.org/10.1149/1.3106112

[165] A. Yuan , Q. Zhang, A novel hybrid manganese dioxide/activated carbon supercapacitor using lithium hydroxide electrolyte, Electrochem. Comm. 8 (2006) 1173–1178. https://doi.org/10.1016/j.elecom.2006.05.018

[166] T. Brousse, M. Toupin, D. Be'langer, A hybrid activated carbon-manganese dioxide capacitor using a mild aqueous electrolyte, J. Electrochem. Soc. 4 (2004) A614-A622. https://doi.org/10.1149/1.1650835

[167] H-Q. Wang, Z-S. Li, Y-G. Huang, Q-Y. Li, X-Y. Wang, A novel hybrid supercapacitor based on spherical activated carbon and spherical MnO2 in a non-aqueous electrolyte, J. Mater. Chem. 20 (2010) 3883-3889. https://doi.org/10.1039/c000339e

[168] S. Nohara, T. Asahina, H. Wada, N. Furukawa, H. Inoue, N. Sugoh, H. Iwasaki, C. Iwakura, Hybrid capacitor with activated carbon electrode, Ni(OH)2 electrode and polymer hydrogel electrolyte, J. Power Sourc. 157 (2006) 605-609. https://doi.org/10.1016/j.jpowsour.2005.07.024

[169] H.B. Li, M.H. Yu, F.X. Wang, P. Liu, Y. Liang, J. Xiao, C.X. Wang, Y.X. Tong, G.W. Yang, Amorphous nickel hydroxide nanospheres with ultrahigh capacitance and energy density as electrochemical pseudocapacitor materials, Nat. Commun. 4 (2013) 1894. https://doi.org/10.1038/ncomms2932

[170] G.-H. Yuan, Z.-H. Jiang, A. Aramata, Y.-Z. Gao, Electrochemical behavior of activated-carbon capacitor material loaded with nickel oxide, Carbon 43 (2005) 2913-2917. https://doi.org/10.1016/j.carbon.2005.06.027

[171] V. Ganesh, S. Pitchumani, V. Lakshminarayanan, New symmetric and asymmetric supercapacitors based on high surface area porous nickel and activated carbon, J. Power Sourc. 158 (2006) 1523–1532. https://doi.org/10.1016/j.jpowsour.2005.10.090

[172] J-W. Lang, L-B. Kong, M. Liu, Y-C. Luo, L. Kang, Asymmetric supercapacitors based on stabilized α-Ni(OH)2 and activated carbon, J Solid State Electrochem. 14 (2010) 1533–1539. https://doi.org/10.1007/s10008-009-0984-1

[173] L.B. Kong, M. Liu, J.W. Lang, Y.C. Luo, L. Kang, Asymmetric supercapacitor based on loose-packed cobalt hydroxide nanoflake materials and activated carbon, J. Electrochem. Soc. 156 (2009) A1000-A1004. https://doi.org/10.1149/1.3236500

[174] L-B. Kong, M. Liu, J-W. Lang, Y-C. Luo, L. Kang, Asymmetric supercapacitor based on loose-packed cobalt hydroxide nanoflake materials and activated carbon, J. Electrochem. Soc. 12 (2009) A1000-A1004. https://doi.org/10.1149/1.3236500

[175] F. Zhou, Q. Liu, J. Gu, W. Zhang, D. Zhang, A facile low-temperature synthesis of highly distributed and size-tunable cobalt oxide nanoparticles anchored on activated carbon for supercapacitors, J. Power Sourc. 273 (2015) 945-953. https://doi.org/10.1016/j.jpowsour.2014.09.168

[176] Z. Lin, X. Yan, J. Lang, R. Wang, L-B. Kong, Adjusting electrode initial potential to obtain high-performance asymmetric supercapacitor based on porous vanadium pentoxide nanotubes and activated carbon nanorods, J. Power Sourc. 279 (2015) 358-364. https://doi.org/10.1016/j.jpowsour.2015.01.034

[177] L-M. Chen, Q-Y. Lai, Y-J. Hao, Y. Zhao, X.Y. Ji, Investigations on capacitive properties of the AC/V2O5 hybrid supercapacitor in various aqueous electrolytes. J Alloy. Comp, 467 (2009) 465–471. https://doi.org/10.1016/j.jallcom.2007.12.017

[178] L-Q. Fan, G-J. Liu, J-H. Wu, L. Liu, J-M. Lin, Y-L. Wei, Asymmetric supercapacitor based on graphene oxide/polypyrrolecomposite and activated carbon electrodes, Electrochimica Acta 137 (2014) 26–33. https://doi.org/10.1016/j.electacta.2014.05.137

[179] Z. Fan, J. Yan , T. Wei, L. Zhi, G. Ning, T. Li, F. Wei, Asymmetric supercapacitors based on graphene/mno2 and activated carbon nanofiber electrodes with high power and energy density, Adv. Funct. Mater. 21 (2011) 2366–2375. https://doi.org/10.1002/adfm.201100058

[180] Y-G. Wang, Z-D. Wang, Y-Y. Xia, An asymmetric supercapacitor using RuO2/TiO2 nanotube composite and activated carbon electrodes, Electrochimica Acta 50 (2005) 5641–5646. https://doi.org/10.1016/j.electacta.2005.03.042

[181] M. Kim, J. Kim, Development of high power and energy density microsphere silicon carbide–MnO2 nanoneedles and thermally oxidized activated carbon asymmetric electrochemical supercapacitors, Phys. Chem. Chem. Phys. 16 (2014) 11323-11336. https://doi.org/10.1039/c4cp01141d

[182] X. Wang, C. Yan, A. Sumboja, P. S. Lee, High performance porous nickel cobalt oxide nanowires for asymmetric supercapacitor, Nano Energy 3 (2014) 119–126. https://doi.org/10.1016/j.nanoen.2013.11.001

[183] Y-g. Wang, Y-y. Xia, A new concept hybrid electrochemical surpercapacitor: Carbon/LiMn2O4 aqueous system, Electrochem. Comm. 7 (2005) 1138-1142. https://doi.org/10.1016/j.elecom.2005.08.017

[184] Y-Y Liang, H-L Li, X-G Zhang, A novel asymmetric capacitor based on Co(OH)2/USY composite and activated carbon electrodes, Mater. Sci. Eng. A 473 (2008) 317–322. https://doi.org/10.1016/j.msea.2007.03.087

[185] L-Q. Fan, G-J. Liu, J-H. Wu, L. Liu, J-M. Lin, Y-L. Wei, Asymmetric supercapacitor based on graphene oxide/polypyrrole composite and activated carbon electrodes, Electrochimica Acta 137 (2014) 26–33. https://doi.org/10.1016/j.electacta.2014.05.137

[186] D. Xuan, W. Chengyang, C. Mingming, J. Yang, W. Jin, Electrochemical performances of nanoparticle Fe3O4/activated carbon supercapacitor using KOH electrolyte solution, J. Phys. Chem. C 113 (2009) 2643–2646. https://doi.org/10.1021/jp8073859

Chapter 4

Ultrasonic Assisted Synthesis of 2D-Functionalized Grapheneoxide@PEDOT Composite Thin Films and its Application in Electrochemical Capacitors

P. Ramyakrishna[a,b], B. Rajender [c,*], G. Sadanandam[d], P. Srinivas[e], Inamuddin[a,f]

[a]Department of Applied Chemistry, Faculty of Engineering and Technology, Aligarh Muslim University, Aligarh 202002, India

[b]College of Chemistry and Chemical Engineering, Hunan University, Changsha 410082, P R China

[c]CAS Key Laboratory of Nanosystem and Hierarchical Fabrication, National Center for Nanoscience and Technology, Beijing 100190, PR China

[d]Department of Civil and Chemical Engineering, University of South Africa, Johannesburg - 1709, South Africa

[e]Department of MaterialsScience and Metallurgical Engineering, Indian Institute of Technology Hyderabad, Sangareddy-502285, India

[f]Chemistry Department, King Abdulaziz University, Jeddah-21589, Saudi Arabia

*research.raaj@gmail.com

Abstract

Poly(3,4-ethylene dioxythiophene) (PEDOT)@functionalized graphene oxide (FGO) hydrogel thin films were fabricated by simple sonochemical polymerization of EDOT with 1 wt% of FGO. The unique properties of the composites were investigated by FT-infrared spectroscopy, X-ray diffraction, scanning electron microscopy, thermogravimetric analysis, cyclic voltammetry and impedance analysis. The potential utility of synthesized composites as electrodes for high-performance supercapacitors has been assessed. Sulfonated graphene oxide (SFGO)-PEDOT showed higher electrochemical performance and long-durability than that of carboxyl functionalized graphene oxide (CFGO)-PEDOT composite and pristine PEDOT. This composite satisfied the prerequisite of elongated cycle life mandatory for a capacitive energy storage device system.

Keywords

Functionalized GO, PEDOT Composite, Two-dimensional, High-performance Supercapacitor Device, Eco-friendly

Contents

1.   Introduction

The dramatic climate change and the restricted accessibility of fossil fuels have spurred universal enthusiasm for creating renewable energy technologies. These days, sustainable and renewable sources from wind power, solar energy, and hydropower are expected to alleviate the heavy burdens on the present vitality and ecological concerns [1]. However, these energy resources are intermittent which vigorously depend on natural conditions. Thus, efficient energy storage systems are critically in demand to turn the off-peak electricity into distinct forms for storage to full the energy shortage throughout the on-peak period. Nowadays, extensive consideration has been a spotlight on electrical double layer capacitors because of their rapid charge-discharge processes, excessive cycle durability, high power density, and low maintenance cost compared to rechargeable batteries [2-4]. Supercapacitors can be successfully utilized in several areas including electric vehicles, consumer electronics, defense and aviation applications. Supercapacitors are frequently hybridized with fuel cells and batteries for large power and energy harvesting applications. They are generally categorized as pseudocapacitors and electrical double-layered capacitors (EDLC) in accordance with their energy storage process. The efficiency of a supercapacitor depends principally on the properties of electrode material. In spite of substantial enhancements in electrode substances for supercapacitors, a few shortcomings which include the low capacitance of carbon materials, the lower electrical conductivity of metal oxides, and poor cycling durability of conducting polymers make the electrode materials less efficient for supercapacitor applications. To overcome these problems, advance endeavours are required to enhance the electrochemical performance of the electrode materials so that higher power and energy densities can be achieved in the case of nanostructured composites through combining the appealing properties of individual nanomaterials and the composites show great potential to utilize those properties for a wide range of applications [4-7]. Synergistic impacts of such multicomponent superior binary systems result in splendid electrochemical overall performance, additionally give positive mechanical strength in addition to fabric.

Graphene oxide (GO), the oxidation profile of graphene, the layered structure contains hydrophilic oxygen functional groups, for example hydroxyl and epoxide on the basal planes and another carbonyl group and carboxylic acid and on the edges of the GO nanosheets and the carbon atoms situated on the edges of the GO sheets are other active sites, which can be utilized for chemical functionalization [8], and GO shows high ionic conductivity and super proton conductivity and promptly dispersed in water as distinctive sheets. The excessive dispersion strength of GO has received attention in applications as functional nanocomposites and nanoelectronic materials. Besides, numerous substances have been integrated GO layers, consisting metal oxides, conducting polymers, and carbon nanotubes to get a synergistic commitment from hybridization to manufacture supercapacitors [9]. GO exhibits extremely long cycle life due to the ultra-high specific surface space and the EDLC mechanism that is the basis of the charge-discharge process. However, the capacitance of pure GO films is often limited by re-stacking of layers. Poly(3,4-ethylene dioxythiophene) [PEDOT] a derivative of polythiophene is one of the best-known π-conjugated polymers and considered a promising material for supercapacitor electrodes mainly because of its superior conductivity, great chemical and electrochemical stability, and excellent dispersibility. Contemporary, researchers have revealed that the synergistic contribution of GO with conducting polymers can offer extra and remarkable electrochemical capacities than every single component. Numerous studies mentioned the aggregate of PEDOT with GO-empowered the material to have higher electrochemical capacitance and a better charge-discharge cycle life [10-14].

During our study, we have synthesized carboxyl and sulfonic functionalized graphene oxide and introduced it into conducting polymer (PEDOT) using sonochemical polymerization pathway (Scheme 1). These composites were characterized by physical, spectral and electrochemical techniques. Functionalized graphene oxide–PEDOT nanocomposite thin film electrodes were fabricated and evaluated for high-performance supercapacitor applications.

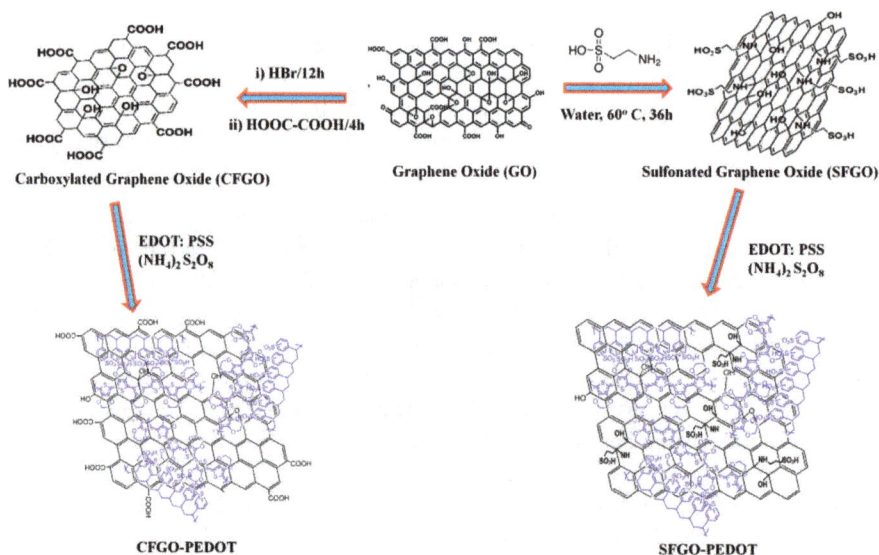

Scheme 1. Schematic illustration of functionalized graphene oxide-PEDOT composite.

2. Experimental

2.1 Formation of functionalized graphene oxide@ PEDOT composites

Carboxylic [15] or sulfonic [16] functionalized graphene oxides (FGO) were synthesized as per the modified reported procedures. In a typical procedure, 1 wt.% of carboxyl or sulfonic acid functionalized graphene oxide was dispersed in a 70 mL aqueous solution containing EDOT (1 mL) and PSS (5mM) and the mixture was sonicated for 0.5 h in an ultrasonic bath (100 W power and 50 kHz). An aqueous solution (30 mL) consisting of ammonium persulfate (2.28 g, 0.1 M) and PSS (5mM) was added all into the above mixture. This mixture was sonicated for 1 h and the starting of polymerization was identified by the appearance of the light blue color in the reaction mixture. The resultant suspension was aged for 24 h. After the required time, the polymer was formed as a hydrogel which was poured into acetone to precipitate out. The resultant precipitation was filtered, washed with acetone and dried overnight in a vacuum oven at 60°C. These composites were labelled as CFGO-PEDOT and SFGO-PEDOT (i.e., C=carboxyl acid groups, S=sulfonic acid groups). For comparison, pristine PEDOT was acquired utilizing this methodology without the addition of FGO.

2.2 Instrumentation

The electrochemical properties (galvanostatic charge-discharge experiments, cyclic voltammetry, and impedance analysis) of all the samples were investigated using prototype supercapacitor device assembled in a symmetrical two-electrode and the electrolytic solution was 1 M H_2SO_4. The electrode was fabricated using PEDOT hydrogel coated on stainless steel mesh and dried in an oven to get a thin film of the electrode. Electrochemical measurements were carried out using Autolab 302N potentiostat/galvanostat (Switzerland) with an impedance analyzer (FRA32M.X). Charge-discharge experiments were recorded from 0 to 0.6 V and cyclic voltammetry was performed between −0.2 to 0.6 V at distinct sweep rates. Impedance spectroscopy measurements were recorded in the frequency range of 40 kHz–10 mHz at various voltages by applying an AC voltage of 5 mV amplitude using three-electrode cell set-up *i.e.*, platinum counter electrode, PEDOT composite as working electrode, and saturated calomel electrode (SCE) as a reference electrode. Samples for FT-IR analysis were mixed with KBr powder and compressed into pellets. The sample powder was evenly dispersed in KBr and spectra's were recorded using gas chromatography-FTIR spectrometer (model 670, Nicolet Nexus, Minnesota). Morphology studies of the composites were performed with the aid of Hitachi S-4300 SE/N field emission scanning electron microscope, Tokyo, Japan. XRD patterns of the composites were obtained using Bruker AXS D8 advance X-ray diffractometer (Karlsruhe, Germany) with Cu-Kα radiation (λ = 1.54 Å) at a scan speed of 0.045° min^{-1}.

3. Results and discussion

The functionalized graphene oxide was incorporated into the PEDOT matrix by ultrasonic irradiation to form a homogeneous binary composite gel. Ultrasonic irradiation assisted method has been widely used because it is fast, facile, low-cost, mild, scalable and environmentally friendly. Ultrasonic assisted or sonication method has numerous chemical and physical effects. The composites were well characterized by electrochemical, spectral, and physical methods and utilized as binder-free electrode materials in the fabrication of symmetric supercapacitor cell.

The FT-IR spectra of functionalized GO-PEDOT composites are shown in Fig. 1. In the Fig. 1a the absorption peaks at 3415 and 3150 cm^{-1} are characteristics to the O-H stretching vibration and aromatic C-H stretching vibration, 2855 and 2930 cm^{-1} are due to the aliphatic C-H stretching vibration of the ethylenedioxy group, 1400 and 1515 cm^{-1} are quinonoid C-C and C=C stretching bands of thiophene ring. The peaks appearing at 1090, 975, 830 and 680 cm^{-1} corresponding to the symmetric stretching of S=O and C–S stretching and bending band of thiophene ring. Fig. 1b and Fig. 1c exhibit similar peaks except 1750 and 1630 cm^{-1}which corresponds to the C=O and C=C stretching vibrations of functionalized GO. In the carboxyl, functionalized GO-PEDOT composite C=O peak at 1750 cm^{-1}clearly shows the high intensity of whereas in case of sulfonic acid functionalized GO-PEDOT composite asymmetric stretching of SO$_2$ peak appeared at 1320 cm^{-1}. This result confirms the presence of functionalized GO present in the PEDOT matrix.

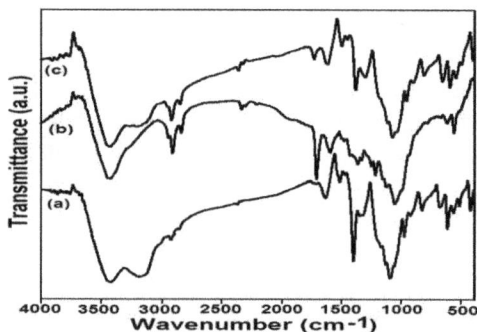

Fig. 1. FT-IR spectra of (a) PEDOT, (b) CFGO-PEDOT, and (c) SFGO-PEDOT.

Fig. 2 depicts the XRD patterns of PEDOT, CFGO-PEDOT, and SFGO-PEDOT. The XRD spectra of pristine PEDOT sample (Fig. 2a) exhibits many peaks at 2θ = 6.0, 16.8, 20.3, 22.7, 26, 28.3, 29.1 and 38.7° with the corresponding d-spacing of 14.3, 5.3, 4.4, 3.9, 3.4, 3.2, 3.1 and 2.3 Å indicates the crystalline nature. The XRD spectra of composites (Figs. 2b and 2c) are identical to pristine PEDOT and the peaks of functionalized GO did not appear in the composites. This result indicates that FGO is wrapped by the PEDOT.

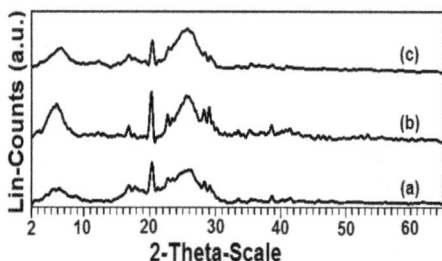

Fig. 2. XRD pattern of(a) PEDOT, (b) CFGO-PEDOT, and (c) SFGO-PEDOT.

Fig. 3 shows the SEM images of PEDOT and its composites. The SEM image of PEDOT displays aggregated densely flake-like structure (Fig. 3a). As evident from Fig. 3b, the most of CFGO surface was wrapped with a smooth thin PEDOT layer and composite shows layer type morphology with two-dimensionality. Incorporation of CFGO into PEDOT (Fig. 3c) exhibits curved and smooth layer like morphology and it seems identical to the CFGO-PEDOT composite. The PEDOT composites reveal two-dimensional layer morphology like graphene oxide.

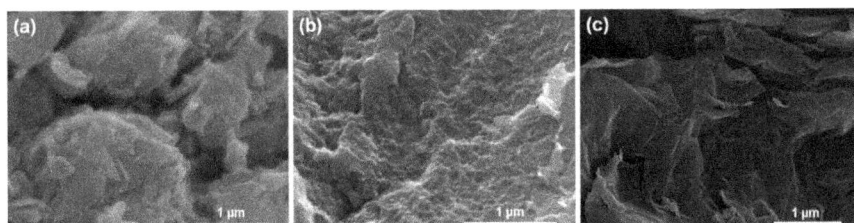

Fig. 3. SEM images of (a) PEDOT, (b) CFGO-PEDOT, and (c) SFGO-PEDOT.

The capacitive performance of the samples was studied by galvanostatic charge-discharge (CD) and cyclic voltammetry (CV) measurements employing two-electrode cell configuration. C supportive investigation of the electrochemical performance and quantitative estimation of the specific capacitance of the PEDOT thin film electrodes. A cyclic voltammogram was constructed between -0.2 and 0.6 V at higher sweep rate 10 mV s^{-1} for the symmetric cell configuration is constructed using PEDOT samples. The cyclic voltammograms of PEDOT, CFGO-PEDOT, and SFGO-PEDOT (Fig. 4) and all electrodes obviously show all the voltammograms have virtually rectangular shape, acknowledging the absolute electrochemical capacitive behavior. All the curves for the electrodes display no obvious peaks, denotes that those electrodes are charged and discharged at a pseudo-constant rate over the full voltammetric curve and that the adsorption-desorption process of ions n the electrode surface. This outcome indicates that the diffusion of ions from the electrolyte can boost entrance to nearly all active sites of the electrode. The value of specific capacitance for the SFGO-PEDOT (156 F g^{-1}) reflects superior capacitance performance than CFGO-PEDOT (119 F g^{-1}) and pristine PEDOT (90 F g^{-1}) in the two-electrode symmetric device. The CV curve of SFGO-PEDOT device exhibited a more rectangular shape and larger area than that of the other devices, indicating a faster charge transfer rate and a higher capacitance. The presence of sulfonic acid functional groups in SFGO-PEDOT boosting the capacitance due to built an electrochemical active larger surface area on the electrodes, increases the conductivity of the composite and provide the rapid Faradaic reaction to facilitating the charge-storage mechanism due to fast electron paths, which is crucial to both the greater specific capacitance and long-cycle life of supercapacitor [17-20].

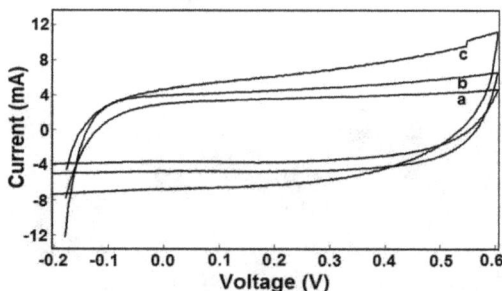

Fig. 4. CV curves of (a) PEDOT, (b) CFGO-PEDOT and (c) SFGO-PEDOT electrodes in symmetric cell device at a sweep rate of 10 mV s^{-1}.

Fig. 5 presents the CV curves of the SFGO-PEDOT composite electrode at different sweep rates from 1 to 100 mV s^{-1} and the value of specific capacitance was obtained as 247, 202, 156, 154, 148, 136, 130, and 100 F g^{-1} with a sweep rate at 1, 5, 10, 20, 30, 40, 50, and 100 mV s^{-1}, correspondingly. As the sweep rate increases the specific capacitance decreases, this loss in capacity can be implied by an ion-interchange mechanism. The reason may be that, at minimum current, the ions have enough time to diffuse into the electrodes inner surfaces while at the maximum current the ions will partly penetrate into the inner surfaces [21]. The CV diagrams at various sweep rates show analogous and symmetric shapes, offering a superb capacitance of this supercapacitor.

Charge-discharge (CD) performance is one of the significant characteristics for the industrial application of supercapacitors. Galvanostatic charge-discharge experiments were observed for the three PEDOT samples. Fig. 6 represents the typical CD curves for the PEDOT thin film electrode at constant specific current density 0.25 A g^{-1} and the values of discharge specific capacitance, energy, and power density were calculated. The charge-discharge capacitance of the SFGO-PEDOT (195 F g^{-1}) was superior to the CFGO-PEDOT (116 F g^{-1}) and PEDOT (73F g^{-1}) samples. The energy densities were 9.8, 5.8, and 1.5Wh kg^{-1}, respectively at a power density of 160 W kg^{-1}. A similar trend is observed in the CV values obtained from cyclic voltammetry analysis. Further, SFGO-

PEDOT composite's encounter for minimum to maximum charge-discharge current densities at 0.25, 0.50, 0.75, 1.25, and 2.5 A g^{-1} (Fig. 7) and the amount of charge-discharge capacitances were 195, 204, 202, 193, and 171 F g^{-1}; energy density was 9.8, 10.2, 10.1, 9.65 and 8.55 Wh kg^{-1} and power densities was 160, 320, 480, 800 and 1600 W kg^{-1}, correspondingly. This outcome shows that the electrode stands for even higher charge-discharge rate of 2.5 A g^{-1}.

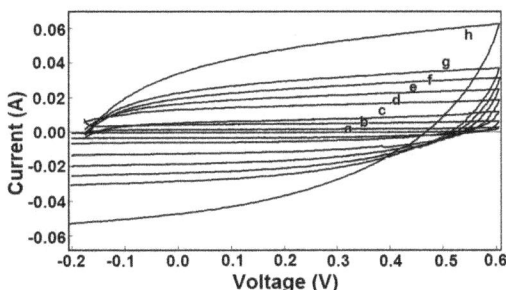

Fig. 5. CV curves of SFGO-PEDOT electrode material in a symmetric cell configuration at scanning rates of (a) 1, (b) 5, (c) 10, (d) 20, (e) 30, (f) 40, (g) 50, and (h) 100 mV s^{-1}.

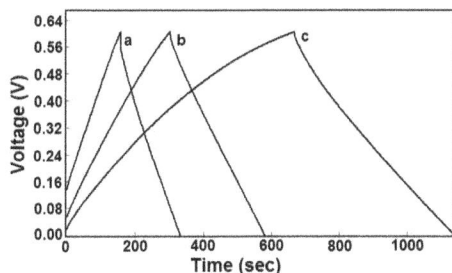

Fig. 6. Charge-discharge curves of (a) PEDOT, (b) CFGO-PEDOT and (c) SFGO-PEDOT electrode materials in symmetric cell configuration at 0.25 A g^{-1} current density.

For commercial applications, supercapacitors should have everlasting cyclic durability and maximum retention of the capacitance. The durability of the thin layer 2D-SFGO-PEDOT thin film was evaluated by galvanostatic charge-discharge experiment at 0.75 A g^{-1} current density for 6000 cycles in 1 M aqueous H_2SO_4 electrolyte (Fig. 8). The symmetrical shaped CD curves represent the electrode has good CD reversibility and the galvanostatic CD curves were observed in distorted Λ-shape attributed to the redox reaction throughout the Faradaic process. The specific capacitance and retention of the SFGO-PEDOT electrode exhibit 202 F g^{-1} and 91% with coulombic efficiency (94-99%) during the 6000 charge-discharge cycles. This result indicates that he SFGO-PEDOT electrode has high electrochemical performance and long cycle life.

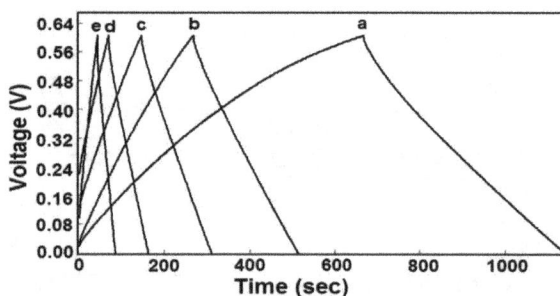

Fig. 7. Charge–discharge curves of SFGO-PEDOT electrode in a symmetric cell at various current densities (a) 0.25, (b) 0.5, (c) 0.75, (d) 1.25, and (e) 2.5 A g^{-1}.

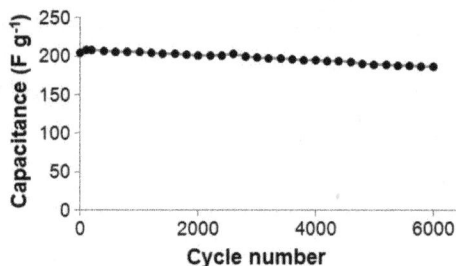

Fig. 8. Specific capacitance with cycle numbers for SFGO-PEDOT electrode symmetric device at current density 0.75 A g^{-1}.

Electrochemical impedance spectroscopy (EIS) is a vital analytical approach utilized to know the information concerning frequency responses of electrochemical capacitors and the capacitance phenomena happening at the electrodes.

EIS test is recorded for the system, the SFGO-PEDOT composite at an applied voltage of 0.8 V in the frequency range of 40 kHz to 10 mHz as shown in Fig. 9. Equivalent circuit obtained for SFGO-PEDOT is shown in Fig. 9a.

This Nyquist plot shows a miserable semicircle in the high-frequency range and a close vertical line in the low-frequency range. Charge-transfer resistance (R_{ct}), solution resistance (R_s), and time constant (τ) values are 0.56 Ω, 0.32 Ω, and 0.1 ms, respectively. The very low value of the time constant, charge-transfer and solution resistance suggests the little equivalent series resistance, rapid charge-discharge progression, and high electrical conductivity of the electrode/electrolyte. Bode diagrams of frequency versus phase angle are shown in Fig. 9b. From the Bode plot, it can be detected that the phase angle value is ~85°; which is nearer to that of a perfect capacitor 90°.

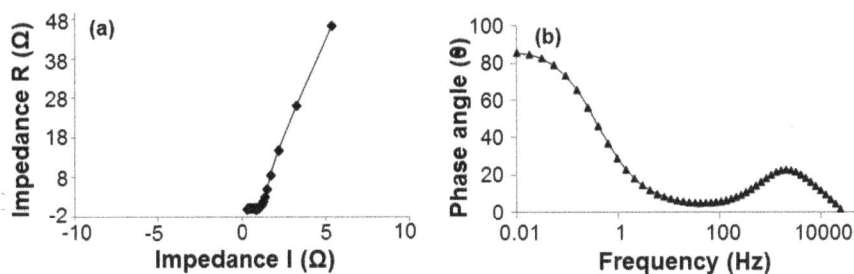

Fig. 9. (a) Nyquist plot and(b) bode plot of frequency versus phase angle of SFGO-PEDOT composite at applied potential 0.8 V.

Conclusions

The PEDOT@functionalized graphene oxide composite electrodes have been successfully fabricated for a prototype supercapacitor. In this study summery, sulfonic acid functionalized graphene oxide introduced to flexible conductive polymer, PEDOT by ultrasonically assisted chemical polymerization pathway. SFGO-PEDOT composite delivers long-cycle life performance with a power density of 480 W kg^{-1}, the specific capacitance of 202 F g^{-1}, and energy density of 10.1 W h kg^{-1} at 0.75 A g^{-1} current density. This composite thin film demonstrates the capacitance retention as 91% after 6000 uninterrupted charge-discharge cycles. The excellent performance of SFGO-PEDOT is attributed to the introduction of the SFGO layer and the PEDOT-PSS layer. The conductive PEDOT-PSS layer can remarkably facilitate the charge transportation by increasing the kinetic inhibition of functional groups to diffusion. Moreover, this methodology will be very useful for wearable electronics.

References

[1]   E.T. Mombeshora, V.O. Nyamori, A review on the use of carbon nanostructured materials in electrochemical capacitors, Int. J. Energy Res. 39 (2015) 1955-1980. https://doi.org/10.1002/er.3423

[2]   A. Burke, Ultracapacitors: why, how, and where is the technology, J. Power Sources 91 (2000) 37-50. https://doi.org/10.1016/S0378-7753(00)00485-7

[3]   JR. Miller, P. Simon, Electrochemical capacitors for energy management, Science 321 (2008) 651-652. https://doi.org/10.1126/science.1158736

[4]   C. Zhong, Y.D. Deng, W.B. Hu, J.L. Qiao, L. Zhang, J.J. Zhang, A review of electrolyte materials and compositions for electrochemical supercapacitors, Chem. Soc. Rev. 44 (2015) 7484-7539. https://doi.org/10.1039/C5CS00303B

[5]   J. Wang, S. Dong, B. Ding, Y. Wang, X. Hao, H. Dou, Y. Xia, X. Zhang, Pseudocapacitive materials for electrochemical capacitors: from rational synthesis to capacitance optimization, Natl. Sci. Rev. 0 (2016) 1-20. https://doi.org/10.1093/nsr/nww072

[6]   H.J. Kim, M. Osada, T. Sasaki, Advanced capacitor technology based on two-dimensional nanosheets, Jpn. J. Appl. Phys. 55 (2016) 1102A3. https://doi.org/10.7567/JJAP.55.1102AA

[7]   M. Shao, R. Zhang, Z. Li, M. Wei, D.G. Evans, X. Duan, Layered double hydroxides toward electrochemical energy storage and conversion: design, synthesis and applications, Chem. Commun. 51 (2015) 15880-15893. https://doi.org/10.1039/C5CC07296D

[8]   D. Sharma, S. Kanchi, M.I. Sabela, K. Bisetty, Insight into the biosensing of graphene oxide: present and future prospects, Arabian J. Chem. 9 (2016) 238-261. https://doi.org/10.1016/j.arabjc.2015.07.015

[9]   H.S. Fan, H. Wang, N. Zhao, J. Xu, F. Pan, Nano-porous architecture of N-doped carbon nanorods grown on graphene to enable synergetic effects of supercapacitance, Sci. Rep. 4 (2014) 7426-7427. https://doi.org/10.1038/srep07426

[10]   I. Cha, E.J. Lee, H.S. Park, J.H. Kim, Y.H. Kim, C. Song, Facile electrochemical synthesis of polydopamine-incorporated graphene oxide/PEDOT hybrid thin films for pseudocapacitive behaviors, Synth. Met. 195 (2014) 162-166. https://doi.org/10.1016/j.synthmet.2014.05.019

[11] M. Wang, R. Jamal, Y. Wang, L. Yang, F.F. Liu, T. Abdiryim, functionalization of graphene oxide and its composite with poly(3,4-ethylenedioxythiophene) as electrode material for supercapacitors, Nanoscale Res. Lett. 10 (2015) 370. https://doi.org/10.1186/s11671-015-1078-x

[12] K. Zhang, X. Duan, X. Zhu, D. Hu, J. Xu, L. Lu, Nanostructured graphene oxide–MWCNTs incorporated poly(3,4-ethylenedioxythiophene) with a high surface area for sensitive determination of diethylstilbestrol, Synth. Met. 195 (2014) 36-43. https://doi.org/10.1016/j.synthmet.2014.05.005

[13]   M. Islam, D. Cardillo, T. Akhter, S.H. Aboutalebi, H.K. Liu, K. Konstantinov, S.X. Dou, Liquid-crystal-mediated self-assembly of porous α-Fe_2O_3nanorods on PEDOT:PSS functionalized graphene as a flexible ternary architecture for capacitive energy storage, Part. Part. Syst. Charact. 33 (2016) 27-37. https://doi.org/10.1002/ppsc.201500150

[14] J. Sun, Y. Huang, C. Fu, Y. Huang, M. Zhu, X.Tao, C. Zhi, H. Hu, A high performance fiber-shaped PEDOT@MnO_2//C@Fe_3O_4 asymmetric supercapacitor for wearable electronics, J. Mater. Chem. A. 4 (2016) 14877-14883. https://doi.org/10.1039/C6TA05898A

[15] Y. Liu, R. Deng, Z. Wang, H. Liu, Carboxyl-functionalized graphene oxide–polyaniline composite as a promising supercapacitor material, J. Mater. Chem. 22 (2012) 13619-13624. https://doi.org/10.1039/c2jm32479b

[16] J. Lu, W. Liu, H. Ling, J. Kong, G. Ding, D. Zhou, X. Lu, Layer-by-layer assembled sulfonated-graphene/polyaniline nanocomposite films: enhanced electrical and ionic conductivities, and electrochromic properties, RSC Adv. 2 (2012) 10537-10543. https://doi.org/10.1039/c2ra21579a

[17] C. Bora, J. Sharma, S. Dolui, Polypyrrole/sulfonated graphene composite as electrode material for supercapacitor, J. Phys. Chem. C. 118 (2014) 29688-29694. https://doi.org/10.1021/jp511095s

[18] W. Feng, Q. Zhang, Y. Li, Y. Feng, Preparation of sulfonatedgraphene/polyaniline composites in neutral solution for high-performance supercapacitors, J. Solid State Electrochem. 18 (2014) 1127–1135. https://doi.org/10.1007/s10008-013-2369-8

[19] B. Ravi, B. Rajender, S. Palaniappan, Improving the electrochemical performance by sulfonation of polyaniline-graphene-silica composite for high performance supercapacitor, Inter. J. Polym. Mater.Polym.Biomater. 65 (2016) 835-840. https://doi.org/10.1080/00914037.2016.1171221

[20]   B. Ravi, B. Rajender, S. Palaniappan, One-step preparation of sulfonated carbon and subsequent preparation of hybrid material with polyaniline salt: a promising supercapacitor electrode material, J. Solid State Electrochem, (2016) 1-10.

[21]   V. Augustyn, P. Simon, B. Dunn, Pseudocapacitive oxide materials for high-rate electrochemical energy storage, Energy Environ. Sci. 7 (5) (2014) 1597-1614. https://doi.org/10.1039/c3ee44164d

Chapter 5

Advanced Supercapacitors for Alternating Current Line Filtering

Rajib Paul1*, Mohd. Khalid[2]

[1]Department of Mechanical and Aerospace Engineering, Case Western Reserve University, Cleveland, OH 44106, USA

[2]Department of Macromolecular Science and Engineering, Case Western Reserve University, Cleveland, OH 44106, USA

* rxp348@case.edu / rajiv2008juniv@gmail.com

Abstract

Ripples in alternating current circuits are very common in domestic and industrial sectors which must be filtered out to increase the lifetime of electronic devices connected in ac-networked circuits. Supercapacitors (SCs) are comparatively more suitable than aluminum electrode capacitors for ac line filtering due to smaller size and volume with 2-5 order higher specific capacitance. Although different types of SCs have been used so far, the proper combination of higher specific capacitance with excellent filtering efficiency is yet to be realized. Here, various SCs for ac line filtering are discussed and summarized to find ways for advancing available technology.

Keywords

Alternating Current Filtering, Supercapacitors, Graphene, Carbon Nanotubes, Polymer, Metal Oxide

Contents

1.   Introduction

Formation of ripples in the alternating current (ac) circuit is a general phenomenon in domestic and industrial sections. Ac electricity has a frequency of either 50 or 60 Hz [1]. When various nonlinear loads, from different electronic devices for domestic requirements, portable electronic, automobiles and medical appliances, are networked together, higher-order harmonics (>120 Hz) of the basic generating frequency are often induced [2]. Consequently, to protect electronic devices from such voltage ripples, aluminum electrolyte capacitors (AECs) are conventionally used for ac line-filtering [3-5]. However, the AECs have very low specific capacitance and they occupy a large volume in electronic circuits. In this context, use of supercapacitors having a specific capacitance of 2-5 order higher in magnitude as compared to that of AECs, are suitable alternatives for filtering ripples in ac-network. Additionally, due to the smaller size of supercapacitors, the ultimate capacitive components remain compact in electronic circuits [6-7].

While mounted on a transmission line operated at 120 Hz, a supercapacitor (SC) behaves like a resistor [5]. The related resistor-capacitor (RC) time constant for supercapacitors is about 1 s which is associated with the high electrochemical series resistance and microporous structure for most of the supercapacitor electrode materials. Such high time constant (~ 1 s) is too long for 120 Hz ac-line filtering which corresponds to a time period of as low as 8.3 ms. This emphasizes in smoothing the leftover ac ripples in common ac line-powered electronics [5]. Such high RC time-constant comes normally from the unsuitable or disordered pore structure of the SC electrodes which impede the high-rate ion diffusion and increases resistance to the effective charge transfer [8]. Therefore, suitable designing and proper fabrication of highly conductive electrodes with optimized micro-/nano-architectures for facilitated electron/ion transportation can improve the performance for ac line-filtering using SCs. In fact, electrode materials with high specific surface area, less inherent porosity and high ordered structural continuity are required for such application. Furthermore, most of the SCs are operated at lower voltages, from 2 to 4 V. Therefore, there is a need to fabricate SCs that are operable at high voltages for easy utilization in ac circuits to filter ripples.

For efficient ac line filtering, the relaxation time constant of a supercapacitor is the minimum time needed to discharge the stored energy with more than 50% efficiency [9]. The corresponding frequency (f) is inversely related to the relaxation time constant (τ_0) ($\tau_0=1/f$), at which the phase angle is 45°. A lower τ_0 indicates a higher power capability (fast charge-discharge) of a supercapacitor which is desirable for efficient ac line filtering. Therefore, high rate capable or ultrafast SCs are very essential to remove undesired high-frequency noises in ac circuits.

Supercapacitors are generally of two types, electrical double layer capacitors (EDLCs) where charge storage occurs through physical adsorption of ions and pseudocapacitors (PCs) where chemical adsorption of ions are attained through using redox active materials like a metal oxide, ion-gel etc. [1]. On the other hand, depending on the size of the SCs, they can be classified as normal SCs or macro-SCs or simply SCs and micro-supercapacitors. Micro-supercapacitors (MSCs) are miniaturized electrochemical energy storage devices which are normally utilized for on-chip electronic circuits. Variety of electrode materials including onion-like carbon [9] and CNTs [10, 11], carbide-derived carbon [12], metal oxides [13], polymers [14] and mesoporous carbons [225], have been utilized to improve the charge-discharge rate capability for ac-line filtering. However, graphene-based materials and graphene/CNT hybrid structures have recently emerged to suppress conventional carbon materials because of graphene's superior electrical conductivity and high specific surface area [8, 15-16]. Graphene and porous carbon composites have also demonstrated excellent ac-line filtering, however, the energy storage capability is not influential in graphene-based SCs yet [17]. Hence, SCs having high energy density and efficient ac-line filtering capability are indeed necessary in this context. Here, the most important SCs (both EDLCs and PSc) and MSCs for efficient ac line filtering application have been discussed. Recently reported flexible and printable SCs are also mentioned for the same purpose. The prospects for efficient SCs and MSCs towards bringing ideal ac-line filtering devices are prescribed with a purpose to make advancement in this technology.

2. Macro-Supercapacitors or Supercapacitors for ac line filtering

For ac line filtering, suitable electrode materials are required to fabricate high rate capable supercapacitors with phase angle closely equal to 90°. The SC performance is also found depend on the fabrication process. In this respect, researchers have used polymers, porous carbon films, carbon nanotubes, graphene and metal oxide based materials to fabricate SCs through various methods. Among polymers, PEDOT:PSS on graphoil (graphite foil) has shown promising line filtering performance. Fig. 1 shows a scalable wet-process to fabricate an electrochemical double layer supercapacitor using sulfuric acid treated commercially available poly(3,4-ethylenedioxythiophene):poly(styrenesulfonate) (AT-PEDOT:PSS) as an electrode material and graphite foil as the current collector. It exhibited high areal (994 $\mu F/cm^2$) and volumetric (16.6 F/cm^3) specific capacitances at 120 Hz, ultrahigh-rate frequency response (phase angle = -83.68° at 120 Hz) with a short resistor-capacitor time constant of 0.15 ms, and an excellent electrochemical stability of 100% after 12000 cycles of charging-discharging at 2 mA/cm^2 [18]. The variance in performance of the device under serious bending conditions, such as, after sharply folding to 180° with a small curvature radius of 0.5 mm, is imperceptible in their EIS and CV curves at a scan rate of 100 V/s as shown in Fig. 1 (j,k).

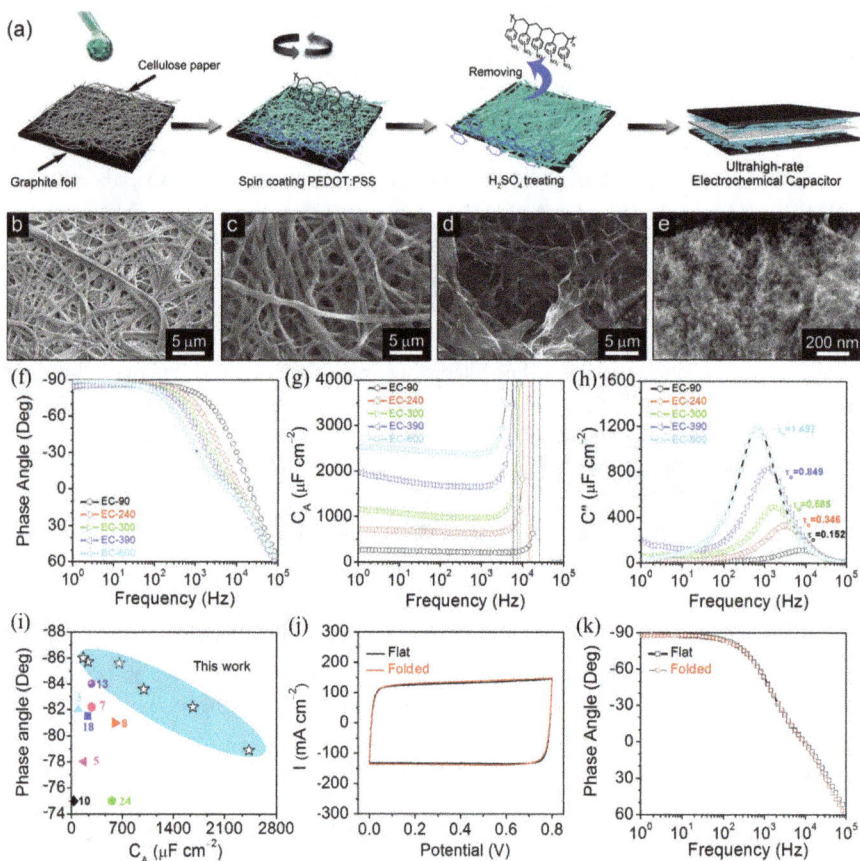

Fig. 1. (a) Schematic illustration of the fabrication of an AT-PEDOT:PSS-based EC; (b–e) top-view SEM images of cellulose paper (b), a PEDOT:PSS coated cellulose paper (c), and an AT-PEDOT:PSS/graphite foil composite at different magnifications (d and e). (f–h) Plots of phase angle (f), specific areal capacitance (g) or imaginary specific capacitance (h) versus frequency for AT-PEDOT:PSS-based ECs-n (n is the thickness of the electrode material, nm); (i) comparison of phase angles and CAs of AT-PEDOT:PSS-based ECs-n with those of reported ECs for ac line-filtering; 3,5,7,8,10,13,18,24 (j) CV

curves of flat and folded all-solid-state EC-300 at 100 V/s; (k) plots of phase angle as a function of frequency for all-solid-state EC-300 in the flat and folded states [18, copyright to Royal Society of Chemistry, 2016].

Carbon nanomaterials grown on the aluminum substrate can also be an interesting strategy to filter the ripples in ac line. The vertically aligned carbon nanostructures are very attractive due to their superior reactivity with surrounding medium through abundant edges. Many attempts have been performed to fabricate supercapacitor electrodes through chemical, electrochemical or electrophoretic methods using graphene oxide flakes. However, those materials demonstrated very poor ac-line filtering capability due to structural complexity. Recently, vertically oriented graphene nanosheets were grown on aluminum foil and used to fabricate electrical double layer capacitors (Fig. 2) [19]. The nanosheets were grown by radio frequency (RF) plasma enhanced chemical vapor deposition (PECVD) from C_2H_2 feedstock at a plasma power of 1100 W. The morphology and density of the sheets were found to depend on the gas flow rate as depicted in Fig. 2a. The surface oxide layer (Al_2O_3) on aluminum

The substrate and its stable nature inhibit good ohmic contact in the device. Therefore, it is important to remove any oxygen content from the substrate surface. The RF-PECVD growth method reduced Al surface oxide and then rapidly covers the surface with carbon to prevent further oxidation before nanosheet growth, thus allowing good ohmic connection between the graphene and the aluminum substrate. The specific capacitance at 120 Hz was approximately 80 mF/cm^2 (using 1.3 mm nanosheet height), which is greater than that achieved from vertically oriented graphene nanosheets grown on nickel foil for similar nanosheet height. The phase angle of capacitors fabricated with these electrodes was almost 90° at 120 Hz with good conductivity (Fig. 2), making these EDLCs suitable for ac filter applications presently dominated by aluminum electrolytic capacitors. Therefore, a combination of nanomaterials and aluminum substrate is an effective technique to fabricate advanced ac-ripple filters. However, in-depth knowledge is still needed to understand such complex graphene networked structures. Moreover, electrode materials grown through plasma-based chemical vapor deposition is superior as compared to other methods. It has been found that the plasma grown nanostructured materials possess superior transport properties due to structural continuity and proper-alignment [20-21].

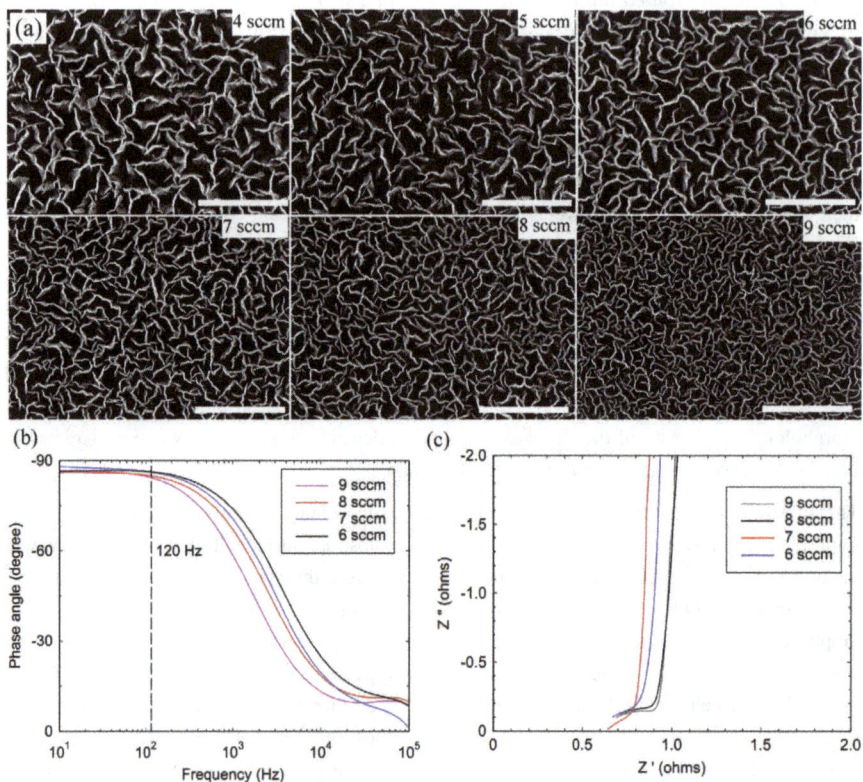

Fig. 2. (a) SEM topography of VOGN/Al growths for 4,5,6,7,8 and 9 sccm flow rates of C2H2 feedstock. Verticality is maintained but the nanosheet density increases significantly with flow rate (all scale bars are 2 mm). (b) Phase angle as a function of frequency for the VOGN on Al capacitors. The phase angle approaches ~90° at low frequency at all flow rates. (c) The complex plane plot shows a 45° intersection with the real axis at high frequency, evidence of porous electrode behavior due to the low conductivity of the organic electrolyte. Note there is no evidence of any semicircles [19, copyright to Elsevier, 2017].

Among other attempts, researchers have adopted heteroatom doping in carbon lattice to attain suitable chemistry for better filtering ac noises. However, heteroatom doping is found not suitable for filtering ac-ripples. This may be due to the fact that heteroatoms induce active scattering centers in carbon lattice which scatters the collected electrons and impedes the charge-discharge rate capability. Some researchers, however, adopted heteroatom doping to modify carbon nanostructures towards efficient ac-line filtering. In this respect, substitutional pyridinic nitrogen dopant sites in carbon nanotubes are found to selectively initiate the unzipping of CNT side walls at a relatively low electrochemical potential (0.6 V) as compared to undoped CNTs [22]. The obtained nanostructures consisted of partially zipped and/or unzipped graphene nanoribbons wrapping around carbon nanotube cores. Such hybrid structure maintains the intact 2D crystallinity with the well-defined atomic configuration at the unzipped edges. The synergistic interaction between the large surface area and robust electrical connectivity of the unique nanoarchitecture led to ultrahigh-power supercapacitor performance, which can serve to filter ac-line with 85° phase angle at 120 Hz which is very close to that of AECs, 83.9° [1]. However, the specific capacitance is around 0.19 mF/cm^2 which needs to be further improved. Therefore, extensive research needs to be performed for fabricating supercapacitors that have both efficient ac-line-filtering capability and excellent charge-storage capability.

The supercapacitors developed so far for ac line filtering have limited for applied voltages which is generally less than or equal to 20 V. To extend the range of applied voltage, it is essential to design electrode materials with a suitable pore structure. In this context, to increase the operating potential, graphitic ordered mesoporous carbon (GOMC) was used as an electrode material in supercapacitors for ac-line filtering applications (Fig. 3, Fig 3g explains the mechanism of ac-line filtering using a SC). By utilizing the open pore structure of GOMC, also known as CMK-3, a 2.5 V supercapacitor with a high areal capacitance (areal ~560 μF/cm^2 at 120 Hz) and a fast frequency response ($\varphi \sim -80°$ at 120 Hz) was obtained. Importantly, the addition of a few CNTs to the CMK-3 results in the formation of an efficient electrical network between the individual CMK-3 particles. the supercapacitors showed excellent performance for ac line filtering, including a rectangular CV, even at a scan rate of 450.

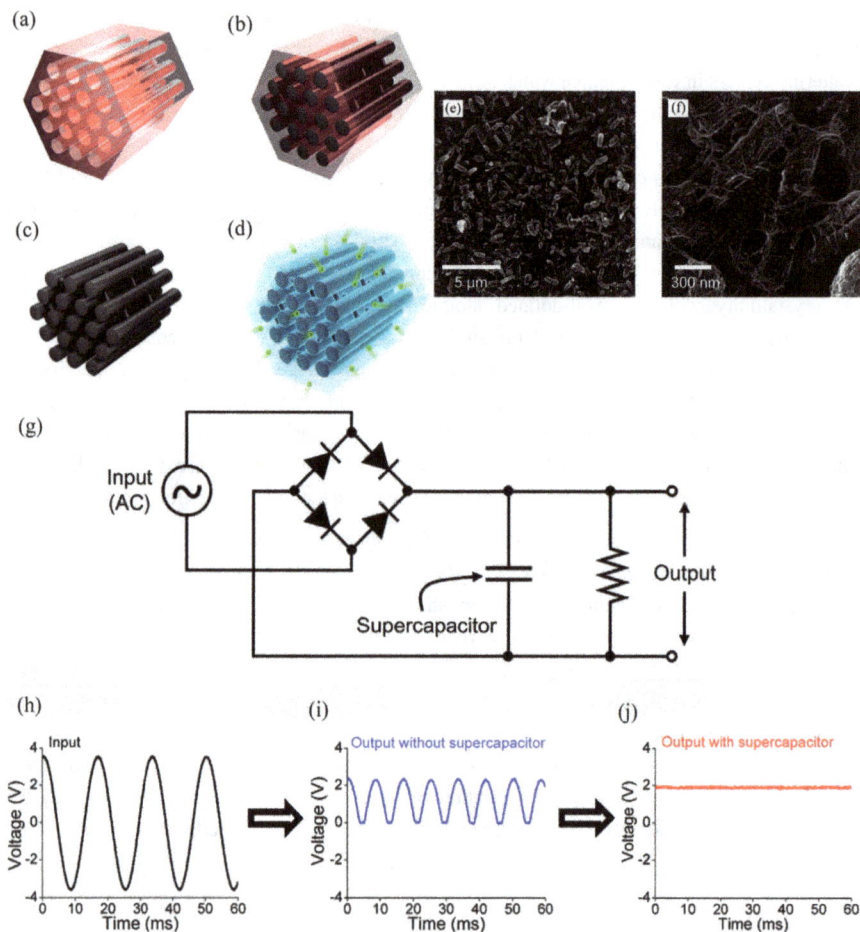

Fig. 3. (a) SBA-15 silica template with hexagonally arranged straight pores. (b) Infiltration and carbonization of iron phthalocyanine in the SBA-15 template. (c) Carbon inverse replica (CMK-3) remaining after template removal. (d) CMK-3 particle with facile ion accessibility to its entire surface. (e) Low- and (f) high-magnification SEM images of a CMK-3/CNT film (CMK-3/CNT ¼ 9 : 1) on Au/Ti/Al foil. (g) Electrical circuit diagram of ac line filtering. (h) ac input signal (60 Hz, Vpeak¼_3.56 V). (i)

Rectified pulsating DC signal (V ¼ 0-2.32 V, 120 Hz). (j) Constant DC output (~1.89 V)
[22, copyright to Royal Society of Chemistry (2016)].

V/s, a low ESR (~0.25), a high areal capacitance (~1 mF/cm^2 at 1 mA/cm^2 and ~560 µF/cm^2 at 120 Hz), and a high negative phase angle (-80° at 120 Hz). The successful conversion of a 60 Hz ac signal to a DC output was demonstrated using this supercapacitor. The advantages of the CMK-3/CNT supercapacitor over a commercial AEC can be extended to applications up to ~40 V. Therefore a porous material with high network-line connectivity would be a suitable candidate for advancement in ac-line filtering by supercapacitors at elevated voltages.

Besides conventional materials like graphene and CNTs, researchers have also developed a simple approach to synthesize porous carbon sponges from cheap material-source like, camphor etc. [23]. It has also been demonstrated that the pore size distribution of carbon sponges can be readily controlled by adding a surfactant in the freeze-drying process. The supercapacitor fabricated from this carbon sponge as electrodes exhibited a phase angle of -78° which was much larger than that of the activated carbon-based capacitor (~0°) but was comparable to that of AEC (85.5°) [24]. In addition, the resistance and reactance of the capacitor reach to an equal value at the phase angle of -45°. This filter had a RC time constant of 0.319 ms at 120 Hz, making it highly promising to replace AECs in ac line-filtering. However, the specific areal capacitance was around 0.172 mF/cm^2 at 120 Hz.

Ultimately, there are several parameters to be considered for fabrication of macro-supercapacitors for effective ac-line filtering. Firstly, the conductivity of the electrode must be excellent which is very crucial for collecting the charge carriers quickly. The charge diffusion time should also be very short for efficient ac ripple filtering applications. Furthermore, the morphology is also an important factor to realize improved specific capacitance of the fabricated supercapacitor. Such purpose requires a porous material which in turn degrades the overall conductivity of the electrode. Therefore, a porous material structure with excellent conductivity would be a game changer for this ac-filtering application.

3. Micro-supercapacitors for ac line filtering

The plethoric demand of portable electronic devices and its continued expansion have foregrounded the requirement for high-performance miniaturized electrochemical storage devices which are able to deliver energy to their host devices. Radio frequency identification (RFID) tags for the development of smart environments is another critical application that requires compact energy storage [25]. In fact, designing efficient miniaturized energy storage devices for energy delivery or harvesting with high-power capabilities remains a challenge. Notably, micro-supercapacitors (MSCs) are highly promising innovations for on-chip applications requiring a high spike of current as well as a high-frequency response for alternating current (ac) line filtering [26]. As such, the capability of retaining high specific capacitance during ultrafast charging/discharging is essential for these high-performance supercapacitor applications. Furthermore, the use of electric double-layer micro-supercapacitors (EDL-MSCs) is preferred due to their rapid and reversible adsorption/desorption of ions at the electrolyte-electrode interface. To achieve high ac line filtering using MSCs, the electrode must facilitate rapid ion transportation and the ion diffusion pathways should be short. Various polymer based MSCs have been fabricated as thin films or as coordination polymer frameworks besides graphene, CNTs and their hybrid structured materials for efficient ac line filtering. Parallelly, to increase the charge storage capability researchers have also opted pseudocapacitors (PSc) for fabricating MSCs to remove ripples in ac lines using metal oxide based materials. We have discussed such attempts briefly here.

Conducting polymers have always been attractive electrode materials for fabrication of MSCs due to their low cost and fabrication ease. Various conducting polymers have been utilized for this purpose, such as poly(3,4-ethylene dioxythiophene) (PEDOT), polypyrrole (PPY) and polyaniline (PANI). MSCs were fabricated using different conducting polymers, like, PEDOT, PPY and PANI and their frequency responses were investigated [27]. It was demonstrated that with proper choice of polymeric material and device structure, miniaturized micro-pseudocapacitors can match the frequency response of commercial bulky electrolytic capacitors. Specifically, PEDOT-based micro-pseudocapacitors exhibited phase angle of -80.5° at 120 Hz which is comparable to commercial bulky electrolytic capacitors, but with an order of magnitude higher capacitance density (3 FV/cm^3) [27]. The trade-off between the areal capacitance (C_A) and frequency response in the 2D architecture (C_A = 0.15 mF/cm^2, the phase angle of -80.5° at 120 Hz) was improved by designing 3D thin film architecture (C_A = 1.3 mF/cm^2, the phase angle of -60° at 120 Hz). Therefore, the morphology and overall conductivity are very important factors in case of conducting polymer-based MSCs.

Recently, a surfactant-mediated process to fabricate flexible micro-supercapacitors (MSCs) combining conventional photolithography and electrochemical deposition has appeared to be very interesting [28]. The anionic surfactant mediated the process of electropolymerization at a lower anodic potential while causing template effects in producing porous conducting PEDOT electrodes as demonstrated in Fig. 4 (a-f). Using this strategy, PEDOT-MSCs with a remarkable performance in terms of tunable frequency response and energy density have been fabricated. Specifically, ultrahigh scan rate capability up to 500V/s has been achieved with a crossover frequency of 400 Hz at a phase angle of 45°. Thus, the micro-supercapacitors exhibited maximum areal cell capacitance of 9 mF/cm^2 with a volumetric stack capacitance of 50 F/cm^3 in 1M H_2SO_4 aqueous electrolyte (Fig. 4).

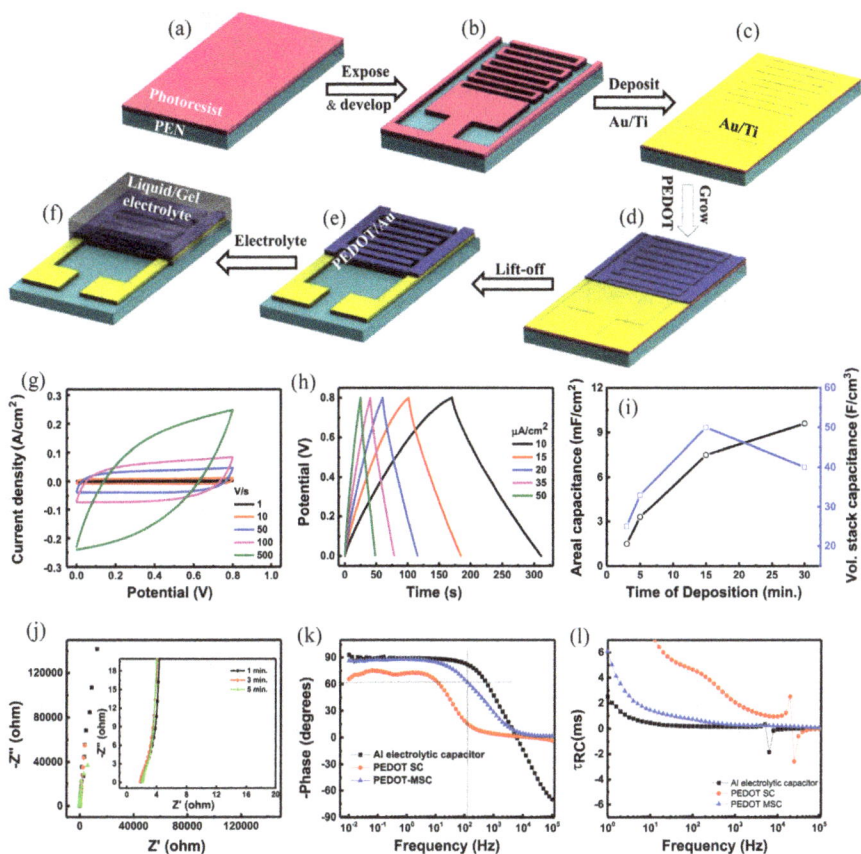

Fig. 4. Schematic illustration for the fabrication of PEDOTMSCs: (a) Spin-coating of photoresist over plastic PEN substrate, (b) Exposure of photoresist film through a mask having interdigitated patterns followed by development, (c) Au/Ti metal layers deposition by sputtering or thermal evaporation, (d) Electrochemical deposition of PEDOT on the metal coated chip before the lift-off process, (e) Lift-off process using acetone to get PEDOT/Au inter digitated finger electrodes, (f) Investigation of electrochemical performance of PEDOTMSCs in liquid and gel electrolyte media. Electrochemical performance of PEDOT-MSC (3min) in 1MH2SO4 electrolyte in two electrode configuration: (g) CV scans at higher scan rates from 1 to 500V/s, (h) Galvanostatic charge-discharge curves at different current densities, (i) Areal and volumetric stack

capacitance with respect to time of deposition of PEDOT at a current density of 35 mA/cm², (j) Impedance spectra of PEDOT-MSC devices for varying deposition times, Inset showing the zoom-in view of the impedance spectra in the high-frequency region, (k) Phase angle and (l) RC time constant vs. frequency of the PEDOT-MSC in comparison with the commercial Al electrolytic capacitor and conventional PEDOTSC [28, copyright to Elsevier (2015)].

The flexibility and stability of these PEDOT-MSCs were tested in aqueous gel electrolyte which showed a capacitance retention up to 80% over 10,000 cycles with a Coulombic efficiency of 100%. The maximum energy density of solid state ion gel based PEDOT-MSCs was found to be 7.7 mWh/cm³, which is comparable to the lithium-based thin film batteries and superior to the current state-of-the-art carbon and metal oxide based MSCs. Furthermore, the tandem configuration of flexible solid-state ion gel based PEDOT-MSCs was employed to demonstrate it as a power source for glowing a red light emitting diode.

The on-chip micro-supercapacitors (MSCs) are also important Si-compatible power-source backups for miniaturized electronics. Despite their tremendous advantages, current on-chip MSCs require harsh processing conditions and typically perform like resistors when filtering ripples from ac supply. In this context, a facile layer-by-layer (LBL) method towards on-chip MSCs based on an azulene-bridged coordination polymer framework (PiCBA) has been reported [29]. The azulene-based building block, 2,2'-diisocyano-1,1',3,3'-tetraethoxycarbonyl-6,6'-biazulene was synthesized. The terminal isocyanide groups of iCBA can coordinate with various metal atoms and metal ions such as Au, Pt, Pd, Co, Ni and Fe. Taking advantage of this feature, a PiCBA monolayer film was prepared through a coordination reaction between iCBA and Co ions ($CoCl_2 \cdot 6H_2O$) at the water-air interface in a Langmuir-Blodgett trough. The cross-section SEM images (prepared by focused ion beam) indicated that the 10-layer PiCBA films had an estimated total thickness of approximately 30 nm, which is slightly larger than the theoretical value, 23.1 nm, due to possible dislocation effect. Hence, the PiCBA layers were directly integrated to fabricate MSCs. First, the iCBA molecules were adsorbed onto the Au surface of the interdigitated Au-SiO_2 substrate through the upright coordination of the terminal isocyanide carbon atoms. Then, the substrate was rinsed in a $CoCl_2$ aqueous solution, deionized water, iCBA monomer solution, and fresh $CHCl_3$ sequentially 10 times to allow the *in-situ* formation of 10-layer PiCBA films on the Au electrode. Finally, the H_2SO_4-polyvinyl alcohol (H_2SO_4-PVA) gel electrolyte was cast on the PiCBA-Au:SiO_2 substrate and allowed to solidify overnight to form a solid-state PiCBA-Au MSC with an in-plane geometry. The fabrication of PiCBA film is shown in Fig. 5. Owing to the good carrier mobility (5×10^{-3} cm²/V/s) of PiCBA, the permanent dipole moment of azulene skeleton, and ultralow band gap of PiCBA, the fabricated MSCs delivered high specific capacitances of up to 34.1 F/cm³ at 50 mV/s and a high volumetric power density of 1323 W/cm^3. Most importantly, such MCSs exhibited ac line-filtering performance (~73° at 120 Hz) with a short resistance-capacitance constant of approximately 0.83 ms which is superior to that of the AECs. Furthermore, the PiCBA-based MSCs showed good cycling stability, with 86% capacitance maintained after 350 cycles [29].

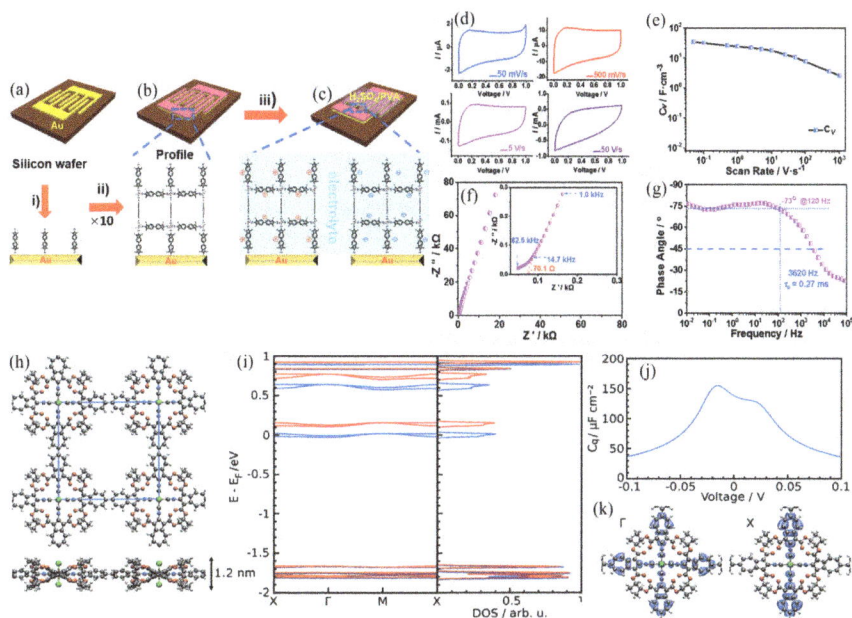

Fig. 5. (a-c) Schematic of LBL fabrication of a PiCBA film on Au interdigital electrodes. (i) The Au interdigital electrode was immersed in iCBA solution for 24 h, followed by rinsing in fresh CHCl₃ and then (ii) immersed in CoCl₂ solution for another 60 s, followed by rinsing in deionized water, iCBA solution for 60 s, and fresh CHCl₃, respectively, to form the 1-layer PiCBA film. This process was repeated nine times to obtain a 10-layer PiCBA-Au electrode. (d) CV curves of PiCBA-based MSCs in the H₂SO₄-PVA gel electrolyte at different scan rates, showing a typical double-layer capacitive behavior even at different scan rates. (e) CV evolution of the MSCs at different scan rates. (f) The complex plane plot of the impedance of the PiCBA-based microdevices. The inset displays a magnification of the high-frequency region. (g) Impedance phase angle on the frequency for the PiCBA-based microdevices. (h) Relaxed unit cell (blue square) of the PiCBA framework. The calculations reveal an almost square Bravais lattice with a=20.948 b, b=20.944 b, and f&90.08. (i) Simulated band structure and density of states (DOS). The Fermi level crosses two spin-up bands, indicating

metallic behavior of the material. (j) Calculated quantum capacitance based on the two bands crossing the Fermi level for zero disorder. (k) The charge density of the lower band crossing the Fermi level at the G point as well as close to the Brillouin zone boundary at X, namely k=(0.4, 0.0, 0.0). The charge density of the second band is almost identical at the G point and complementarily delocalized horizontally at (0.4, 0.0, 0.0) [29, copyright Wiley (2017)].

Density Functional Theory (DFT) study revealed that the optimization of the unit cell yielded an approximately square Bravais lattice, with the symmetry of the structure slightly broken by the torsion of the azulene units (Fig. 5h) [29]. Fig. 5i shows the band structure and density of states (DOS). Interestingly, two spin-up bands cross the Fermi level, indicating the metallic behavior of the PiCBA framework in this configuration, while the narrow bandwidth is consistent with the observed mobility. The quantum areal capacitance (Cq) extracted from the DFT calculations reaches 150 mF/cm^2 and is clearly larger than that obtained from experiments at standard scan rates (Fig. 5j); this was explained by the limiting scenario of a disorder free system. To include the effects of disorder on Cq, the DOS of the four bands closest to the Fermi level was broadened. With DOS broadening, the quantum capacitance decreased below 50 mF/cm^2, which was closer to experimental observations. The calculations indicated that the charge densities of these bands are spread over the azulene units (Fig. 5k) and the dispersion of these bands is rather small, which appears to be mainly because of the metal centers as well as partly because of the torsion between the azulene units [29]. Such theoretical support of the experimental observation not only helps to understand the mechanism of the functionality of the electrochemical device but it also helps in further advancements through proper knowledge and accurate pathway.

Graphene-based MSCs are also very interesting owing to their excellent physical and electrochemical properties. A flexible reduced graphene oxide (rGO)/Au based MSC was fabricated through *in-situ* femtolaser writing [26]. Without the need for complicated multi-step photolithography processes, the femtolaser writing allowed for the simultaneous and direct reduction of GO/HAuCl$_4$ composites into a combination of rGO electrodes and Au collectors (Au NPs) which is indeed interesting approach for fabricating MSCs. It was demonstrated that the fused Au nanoparticle in-plane network significantly increased the conductivity meanwhile enhancing the capacitance of porous rGO. These flexible MSCs showed two-orders-of-magnitude increase in electrode conductivity of 1.1 x 10^6 S/m due to the highly conductive pathways created by Au NPs, with short ion diffusion pathways due to the in-plane design, superior rate capabilities (71% and 50% of capacitance retention for the charging rate increase from 0.1 V/s to 10 V/s and to 100 V/s, respectively), sufficiently high frequency responses (362 Hz, 2.76 ms time constant), phase angle of ~60° at 120 Hz, large specific capacitance (0.77 mF/cm^2 at 1 V/s), and outstanding cycle stability (10000 cycles). Furthermore, a proof-of-concept for 3D MSCs was demonstrated by the enhanced areal capacitance (up to 4.92 mF/cm^2 at 1 V/s) while retaining high rate capabilities, which allow for the more compact on-chip design. Such high performance of rGO/Au-FS MSCs promotes their applications in flexible, portable and wearable electronic devices that require fast charge/discharge rates and high power densities with ac-line filtering capability to some extent. This new femtolaser-written MSCs can be potentially integrated with other electronic devices such as solar cells, gas sensors, field-effect transistors, or chip-level display devices etc. [26].

Furthermore, with an aim to improve both specific capacitance and ac line filtering capability in MSCs, three-dimensional (3D) graphene-carbon nanotube carpets hybrid materials (G/CNTCs) were used which demonstrated excellent electrochemical performance as presented in Fig. 6 [30]. The microdevices constructed with short CNTCs exhibited a much higher energy capacity than AECs while having comparable ac line filtering performances. With the increased heights of CNTCs, the as-produced microdevices showed improved capacitances while maintaining satisfactory ac response. The most dramatic materials difference between these and other related structures are that the G/CNTCs have a seamless transition structure, maximizing the electrical conductivity. The 3D G/CNTCs-based micro supercapacitors (G/CNTCs-MCs) were fabricated *in-situ* on nickel electrodes using catalysts (Fig. 6). The G/CNTCs-MCs showed impedance phase angle of -81.5° at a frequency of 120 Hz, comparable to commercial aluminum electrolytic capacitors (AECs) for alternating current (ac) line filtering applications (Fig. 6). In addition, G/CNTCs-MCs delivered a high volumetric energy density of 2.42 mWh/cm^3 in the ionic liquid which is more than 2 orders of magnitude higher than that of AECs. The ultrahigh rate capability of 400 V/s enabled the microdevices to demonstrate a maximum power density of 115 W/cm^3 in aqueous electrolyte. The high-performance electrochemical properties of G/CNTCs-MCs can provide more compact ac filtering units and discrete power sources in future electronic devices. The elevated electrical features were likely offered by the seamless nanotube/graphene junctions at the interface of the differing carbon allotropic forms. Given these performance characteristics, the G/CNTCs-MCs would provide a route to addressing the demands of the future microscale energy storage devices.

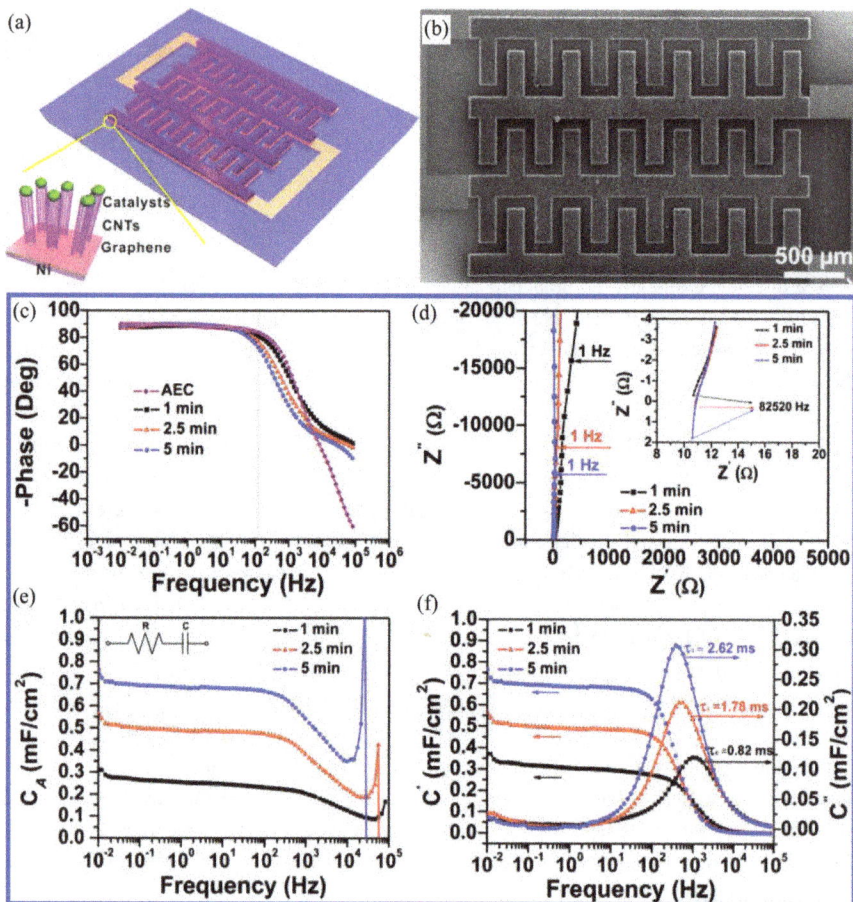

Fig. 6. Design of micro-supercapacitors and material characterizations of CNTCs. (a) Schematic of the structure of G/CNTCs-MCs. Inset: enlarged scheme of Ni-G-CNTCs pillar structure that does not show the Al_2O_3 atop the CNTCs; (b) SEM image of a fabricated G/CNTCs-MC. The ac impedance characterizations of micro-supercapacitors with CNTCs grown for various durations (1, 2.5, and 5 min) using 1 M Na_2SO_4. (c) Impedance phase angle versus frequency. The phase angles occurring at 120 Hz are 81.5°, 77.2°, and 73.4° for 1, 2.5, and 5 min growth, respectively; the phase angle at 120 Hz for an AEC is 83.9°; (d) Nyquist plots of impedance from the three different growth-time structures. The inset is the expanded view in the high-frequency region; (e) C_A versus f using series-RC circuit model; τ_{RC} of 195, 325, and 402 µs were obtained for 1,

2.5, and 5 min growth; a Randels equivalent circuit representing the RC circuit elements is shown in the inset; (f) C' and C" versus f. The extremely low τ0 of 0.82, 1.78, and 2.62 ms were extracted from 1, 2.5, and 5 min growth [30, copyright to American Chemical Society (2012)].

Among pseudocapacitors for ac-line filtering, titanium oxide (TiO_2) and ruthenium oxide (RuO_2) based MSCs have performed promisingly. The electric capacitance of an amorphous TiO_2-x surface was found to increase proportionally to the negative sixth power of the convex diameter [31]. This occurred because of the van der Waals attraction on the amorphous surface of up to 7 mF/cm^2, accompanied by extremely enhanced electron trapping resulting from both the quantum-size effect and an offset effect from positive charges at oxygen-vacancy sites. The capacitive behavior (near the -90° phase angle) throughout the frequency region (Fig. 7) was a clear evidence for the series-RC circuit. Thus, the ATO offered a nearly ideal electric distributed-constant structure for enhancing electrical power storage. Although the electrical storage was not always the same as the capacitance, the value of the series capacitance was 4.17 μF (2.085 F/cm^3, 537.6 μ F/kg) at 0.1 Hz. The supercapacitor, constructed with a distributed constant-equipment circuit of large resistance and small capacitance on the amorphous TiO2-x surface, illuminated a red LED for 37 ms after it was charged with 1 mA at 10 V [31]. The fabricated device showed no dielectric breakdown up to 1100 V which is the very attractive property for a metal oxide based electrode. However, further research work in further developing the amorphous titanium-dioxide based supercapacitors might be attained by integrating oxide ribbons with a micro-electro-mechanical system [31].

Recently, hydrated ruthenium oxide ($hRuO_2$) was used for realizing high energy MSCs. In this effort, ultra-high power micro-supercapacitors were fabricated based on multi-walled carbon nanotubes using a simple and integrated circuit-compatible approach for the deposition of the active material ($hRuO_2$), with extremely high resolutions [32]. The fabrication process is illustrated in Fig. 8. The resulting MSCs exhibited excellent electrochemical performances with specific cell capacitance up to 3 mF/cm^2 for $hRuO_2$ and life cycle exceeding 30000 cycles for MWCNTs with a maximum specific power of 1.3 W/cm^2. All micro-devices displayed impressive scan rate abilities, up to 10000 V/s for MWCNTs, which were more than 5 orders of magnitude higher than those of conventional supercapacitors, the specific capacitances were 0.2 and 3.2 mF/cm^2 at 10 V/s for the MWCNTs and the $hRuO_2$-based MSCs, respectively. These values are comparable to the specific capacitances reported in the literature at much lower scan rates for micro-EDLCs (0.4-1.5 mF/cm^2) and pseudo-capacitive MSCs (0.4–12.6 mF/cm^2). The MWCNT-based MSC was characterized to possess extremely low time constant of 32 ms (512 ms for the $hRuO_2$-based MSC), which is in complete agreement with the theoretically determined values. The proposed strategy is promising for batch fabrication and integration of operational on-chip MSCs into existing microsystems. With electron beam or nano-imprint lithography, the process could also be used for the realization of nano-supercapacitors with sub-micrometer pattern resolution. This process can be applied to other materials or deposition techniques. The use of conventional UV-photolithographytechniques (1 mm resolution) currently limits the size of the pattern, but this work considered the stepper projection lithography (350 nm resolution) and electron beam lithography (sub-10 nm resolution) for further reduction of the interspace (Fig. 8e). In turn, these improvements in electrode design are hoped to lead unprecedented performances of micro-supercapacitors.

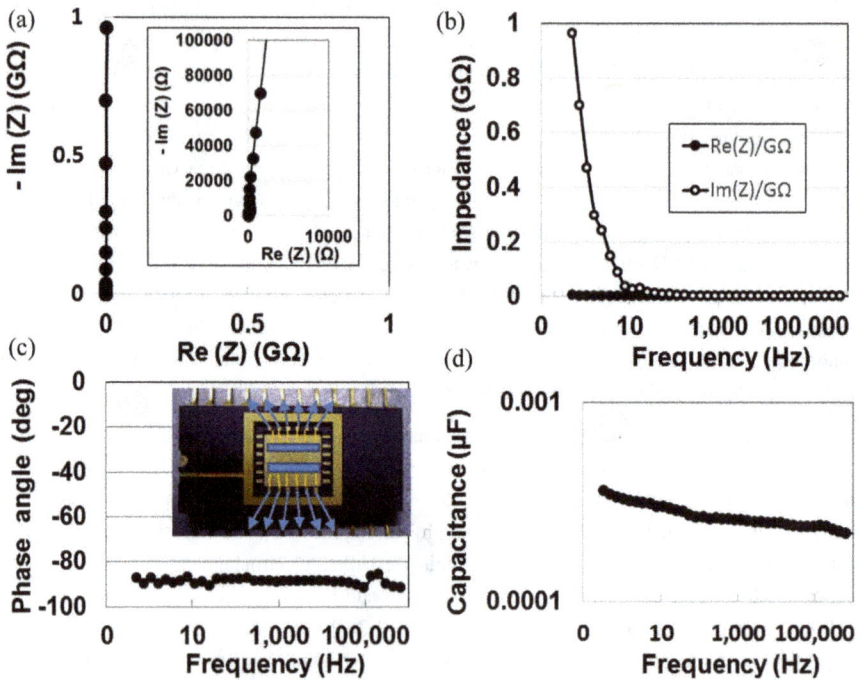

Fig. 7. Non-destructive analysis of the electrostatic contribution of an amorphous titanium oxide (ATO) based MSC. (a) The Nyquist plot as a function of frequency for ATO device. (b) Real and imaginary impedances. (c) Phase angle. The inset of (c) depicts a device fabricated by MEMS. (d) Series capacitance [31, copyright to Macmillan publishers (2016)].

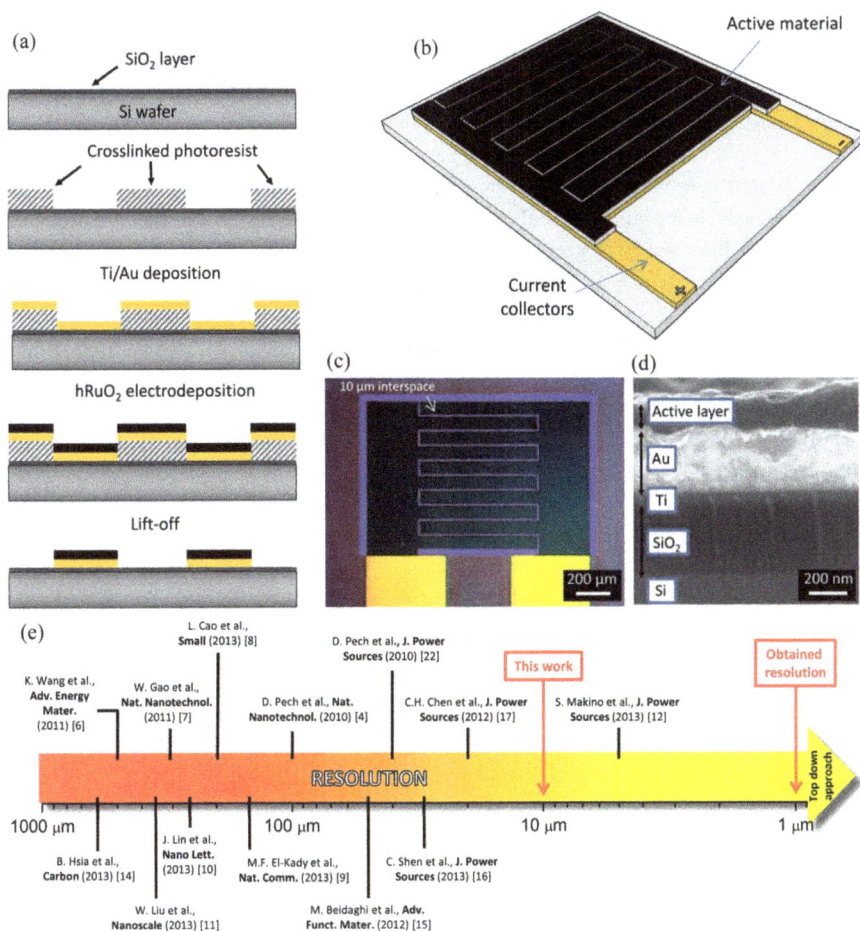

Fig. 8. Micro-supercapacitor fabrication and design. (a) A simplified illustration of the fabrication process using the lift-off process of the active material for a high-resolution pattern (not to scale). Schematic (b) and an optical image (c) of the micro-device. (d) Scanning electron microscopy image of the cross-section of the electrode. (e) Comparison of the resolution of state-of-the-art micro-supercapacitors reported in the literature [32, copyright to The Royal Society of Chemistry (2014)].

4. Flexible-supercapacitors for ac line filtering

Recent requirement and development of flexible and wearable electronics as well as flexible and wearable supercapacitors, in either a thin film or fiber-shaped (coating, fabric/ cloth, paper, textile, etc.), have led to encroaching research interest towards advanced flexible electronics and power sources. In this context, the fabrication of flexible supercapacitors for efficient ac-line filtering is essential. Due to large surface area, excellent mechanical and electrical properties, and high electrochemical stability, carbon nanomaterials have been primarily promising as electrode materials for this purpose. For example, flexible solid-state supercapacitors based on ultrathin CNT films and ion gel for high-frequency applications have been reported (Fig. 9). The gel electrolytes were prepared by mixing poly(styrene-block-ethylene oxide-block styrene) triblock copolymer (50 mg) and 1-ethyl-3-methylimidazolium bis-(trifluoromethysulfonyl)imide ([EMIM][NTf2], C-TRI, 0.7 mL) in acetonitrile (1.4 mL) [33]. The mixture was stirred magnetically for 12 h in a glovebox under an argon atmosphere. Because of the appropriate pore structure of the CNT films and nanochannel structure in gel electrolytes and the high compatibility between the pores of CNT films and ion-sizes in the gel, the CNT/ion gel supercapacitor exhibited a fast response speed and high cell voltage, enabling its application for ac-line filtering. Thus, the energy density of the supercapacitor was found to be improved by more than 20 times than previously reported values. The phase angle, the areal capacitance at 120 Hz, and τ_0 were -78.1°, 106 $\mu F/cm^2$, and 1.00 ms, respectively (Fig. 10). Moreover, an ac-line filtering circuit implemented with the flexible supercapacitor successfully converted a 60-Hz ac signal into a smooth DC signal. Furthermore, the electrochemical properties of the supercapacitors were well-retained over 1200 cycles of bending [33]. The flexible supercapacitors with ultrahigh power performance and high energy density may contribute to the miniaturization of wearable and portable electronics by replacing bulky AECs.

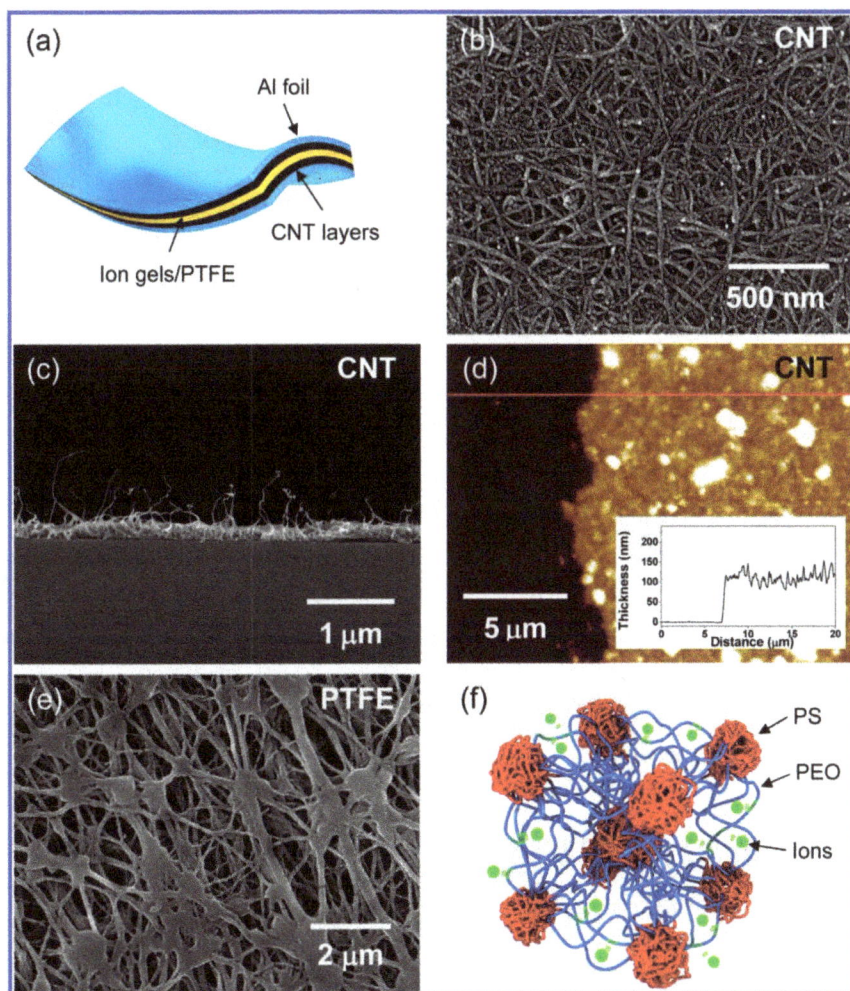

Fig. 9. (a) Schematic of an all-solid-state supercapacitor. SEM (b) overview and (c) cross-sectional images of a CNT film deposited on a Si substrate. (d) AFM image of a CNT film. (e) SEM images of PTFE. (f) Schematic of ion gels [33, copyright to American Chemical Society (2016)].

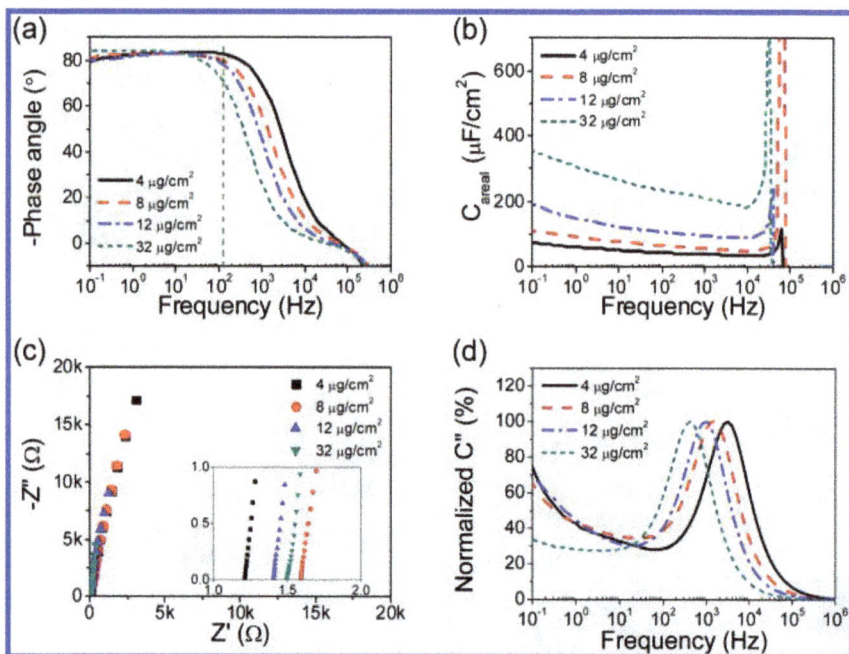

Fig. 10. CNT-mass dependent frequency response of all-solid-state supercapacitors with a gel electrolyte (~30 μm): (a) phase angle, (b) areal capacitance, (c) Nyquist plot, and (d) normalized capacitance [33, copyright to American Chemical Society (2016)].

Hybrid carbon films composed of graphene and porous carbon films may also give full play to the advantages of both carbon materials, and have great potential for application in energy storage and conversion devices. The carbon film can provide porous channels while the graphene can increase the transport and charge diffusion into carbon film and the hybrid structure could be fruitful for ac line filtering application. Unfortunately, there are limited reports on the fabrication of hybrid carbon films. It can be fabricated adopting a very simple approach as reported to achieve a free-standing three-layered sandwich-structure of hybrid carbon film (HCF) composed of porous amorphous carbon film in the middle and multilayer graphene films on both sides by chemical vapor deposition in a controllable and scalable way [34]. The fabrication steps have been presented in Fig. 11. The hybrid carbon films revealed very good electrical conductivity, excellent flexibility, and good compatibility with the substrate. Supercapacitors assembled by such hybrid carbon films exhibited ultrahigh rate capability, wide frequency range, good capacitance performance, and high-power density. Moreover, this approach may provide a general path for fabrication of hybrid carbon materials with different structures by using different metals with high

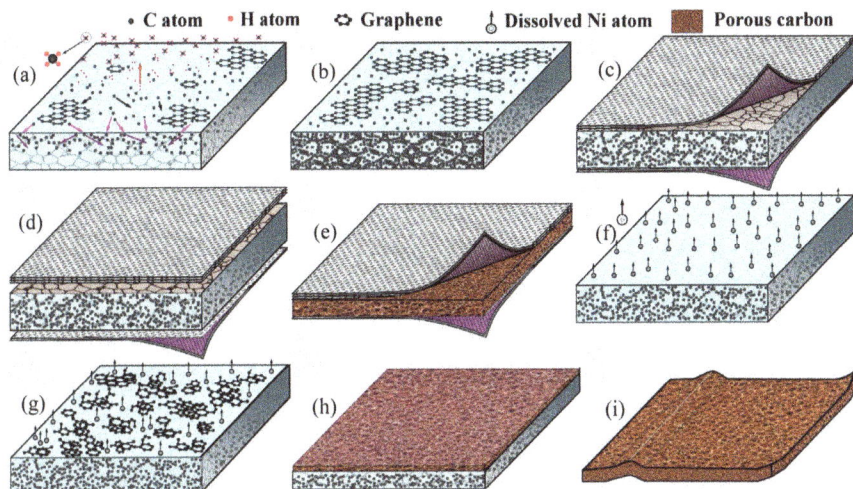

Fig. 11. Design of free-standing carbon films. Schematic illustration of the fabrication process of HCF and PACF, (a) decomposition of methane, carbon atom adsorption, diffusion and dissolution at growth temperature, (b) graphene growing and a large number of carbon atoms dissolving in Ni foil, (c) part of the carbon atoms segregating/precipitating on Ni surface to form MGF, and another part of the carbon atoms trapped in Ni foil, (d) MGFs detaching from both sides of the Ni foil in ferric chloride solution, (e) free-standing HCF after Ni foil completely etched, (f) etching Ni foil in ferric chloride solution after the removal of MGFs, (g) and (h) etching Ni foil and formation of the PACF, (i) free-standing PACF after Ni foil completely etched [34, copyright to Macmillan publishers (2014)].

carbon solubility. The area-specific capacitances (Cs) of the HCF (1.12 mm) electrodes reached 0.48 mF/cm^2 at a scan rate of 50 mV/cm^2, which is larger than the area specific capacitances of the multilayer graphene sheets (0.39 and 0.28 mF/cm^2) [5, 35]. The characteristic time constants were 143 and 472 ms for the SC based on the HCF (1.12 mm) and the HCF (1.83 mm) electrodes, respectively, indicating the HCF-based SC possesses ultra-high rate performance. At 120 Hz, the impedance phase angles of the HCF-based SCs were approximately 67° and 64° for the SC based on HCF (1.12 mm) and HCF (1.83 mm) electrodes, respectively, as compared with, 0° for the activated carbon SC [5], -15° for the VOG-bridged coated nickel-foam SC [36], and -82° for the vertically oriented graphene SCs [5]. The advantage of carbon film based SCs lays in their robust mechanical properties. Therefore, flexible supercapacitors for ac line filtering needs to be explored more for advanced flexible electronics device applications operated by alternating voltages.

5. Printable supercapacitors for ac line filtering

Printed electronics is exponentially emerging as an interesting technology representing greatly integrated functionality of thinness, flexibility, scalability, and lightweight, and it has enormous potential applications, such as, printable transistors, photovoltaic cells, flexible displays, sensors, radio frequency identification, and organic light-emitting diodes, which have ultimately stimulated the rapid development of new-concept, miniaturized printable energy-storage systems that can be compatible with them [37-41]. In this respect, printable SCs, commonly fabricated using printed films on plastic or paper substrates, have been widely acknowledged as an ideal power source for printed electronics because they can incorporate multiple functions into a single device. In such context, it is crucial to have printed SCs capable of filtering ripples in ac line supply for successful printed electronics applications. There are very few reports in this regard.

Among such research works, SCs with good ac line filtering capability have been realized through the fabrication of ultrathin printable supercapacitors (UPSCs) based on graphene-conducting polymer (EG/PH1000) hybrid films [42]. The conductive hybrid ink of electrochemically exfoliated graphene (EG) and poly(3,4-ethylene dioxythiophene):poly(styrenesulfonate) (PEDOT:PSS) (Clevios PH1000) was spray coated on a gold-deposited 2.5-μm-thick PET substrate with a tunable thickness ranging from nanometers to micrometers (Fig. 12). Thus, the UPSCs constructed with two EG/PH1000 hybrid films sandwiching a thin polymer gel electrolyte of poly(vinyl alcohol)/H_2SO_4 exhibited an unprecedented volumetric capacitance of 348 F/cm^3 and excellent cycling stability with no capacitance loss even after 50000 cycles. Such UPSCs could be operated at an ultrahigh rate of up to 2000 V/s. More importantly, the devices delivered ac line-filtering performance with a phase angle of -75° at 120 Hz and extremely short RC time constant of <0.5 ms. Notably, at 120 Hz, the UPSCs-25, UPSCs-77, and UPSCs-128 exhibited capacitances of 75, 127, and 179 μF, respectively, and the measured resistances about 6.3, 4.7, and 3.6 Ω, respectively. Thus, RC time constants of 472, 597, and 644 μs were achieved for UPSCs-25, UPSCs-77, and UPSCs-128, respectively. These values are well comparable to those of the state-of-the-art EDLCs based on vertically oriented graphene (<200 μs) [5], ErGO (1.35 ms) [24], and thermally reduced GO (2.3 ms) [43]. This result suggests that our printable supercapacitors

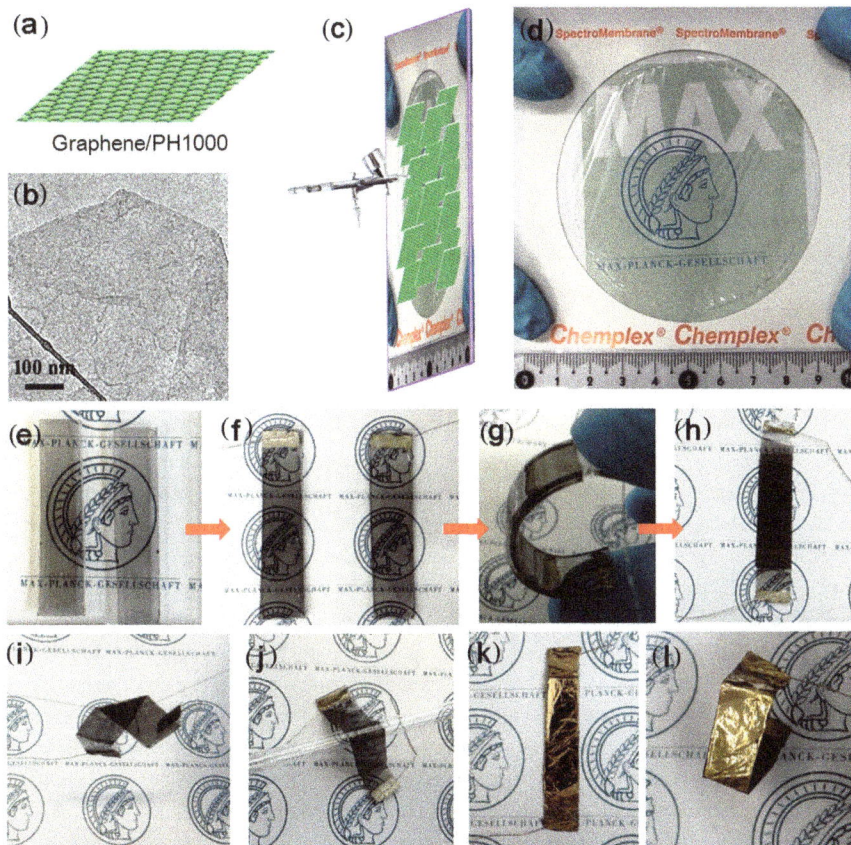

Fig. 12. Schematic illustration of UPSCs: (a) Schematic description of 2D EG/PH1000 nanosheets. (b) TEM image of the PH1000 adsorbed EG nanosheets. (c) Fabrication scheme of EG/PH1000 film on 2.5-µm-thick PET substrate with an air brush pistol. (d) Optical image of the EG/PH1000 film on a 2.5-µm-thick PET substrate with a thin-film diameter of 76 mm. (e–h) Optical images showing the fabrication of UPSCs, including: (e) spray coating of EG/PH1000 film on 2.5-µm-thick PET/PDMS substrates, with lengths of 4.5 cm and widths of 1 cm; (f) coating of gel electrolyte on the surface of film electrodes and connection of the Pt wire on the film by silver paste glue; (g) assembly of UPSCs sandwiched between two gel-electrolyte/film electrodes on PET/PDMS substrates; (h) all-solid-state UPSCs on 2.5-µm-thick PET substrates (after peeling off the PDMS substrates). (i-l) UPSCs with excellent flexibility, e.g., naturally bending state when: (i) hung on a hair and j) placed on an NMR glass tube. (k,l) A UPSC based on

EG/PH1000 film on Au-coated PET substrates (with the same procedure as (e-g)), showing: (k) flat and (l) folded bending states [42, copyright to Wiley (2015)].

hold great potential to replace AECs (that need 8.3 ms) for ac-line-filtering applications [5, 24]. Additionally, the Ragone plot revealed that the energy density and power density (based on two electrodes) were 12 mWh/cm^3 and 4386 W/cm^3 for our UPSCs, respectively, which could potentially fill the gap between high-energy lithium thin-film batteries (10 mWh/cm^3) [9] and high-power electrolytic capacitors (10^3 W/cm^3) [44]. Table 1 summarizes different types of supercapacitors for ac line filtering applications and provides the detailed physical parameters.

6   Conclusion and perspectives

There have been numerous efforts to fabricate various types of supercapacitors (EDLC and PC) in two sizes (macro and micro) sometimes with excellent flexibility of the electrodes and sometimes on-chip and/or printable capability for the same purpose of efficient filtering of ripple voltages from ac supply within complex electrical circuits conventionally used in domestic and industrial sections. The excellent conductivity of electrode material and its proper pore structure helps in fast charge diffusion and transport making the SCs ultrafast. Graphene, carbon nanotube, their hybrid composite structures, polymer films, coordination conducting polymer frameworks, porous carbon and metal oxide electrodes have been utilized for this purpose which demonstrated phase angles up to very close to 90° (65 to 89°) corresponding to 120 Hz frequency. Most of the carbon and polymer-based SCs demonstrated phase angle around 80 to 85°. The main limitation in ac line filtering using different types of SCs is that when the phase angle is close to 90°, the specific capacitance is low and vice versa. In fact, some of the metal oxide based MSCs exhibit higher capacitances, however, the ac line filtering capability is inferior. Generally, the areal capacitance for most of the carbon-based SCs are low, although, vertically oriented graphene grown on an alumina substrate and carbon nanotube-based ion get SCs demonstrated excellent areal capacitances around 80 and 106 mF/cm^2 together with appreciable phase angles (87 and 78.1° respectively). In contrary, the areal capacitances of various polymer based SCs are also very low and their phase angles are mostly low.

Therefore, intense research focus must be directed to fabricate supercapacitors with high areal capacitance with phase angle very close to 90°. Such SCs can bring a breakthrough in the technology related to ac line filtering using supercapacitors. For this purpose electrode materials with ordered morphology and porosity having excellent conductivity must be synthesized. Better understanding the suitable pore structure and their manipulation can reduce the ion diffusion time to realize ultra-fast supercapacitors. In addition, proper electrolyte and redox active materials should also be selected to obtain high areal capacitance. A proper combination of electrode material, its morphology, and crystallinity together with a suitable electrolyte and redox active material can bring forward advanced SCs for efficient ac line filtering capability. As such, flexible and wearable SCs and MSCs with excellent ac-line filtering efficiency would not only be an interesting but essential approach for supporting multifunctional and portable electronic devices for next generation.

Table 1. Comparison of Electrochemical Performance of Various Capacitors[a]

Supercapacitor type	Material	Phase at 120 Hz (°)	F (Hz) at -45°	t_{RC} at 120 Hz (ms)	t_0 (ms)	R_c (Ω)	C_A (mF/cm^2)	P_v (W/cm^3)	E_v (mWh/cm^3)	Ref.
Macro	AEC	83.9	1600	0.14	-	>400	0.3	>100	<0.01	1
	Unzipped NCNT	85	8150	0.21	0.25	1.15	0.19	-	-	16
	LSG	<20	30	-	-	10	3.67	<10	<0.1	44
	VG	82	15000	0.2	-	-	<0.2	-	-	5
	ErGO	85.5	4200	1.35	0.24	3.5	<1	-	-	8
	CNTs	<75	636	-	1.5	1	-	-	-	47
	AT-PEDOT:PSS	83.68	~8000	-	0.15	-	0.994	-	-	18
	VG/Al$_2$O$_3$	~87	~4000	-	-	-	80	-	-	19
	GOMC	80	-	-	-	-	1	-	-	22
	Camphor-C	78	4200	0.319	-	1.86	0.172	-	-	23
Micro	G-CNTCs	81.5	1343	0.195	0.82	400	2.16	115	0.16	30
	AC	~1	<5	-	700	1	-	-	-	9, 46
	OLC	-	<100	-	26	100	-	-	-	9
	LRG	-	<5	-	-	<0.1	~1	100	1	45
	PEDOT	80.5	-	-	-	-	0.15	-	-	27
	PEDOT	~60	400	-	-	-	9		7.7	28
	PiCBA	73	3620	0.83	0.27	-	34.1 F/cm^3	1323	4.7	29
	rGO/Au	~60	362	-	2.76	-	4.92	-	-	26
	ATO	~89	-	-	-	-	0.004	-	-	31
	hRuO$_2$	<50	-	-	512	-	3.2	-	-	32
Flexible	CNT/ion gel	78.1	~6000	-	1	-	106	-	-	33
	HCF	67	-	-	143	-	0.48	-	-	34
Printable	G-PEDOT:PSS	75	-	<0.5	-	-	0.179	-	-	42

[[a]AEC: aluminum electrolytic capacitor; LSG: laser-scribed graphene; VG: vertical graphene; ErGO: electrochemical reduced graphene oxide; CNTs: carbon nanotubes; AT-PEDOT-PSS: sulfuric acid treated commercially available poly(3,4-ethylenedioxythiophene)-poly(styrenesulfonate); GOMC: graphitic ordered mesoporous carbon; G/CNTCs: graphene/carbon nanotube carpets; OLC: onion-like carbon; LRG: laser reduced graphene; AC: activated carbon; PiCBA: azulene-bridged coordination polymer framework; ATO: amorphousTiO_2; hRuO_2: hydrated ruthenium oxide; HCF: hybrid carbon film (multilayer graphene/carbon film/ML-graphene)]

References

[1] X. Chen, R. Paul, L. Dai, Carbon-based supercapacitors for efficient energy storage, Nat. Sci. Rev. 4 (2017) 453-489. https://doi.org/10.1093/nsr/nwx009

[2] A. Dale, C. Brownson, C.E. Banks, Fabricating graphene supercapacitors: highlighting the impact of surfactants and moieties, Chem. Commun. 48 (2012) 1425-1427. https://doi.org/10.1039/C1CC11276G

[3] L.L. Zhang, X.S. Zhao, Carbon-based materials as supercapacitor electrodes, Chem. Soc. Rev. 38 (2009) 2520-2531. https://doi.org/10.1039/b813846j

[4] A.G. Pandolfo, A.F. Hollenkamp, Carbon properties and their role in supercapacitors, J. Power Sources 157 (2006) 11-27. https://doi.org/10.1016/j.jpowsour.2006.02.065

[5] J.R. Miller, R.A. Outlaw, B.C. Holloway, Graphene double-layer capacitor with ac line-filtering performance, Science 329 (2010) 1637-1639. https://doi.org/10.1126/science.1194372

[6] C. Niu, E.K. Sichel, R. Hoch et al, High power electrochemical capacitors based on carbon nanotube electrodes, Appl. Phys. Lett. 70 (1997) 1480-1482. https://doi.org/10.1063/1.118568

[7] J. Joseph, A. Paravannoor, S.V. Nair, Supercapacitors based on camphorderived meso/macroporous carbon sponge electrodes with ultrafast frequency response for ac line-filtering, J. Mater. Chem. A 3 (2015) 14105-14108. https://doi.org/10.1039/C5TA03012A

[8] K.X. Sheng, Y.Q. Sun, C. Li, W.J. Yuan, G.Q. Shi, Ultrahigh-rate supercapacitors based on eletrochemically reduced graphene oxide for ac line-filtering, Sci. Rep. 2 (2012) 247. https://doi.org/10.1038/srep00247

[9]   D. Pech, M. Brunet, H. Durou, P. Huang, V. Mochalin, Y. Gogotsi, P. Taberna, P. Simon, Ultrahigh-power micrometre-sized supercapacitors based on onion-like carbon, Nat. Nanotechnol. 5 (2010) 651-654. https://doi.org/10.1038/nnano.2010.162

[10]   D. N. Futaba, K. Hata, T. Yanada et al, Shape-engineerable and highly densely packed single-walled carbon nanotubes and their application as super-capacitor electrodes, Nat. Mater. 5 (2006) 987-994. https://doi.org/10.1038/nmat1782

[11]   Y. Rangom, X. Tang, L.F. Nazar, Carbon nanotube-based supercapacitors with excellent ac line filtering and rate capability via improved interfacial impedance, ACS Nano 9 (2015) 7248-7255. https://doi.org/10.1021/acsnano.5b02075

[12]   Y. Korenblit, M. Rose, E. Kockrick et al, High-rate electrochemical capacitors based on ordered mesoporous silicon carbide-derived carbon, ACS Nano 4 (2010) 1337-1344. https://doi.org/10.1021/nn901825y

[13]   X. Lang, A. Hirata, T. Fujita et al, Nanoporous metal/oxide hybrid electrodes for electrochemical supercapacitors, Nat. Nanotech. 6 (2011) 232-236. https://doi.org/10.1038/nnano.2011.13

[14]   Y. Hou, Y. Cheng, T. Hobson et al, Design and synthesis of hierarchical MnO2 nanospheres/carbon nanotubes/conducting polymer ternary composite for high performance electrochemical electrodes, Nano Lett. 10 (2010) 2727-2733. https://doi.org/10.1021/nl101723g

[15]   Z.S. Wu, X. Feng, H.M. Cheng, Recent advances in graphene-based planar micro-supercapacitors for on-chip energy storage, Nat. Sci. Rev. 1 (2014) 277-292. https://doi.org/10.1093/nsr/nwt003

[16]   J. Lim, U.N. Maiti, N.Y. Kim et al, Dopant-specific unzipping of carbon nanotubes for intact crystalline graphene nanostructures, Nat. Commun. 7 (2016) 10364. https://doi.org/10.1038/ncomms10364

[17]   H. Wei, S. Wei, W. Tian et al, Fabrication of thickness controllable freestanding sandwich-structured hybrid carbon film for high-rate and high-power supercapacitor, Sci. Rep. 4 (2014) 7050. https://doi.org/10.1038/srep07050

[18]   M. Zhang, Q. Zhou, J. Chen, X. Yu, L. Huang, Y. Li, C. Li, G. Shi, An ultrahigh-rate electrochemical capacitor based on solution-processed highly conductive PEDOT:PSS films for AC line-filtering, Energy Environ. Sci. 9 (2016) 2005-2010. https://doi.org/10.1039/C6EE00615A

[19] D. Premathilake, R.A. Outlaw, S.G. Parler, S.M. Butler, J.R. Miller, Electric double layer capacitors for ac filtering made from vertically oriented graphene nanosheets on aluminum, Carbon 111 (2017) 231-237. https://doi.org/10.1016/j.carbon.2016.09.080

[20] R. Paul, V. Etacheri, V.G. Pol, J. Hu, T.S. Fisher, Highly porous three-dimensional carbon nanotube foam as a freestanding anode for a lithium-ion battery, RSC Adv. 6 (2016) 79734-79744. https://doi.org/10.1039/C6RA17815D

[21] R. Paul, D. Zemlyanov, A.A. Voevodin, A.K. Roy, T.S. Fisher, Methanol wetting enthalpy on few-layer graphene decorated hierarchical carbon foam for cooling applications, Thin Solid Films 572 (2014)169-175. https://doi.org/10.1016/j.tsf.2014.08.020

[22] Y. Yoo, M.S. Kim, J.K. Kim, Y.S. Kim, W. Kim, Fast-response supercapacitors with graphitic ordered mesoporous carbons and carbon nanotubes for AC line filtering, J. Mater. Chem. A 4 (2016) 5062-5068. https://doi.org/10.1039/C6TA00921B

[23] J. Joseph, A. Paravannoor, S.V. Nair, Z.J. Han, K. Ostrikov, A. Balakrishnan, Supercapacitors based on camphor-derived meso/macroporous carbon sponge electrodes with ultrafast frequency response for ac line-filtering, J. Mater. Chem. A 0 (2015) 1-5. https://doi.org/10.1039/C5TA03012A

[24] K. Sheng, Y. Sun, C. Li, W. Yuan, G. Shi, Ultrahigh-rate supercapacitors based on eletrochemically reduced graphene oxide for ac line-filtering, Sci. Rep. 2 (2012) 247 1-5. https://doi.org/10.1038/srep00247

[25] P. Huang, C. Lethien, S. Pinaud, K. Brousse, R. Laloo, V. Turq, M. Respaud, A. Demortière, B. Daffos, P.L. Taberna, B. Chaudret, Y. Gogotsi, P. Simon, On-chip and freestanding elastic carbon films for micro-supercapacitors, Science 351 (2016) 691-695. https://doi.org/10.1126/science.aad3345

[26] R.Z. Li, R. Peng, K.D. Kihm, S. Bai, D. Bridges, U. Tumuluri, Z. Wu, T. Zhang, G. Compagnini, Z. Feng, A. Hu, High-rate in-plane micro-supercapacitors scribed onto photo paper using in situ femtolaser-reduced graphene oxide/Au nanoparticle microelectrodes, Energy Environ. Sci. 9 (2016) 1458-1467. https://doi.org/10.1039/C5EE03637B

[27] N. Kurra, Q. Jiang, A. Syed, C. Xia, H.N. Alshareef, Micro-pseudocapacitors with Electroactive Polymer Electrodes: Towards Ac-Line Filtering Applications, ACS Appl. Mater. Interfaces 8 (2016) 12748-12755. https://doi.org/10.1021/acsami.5b12784

[28] N. Kurra, M.K. Hota, H.N. Alshareef, Conducting polymer micro-supercapacitors for flexible energy storage and Ac line-filtering, Nano Energy 13 (2015) 500-508. https://doi.org/10.1016/j.nanoen.2015.03.018

[29] C. Yang, K.S. Schellhammer, F. Ortmann, S. Sun, R. Dong, M. Karakus, Z. Mics, M. Lçffler, F. Zhang, X. Zhuang, E. Canovas, G. Cuniberti, M. Bonn, X. Feng, Coordination polymer framework based on-chip micro-supercapacitors with ac-line filtering performance, Angew. Chem. Int. Ed. 56 (2017) 3920-3924. https://doi.org/10.1002/anie.201700679

[30] J. Lin, C. Zhang, Z. Yan, Y. Zhu, Z. Peng, R.H. Hauge, D. Natelson, J.M. Tour, 3-Dimensional graphene carbon nanotube carpet-based microsupercapacitors with high electrochemical performance, Nano Lett. 13 (2013) 72-78. https://doi.org/10.1021/nl3034976

[31] M. Fukuhara, T. Kuroda, F. Hasegawa, Amorphous titanium-oxide supercapacitors, Sci. Rep. 6 (2016) 35870. https://doi.org/10.1038/srep35870

[32] T.M. Dinh, K. Armstrong, D. Guay, D. Pech, High-resolution on-chip supercapacitors with ultra-high scan rate ability, J. Mater. Chem. A 2 (2014) 7170-7174. https://doi.org/10.1039/C4TA00640B

[33] Y.J. Kang, Y. Yoo, W. Kim, 3-V Solid-State flexible supercapacitors with ionic-liquid-based polymer gel electrolyte for ac line filtering, ACS Appl. Mater. Interfaces 8 (2016) 13909-13917. https://doi.org/10.1021/acsami.6b02690

[34] H. Wei, S. Wei, W. Tian, D. Zhu, Y. Liu, L. Yuan, X. Li, Fabrication of thickness controllable free-standing sandwich-structured hybrid carbon film for high-rate and high-power supercapacitor, Sci. Rep. 4 (2014) 7050. https://doi.org/10.1038/srep07050

[35] D.W. Wang, F. Li, Z.S. Wu, W. Ren, H.M. Cheng, Electrochemical interfacial capacitance in multilayer graphene sheets: Dependence on number of stacking layers, Electrochem. Commun. 11 (2009) 1729-1732. https://doi.org/10.1016/j.elecom.2009.06.034

[36] Z. Bo, W. Zhu, W. Ma, Z. Wen, X. Shuai, J. Chen, J. Yan, Z. Wang, K. Cen, X. Feng, Vertically oriented graphene bridging active-layer/current-collector interface for ultrahigh rate supercapacitors, Adv. Mater. 25 (2013) 5799-5806. https://doi.org/10.1002/adma.201301794

[37] J.H. Ahn, H.S. Kim, K.J. Lee, S. Jeon, S.J. Kang, Y.G. Sun, R.G. Nuzzo, J.A. Rogers, Heterogeneous three-dimensional electronics by use of printed

semiconductor nanomaterials, Science 314 (2006) 1754-1757.
https://doi.org/10.1126/science.1132394

[38] J.S. Shi, C.X. Guo, M.B. Chan-Park, C.M. Li, All-printed carbon nanotube finFETs on plastic substrates for high-performance flexible electronics, Adv. Mater. 24 (2012) 358-361. https://doi.org/10.1002/adma.201103674

[39] S.H. Kim, K. Hong, W. Xie, K.H. Lee, S.P. Zhang, T.P. Lodge, C.D. Frisbie, Electrolyte-gated transistors for organic and printed electronics, Adv. Mater. 25 (2013) 1822-1846. https://doi.org/10.1002/adma.201202790

[40] M. Kaltenbrunner, M.S. White, E.D. Glowacki, T. Sekitani, T. Someya, N.S. Sariciftci, S. Bauer, Ultrathin and lightweight organic solar cells with high flexibility, Nat. Commun. 3 (2012) 770 1-7. https://doi.org/10.1038/ncomms1772

[41] Z.S. Wu, X.L. Feng, H.M. Cheng, Recent advances in graphene-based planar micro-supercapacitors for on-chip energy storage, Natl. Sci. Rev. 1 (2014) 277-292. https://doi.org/10.1093/nsr/nwt003

[42] Z.S. Wu, Z. Liu, K. Parvez, X. Feng, K. Müllen, Ultrathin printable graphene supercapacitors with ac line-filtering performance, Adv. Mater. 27 (2015) 3669-3675. https://doi.org/10.1002/adma.201501208

[43] T.N. Walleser, I.M. Lazar, M. Fabritius, F.J. Tolle, Q. Xia, B. Bruchmann, S.S. Venkataraman, M.G. Schwab, R. Mulhaupt, Adv. Funct. Mater. 24 (2014) 4706. https://doi.org/10.1002/adfm.201304151

[44] M.F. El-Kady, V. Strong, S. Dubin, R.B. Kaner, Laser scribing of high-performance and flexible graphene-based electrochemical capacitors, Science 335 (2012) 1326. https://doi.org/10.1126/science.1216744

[45] W. Gao, N. Singh, L. Song, Z. Liu, A.L.M. Reddy, L.J. Ci, R. Vajtai, Q. Zhang, B.Q. Wei, P.M. Ajayan, Direct laser writing of micro-supercapacitors on hydrated graphite oxide films, Nat. Nanotechnol. 6 (2011) 496-500. https://doi.org/10.1038/nnano.2011.110

[46] D. Pech, M. Brunet, P.L. Taberna, P. Simon, N. Fabre, F. Mesnilgrente, V. Conedera, H.J. Durou, Elaboration of a microstructured inkjet-printed carbon electrochemical capacitor, J. Power Sources 195 (2010) 1266-1269. https://doi.org/10.1016/j.jpowsour.2009.08.085

[47] C.S. Du, N. Pan, Supercapacitors using carbon nanotubes films by electrophoretic deposition, J. Power Sources 160 (2006) 1487-1494. https://doi.org/10.1016/j.jpowsour.2006.02.092

Chapter 6

High-Performance Polymer Type Capacitors

Tulay Y. Inan

SAUDI ARAMCO, Research and Development Center, PO Box 5074, 31311 Dhahran, Saudi Arabia

Abstract

This book chapter concentrates on energy storage applications of thin film containing electrochemical capacitors (supercapacitors) with specifically high-temperature polymers (fluorinated polymers, polyimides, polyarylenes, polyvinyl pyrollidone, etc.) for their good thermal and hydrolytic stability, excellent mechanical and chemical stability, low cost, flexibility and commercial availability. It consists of three main parts as given in the table of contents related to the 1) fluorinated high-performance polymers (FHPP) 2) nonfluorinated high-performance polymers (NFHPP) and 3) fluorinated and nonfluorinated high-performance polymers (FNFHPP) with their composites. A total of 141 literature reviewed references are cited.

Keywords

Thin Film, Electrochemical Capacitors, Supercapacitor, Dielectric Material, Separator, Solid Separators, High-Performance Polymers, Fluorinated Polymers, Polyimides, Polyarylenes, Polyvinyl Pyrollidone

Contents

1. Introduction

Electrochemical energy systems can be classified as (i) energy conversion and (ii) energy storage electrochemical devices as shown in Fig. 1.

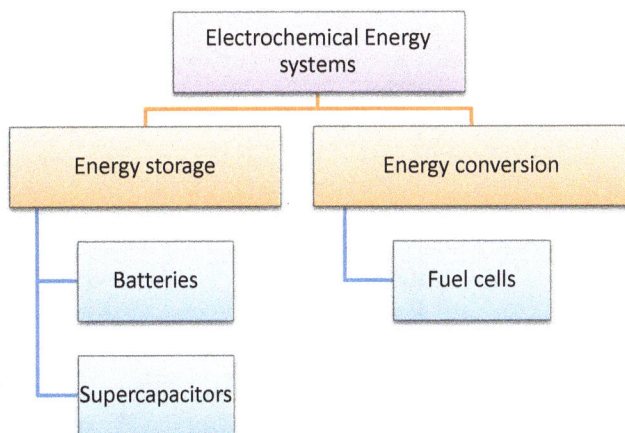

Fig. 1. Electrochemical energy systems.

Although there are some differences in the mechanisms of such devices, most of these depend on polymer electrolyte membranes/separators and have two main functions i) to separate the electrochemically active masses (electrodes) and ii) to intercede the electrochemical reactions occurring at the anode and cathode by conducting specific ions [1].

1.1   Electrochemical capacitors (Supercapacitor)

Supercapacitors are energy storage devices and like batteries, supercapacitors consist of two conducting electrodes separated by an insulating dielectric material or separator. When a voltage is applied to a capacitor, opposite charges accumulate on the surfaces of each electrode as shown in the schematic diagram (Fig. 2). The charges are kept separated by the dielectric separator, thus producing an electric field that allows the capacitor to store energy. They combine the high energy storage capability of batteries with the high power delivery capability of the capacitors, as indicated by the high charging-discharging rate.

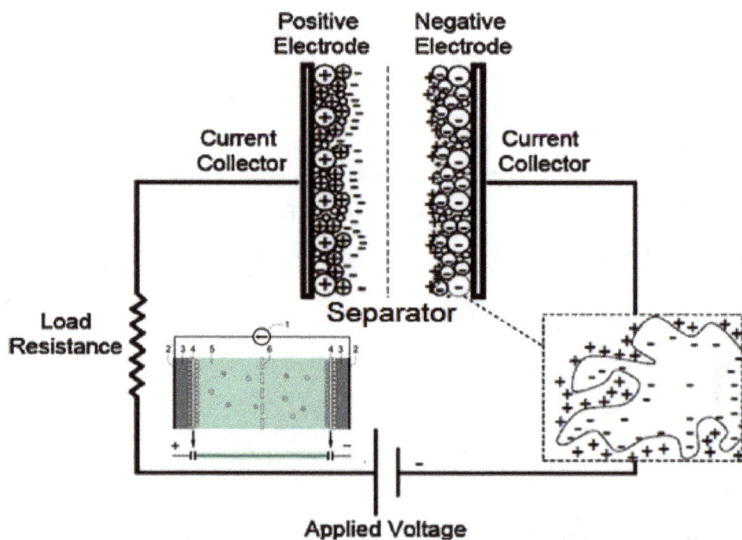

Fig. 2. Typical construction of a supercapacitor: 1) Power source, 2) Collector, 3) Polarized electrode, 4) Helmholtz double layer, 5) Electrolyte having positive and negative ions, 6) Separator [1,3].

Also, depending on electrode material and surface shape, some ions become specifically adsorbed after permeating to double layer. As a result, they contribute with pseudocapacitance to the total capacitance of the supercapacitor [3] (Fig. 3.).

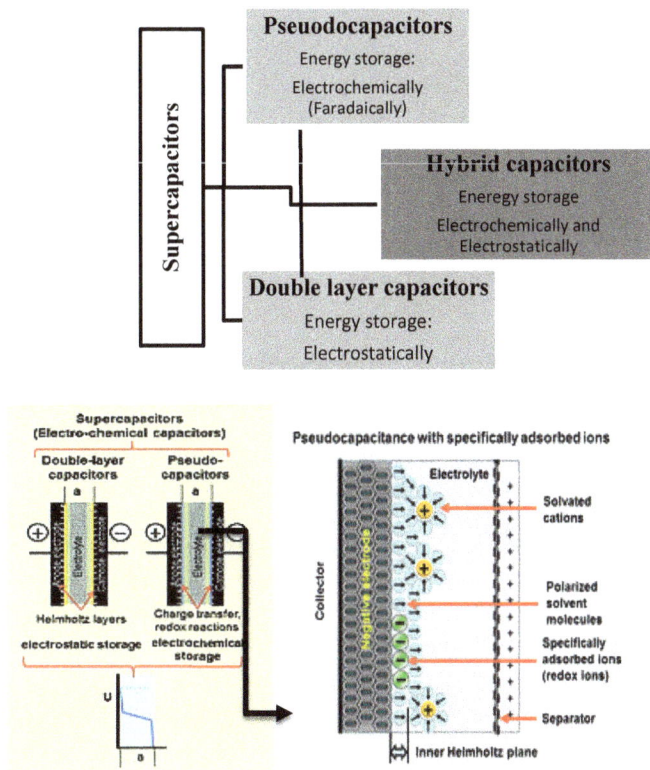

Fig. 3. Charge storage principles of different capacitor types and their internal potential distribution [2].

The progress in these devices showed that electrodes with much higher surface areas and thinner dielectric materials compared to those used in conventional capacitors are leading to an increase in both capacitance and energy [1,2]. This provides several advantages to supercapacitors over electrochemical batteries and fuel cells, including higher power density, shorter charging times, and longer cycle life and shelf life. The relative performance of energy storage devices in terms of the power density versus energy density is represented in a Ragone diagram as shown in Fig. 4 [3].

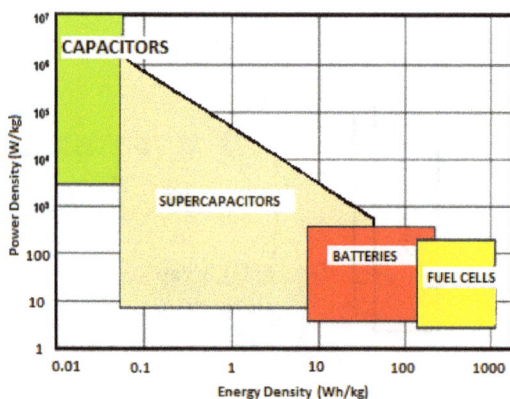

Fig. 4. Ragone diagram for energy storage devices [3].

Lithium-ion batteries and supercapacitors are receiving high attention nowadays as energy storage devices. These have great potential in different applications, such as power source for a hybrid electric vehicle, laptop or mobile phone power backup, portable electronics, medical electronics and military devices.

The materials used for producing solid separators can be divided into three basic groups: cellulose (paper), textiles and polymers. The use of polymer separators in supercapacitors is very promising considering the high chemical and mechanical resistance of these materials, low price and ease of processing.

Efforts have been made to develop low-cost materials with (a) high ionic conductivity, (b) high chemical and thermal stability in the operating temperature range to increase the efficiency and (c) reliability and service lifetime, in addition to reducing the capital costs of energy systems. Various polymerization methods have been used to prepare this fundamental part of supercapacitors of polymer electrolyte membranes/separators for different applications. Supercapacitor polymeric materials are found in two forms:

(i) Electrolyte gel, ion conduction takes place through highly porous (40–80% porosity) or swollen type separators by using an entrapped liquid electrolyte either aqueous (e.g. H_2SO_4 or KOH) or organic (e.g. acetonitrile or propylene carbonate) solvents.

(ii) Solvent-free solid electrolytes; ion conduction takes place by fixed ionic groups.

None of the separators participates in the electrochemical reactions but plays an important role in terms of safety of the devices. Specifically, the main function of separators is to separate the anode and the cathode to prevent internal short circuit without blocking the ionic conductivity. That's why the mechanical properties of the polymer become an important issue for the safety of the device.

In the last decades, many researchers focused on the synthesis and development of high-performance aromatic type hydrocarbon polymers, blending, organic and inorganic composite applications and also crosslinking reactions like radiation-induced crosslinking, thermal crosslinking and sol-gel method in order to modify the properties of polymer separators for application in the supercapacitors. Specifically, high-temperature polymers were chosen for this purpose because of their good thermal and hydrolytic stability, excellent mechanical and chemical stability, low cost, flexibility and commercial availability. This chapter focused on 1) fluorinated high-performance polymers (FHPP)2) nonfluorinated high-performance polymers (NFHPP) and 3) fluorinated and nonfluorinated high-performance polymers (FNFHPP)with their composites for energy storage applications of supercapacitors.

1.2 High-performance polymers (HPP) [4]

High-performance polymers (HPP), also known as non-metallic, high-temperature polymers, advanced engineering materials, and heat resistant polymers are the polymers that can hold their outstanding properties of mechanical, dimensional, and chemical stabilities under high temperature and/or high-pressure corrosive environments.

HPP type polymers can be developed using high bond strength monomers. Because of increasing bond dissociation energy the strength and resistance of the polymer to the harsh environments is increased. It is difficult to break a C=C as compared to C-C since the bond energy of C-C is 83 kcalmol^{-1}, C=C is 145 kcal mol^{-1}. Therefore, most of the HPP contained a lot of C=C relative to C-C. Adding aromatic components along the backbone also improves resistance and stability of the HPP. Polymers with high aromatic contents, when exposed to elevated temperature do not produce volatile fragments due to their resonance-stabilized bonds. These bonds add 40–70 kcal mol^{-1} in the bond strength and as a result at high temperature, radical species are produced at the onset of the degradation of the polymer. These radical species then react with the polymer chain and, as a result, breaks it. These radical species if generated from the resonance stabilized units, become less reactive as the radicals are delocalized via the π system.

Having very high aromatic content of HPP gives polymers outstanding properties but at the same time, it causes one major disadvantage for processing due to the stiff and very hard polymeric structures. Heteroatoms such as nitrogen, oxygen, and sulfur have been introduced into the aromatic units to solve this problem. However, the thermal stability, as well as the chemical resistance, are decreased because of the decrease in bond strength. But at the same time, secondary forces such as hydrogen bonding, polar interaction, and van der Waals forces – all of which can drastically improve the thermal stability of the HPPs due to the intermolecular attraction among the polymer chains. Polymer molecular weight and its distribution, crosslinking density, crystallinity, and the presence of additives and reinforcements are the other factors known to improve the thermal stability of the HPP.

Polymers can be classified as HPP, engineering polymers (EP) and commodity polymers (CP). HPP can be divided into amorphous and crystalline (crystalline and semi-crystalline) polymers. Examples include amorphous polymers: polysulfone (PSU), polyimide (PI), polyetherimide (PEI) and polyamide-imide (PAI); semi-crystalline polymers: poly(phenylene sulfide) (PPS), polyetheretherketone (PEEK), or poly(ether ketone ketone) (PEKK), poly(ether sulfone) (PSF) and various liquid crystalline polymers (Fig. 5.)

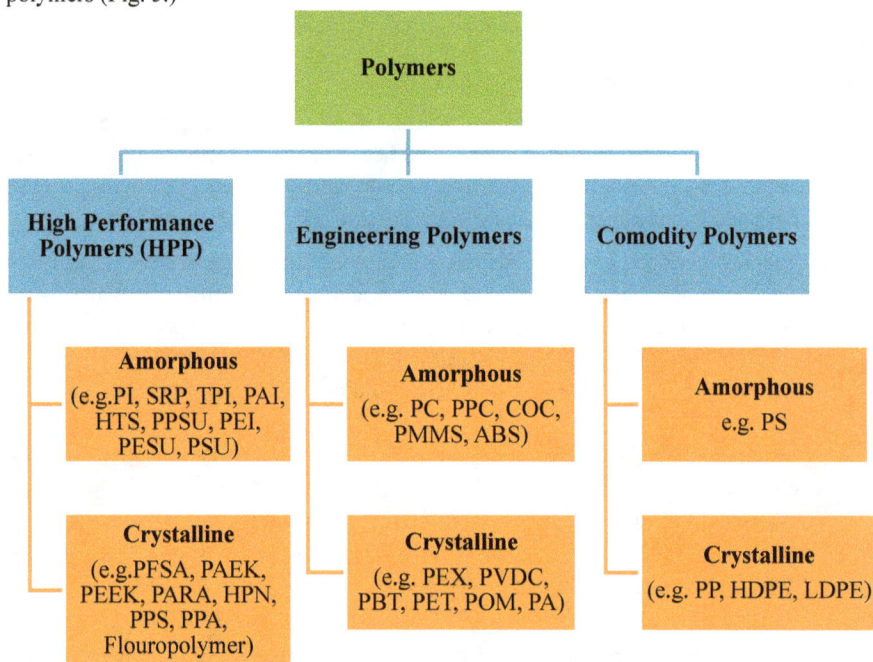

Fig. 5. Classification of polymers.

2. HPP Type separators/electrolytes/substrates in supercapacitors

HPP type solid supercapacitor separators with high dielectric constant, low dielectric loss, good breakdown strength, and high-temperature capability are charming capacitive energy-storage systems. Biaxially oriented polypropylene (BOPP), poly(ethylene terephthalate) (PET), poly(ethylene naphthalate) (PENt), polycarbonate (PC), and poly(vinylidene) fluoride (PVDF) type dielectrics are commercially available polymer separators that can be operated at just below 200 °C [5]. The state-of-the-art polymer of biaxially oriented polypropylene (BOPP) film dielectric has a maximal energy density of 5 J/cm^3 and a high breakdown field of 700 MV/m, but with a limited dielectric constant (~2.2) and a reduced breakdown strength above 85 °C [6]. Sincere efforts have been put into exploring high-temperature polymer dielectrics to fulfill the demand of high-temperature applications, such as aerospace and military power supply. Mostly used solid type polymer electrolyte separators are polyimide, fluorinated type homo and copolymers like PVDF, PVDF-HFP, poly(vinylidene fluoride-co-trifluoroethylene) (PVDF-TrFE), poly(tetrafluoroethylene) (PTFE), perfluoroalkoxy (PFA), polyarylene ethers (PAEK, PEEK, PEKK), polyarylene ether nitrile (PEN), polybenzoxazoles, their sulfonated type structures like sulphonatedpolyetherether ketone (SPEEK), and also their different combinations of organic (blends) and inorganic added composites.

For polymeric gel electrolytes, solid polymer electrolytes/separators are alternative materials that are preferred over crystalline solids for their different advantages like a wide range of compositions, and hence control of properties, good interfacial contacts between the electrode and electrolyte and ease of preparation in thin and bulk forms. In recent years, considerable progress has been made to enhance the electrical conductivity as well as electrochemical and mechanical stabilities of these polymer electrolyte materials to be utilized for various applications. For example, Armand et al. used solid polymer electrolyte such as poly(ethylene oxide) (PEO) without any modification that may transport ions [1]. However, the electrolyte conductivity is lower than liquid electrolyte-based systems. Grafting non-conducting polymers with conducting polymers could be the other option to obtain a conducting system [1]. Polymer modification with chemical and/or radiation-induced techniques (grafting or crosslinking) may also be applied to overcome the electrolyte limitation within the system.

2.1 Fluorinated type polymers and their composites

The commercial ionomer of Nafion received considerable attention as a proton conductor for PEM fuel cells, battery and also supercapacitor applications because of its excellent thermal and mechanical stability. Nafion is a perfluorosulfonic acid (PFSA) membrane and needs water to become proton conductive. Nafion has a polymer with a polytetrafluoroethylene main chain and a perfluorinated side chain ending with a sulfonic acid group $-SO_3H$ (Fig. 6.). The proton conductivity arises from the sulfonic acid groups. Because a C–F bond is stronger than a C–H bond (their bond strengths, at 298K, in diatomic molecules, are 552 and 338 kJ mol^{-1}, respectively) and the C–C bond is well shielded by the F atoms [7]. It is mainly used in electrode applications [8,9,10,11] as a binder but it is also used in solid polymer electrolyte applications. For example, Staiti and Lufrano [12] prepared more compact, flexible and safer solid polymer electrolytes for solid-state supercapacitors. They used two different perfluorinated sulfonic acids (PFSA) type solid polymer electrolytes specifically Nafion® 115 (purchased from DuPont™) and Fumapem® F-950 (received from FuMA-Tech GmbH (Germany) between the electrodes of supercapacitors. They also used 5 wt% Nafion (Aldrich Chemistry) and Fumion solution (FuMA-Tech) in the preparation of the electrodes and produced four different supercapacitors to determine their respective specific capacitances, resistances, specific energies and power densities. The results of the cyclic voltammetry, galvanostatic charge/discharge, and electrochemical impedance spectroscopy analysis, showed the superior performance of supercapacitor based on Nafion electrolyte membrane and Nafion/carbon electrodes in terms of specific capacitance and electrical series resistance.

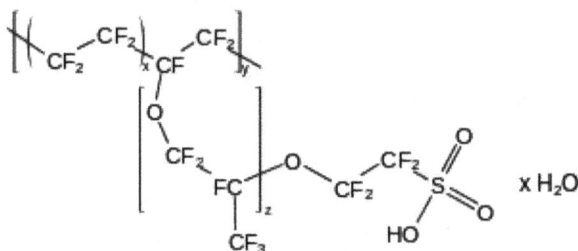

Fig. 6. Schematic representation of Nafion.

There are also other types of fluorinated polymers used in supercapacitor applications. Hibino et al. [13] fabricated a proton-conducting composite solid-state supercapacitor electrolyte using $Sn_{0.95}Al_{0.05}H_{0.05}P_2O_7$ (SAPO)-polytetrafluoroethylene (PTFE) with a highly condensed H_3PO_4 electrode ionomer. The resulting supercapacitor exhibited proton conductivity of 0.02 S cm^{-1} and a wide withstanding voltage range of ±2 V with an energy density of 32 Wh kg^{-1} at 3 A g^{-1}. It was also shown that up to 7000 cycles stability was achieved from room temperature to 150°C. The H_3PO_4 ionomer had good wettability with micropore-rich activated carbon, resulting in a capacitance of 210 F g^{-1} at 200°C.

Park et al. [14,15] proposed thin field-activated electroactive polymer (FEAP) films including poly(vinylidene fluoride) (PVDF), ferroelectric poly(vinylidene fluoride-co-trifluoroethylene) (PVDF-TrFE) and poly(vinylidene fluoride-co-trifluorocthylene-co-chlorotrifluoroethylene) (PVDF-TrFE-CTFE) for high performance organic electronic device applications. They developed highly ordered semicrystalline domains of FEAPs on a molecularly ordered poly(tetrafluoroethylene) (PTFE) substrate, with the combination of the spin coating method on a few centimeter square area. They demonstrated that their epitaxial (refers to the deposition of a crystalline overlayer on a crystalline substrate grown) FEAP films on PTFE surface were successfully incorporated into various organic electronic devices for high device performance, including non-volatile memory capacitor, metal-ferroelectric-insulator-semiconductor memory element, field effect transistor type memory, and high energy storage capacitor. They also prepared highly ordered poly(vinylidene fluoride-co-trifluoroethylene) (PVDF-TrFE) ultrathin films epitaxially grown on friction-transferred polytetrafluoroethylene (PTFE) surface which were incorporated in the metal-ferroelectric-metal (MFM) and metal-ferroelectric-insulator-semiconductor (MFIS) memory structure. Ultrathin ferroelectric polymer films grown by epitaxy were microimprinted with a silver coated polydimethylsiloxane (PDMS) mold at 170 °C with excellent quality and the simultaneous transfer of silver electrodes on the imprinted PVDF-TrFE. Another epitaxially grown PVDF-TrFE film incorporated for arrays of ferroelectric capacitors that exhibited not only the significant reduction of ferroelectric thermal hysteresis and descent remnant polarization at very low effective operating voltage of ±5 V maintained 88% of its initial value after a number of fatigue cycles of 5 × 108 in the mode of bipolar pulse switching. A ferroelectric field effect transistor memory with the epitaxially grown PVDF-TrFE layer as gate dielectric showed saturated I-V hysteresis with bistable on/off ratio of approximately 102.

Because of the attractive packaging characteristics, applications in high performance military data processors as well as commercial telecom switches are being qualified for high volume production and assembly, with more than one application currently in high volume ramp multichip applications to go from paper concept to use a thin PTFE composite substrate due to the electrical and mechanical advantages to create very high performance applications with this polymer. Murali et al. [16] incorporated micron and nanosize silica fillers in the PTFE matrix to prepare flexible composite substrates. A proprietary process comprising of sigma mixing, extrusion, calendaring followed by hot pressing (SMECH process) to obtain nearly isotropic and dimensionally stable filled

PTFE substrates has been proposed. They also employed theoretical modeling to predict the effective dielectric constant of the composite system and validated the results with experimental data. Their findings were in good agreement with theoretical modeling.

Xiang et al. [17] obtained composite films by mixing a series of polytetrafluoroethylene (PTFE) with different amount of ceramic filler. They reported that as the amount of ceramic filler increased, both of the relative permittivity and dissipation factor of composites increased. They examined the dielectric and thermal expansion properties of Bi_2O_3-ZnO-Nb_2O_3-based (BZN) ceramics composites as a function of temperature within the range from -50 to 175 °C.

Nowak et al. [18] presented large fluorinated organic substrates capable of interconnecting one or a few application-specific integrated circuit (ASIC) semiconductor devices with packaged memory devices. They used a a 0.47mm thick PTFE substrate in a 65mm size format, having 8 memory devices surrounding a central ASIC, with numerous capacitors, in double-sided format. On the other hand, Nowogrodzki et al. [19] proposed five granulation PTFE exhibiting dielectric strength values approximately twice to that of the general purpose pelletized PTFE. They concluded that it is possible to design high-voltage capacitors of superior electrical characteristics when this high-electric-strength material is combined with the technology of electroplating PTFE.

Su et al. [20] developed a free-standing-mode based triboelectric nanogenerator (F-TENG) which may have potential applications of the triboelectric generator in gas flow harvesters, self-powered air navigation, self-powered gas sensors and wind vector sensors. This F-TENG is composed of indium tin oxide (ITO) foils and a polytetrafluoroethylene (PTFE) thin film. They suggested that by utilizing the wind-induced resonance vibration of a PTFE film between two ITO electrodes, the F-TENG delivering an open-circuit voltage up to 37 V and a short-circuit current of 6.2 μA, can be used as a sustainable power source to simultaneously and continuously light up tens of light emitting diodes (LEDs) and charge capacitors. They also reported that uniform division of the electrode into several parallel units efficiently suppressed the inner counteracting effect of undulating film and enhanced the output current by 95%. The F-TENG demonstrated prominent durability and an excellent linear relationship between output current and flow rate, revealing its feasibility as a self-powered sensor for detecting wind speed.

Kim et al. [21] used polytetrafluoroethylene (PTFE) film as a hybrid nanogenerator to integrate a triboelectric generator and a thermoelectric generator (TEG) for harvesting both the kinetic friction energy and heat energy that would otherwise be wasted. The triboelectric part consists of a PTFE film with nanostructures and a movable aluminum panel. The thermoelectric part is attached to the bottom of the PTFE film by an adhesive phase change material layer. They confirmed that the hybrid nanogenerator can generate higher output power than that generated by a single triboelectric nanogenerator or a TEG. The hybrid nanogenerator is capable of producing a power density of 14.98 mW cm^{-2}. The output power, produced from a sliding motion of 12 cm s^{-1}, was capable of instantaneously lighting up 100 commercial LED bulbs. The hybrid nanogenerator is capable to charge a 47 μF capacitor at a charging rate of 7.0 mV s^{-1}, which is 13.3% faster than a single triboelectric generator. Furthermore, the efficiency of the device was significantly improved by the addition of a heat source.

In supercapacitors, poly (vinylidene fluoride), poly(vinylidene fluoride-co-hexafluoropropylene) and poly(methyl methacrylate) (PVDF, PVDF-HFP, and PMMA) based polymers with cellulose and polypropylene (Celgard®) [1] have been mostly used as gel type macroporous separators/polymer electrolytes. PVDF type polymers and copolymers have relatively high melting points, very good mechanical and dielectric properties and strong affinity with electrolyte solvents [1] and hence are very useful to be used as separators/polymer electrolyte substrates in electrochemical devices. However, PVDF can easily swell and plasticized in solvents and its mechanical strength is reduced. Such effect is more dominant in PVDF copolymers due to their lower crystallinity. Different methods like organic and inorganic composite applications, chemical cross-linking, etc. have been used to overcome this problem. One of the earliest studies addressing crosslinking of PVDF with an electron beam (EB) was reported by Bhateja et al. [1]. In this study, crosslinking of PVDF films (25 μm thick) was investigated after irradiation at various doses in the range of 20–500 kGy. Their results showed that the crystallinity increased slightly with the increase of the dose without changing the structure of the crystal phase.

Sivaraman et al. [22] prepared a polymer electrolyte membrane for supercapacitor applications by radiation-induced grafting of a mixture of styrene and acrylic acid (AAc) monomers onto fluorinated ethylene propylene (FEP) film followed by sulfonation. FEP and some additives were mixed and the mixture was exposed to radiation (50 kGy). After washing and regeneration of the grafted film, it was sulfonated. The resultant membrane showing the good tensile strength of 7.5 MPa, and 125% elongation, was very suitable to separate the two electrodes when tested in the supercapacitor. The combination of polyaniline (PANI) based electrode and poly(vinyl sulfonic acid) (PVSA) as were reached proton-conducting solid electrolyte with the FEP-g-AA-SO$_3$H membrane, 1500 cycles being achieved with a capacitance of 98 F g^{-1}. The reduction in the capacitance after 1500 cycles was less than 20%.

The wettability of a separator is an important property and may be enhanced by modification of the surface of the polymer. Modification of the PVDF surface by radiation-induced grafting of another fluoropolymer and further hydrophilization by thiol-para fluoro coupling was suggested by Dumas et al.[23]. They induced irradiation (60Co, the total dose of 150 kGy) in order to graft 2,3,4,5,6-pentafluorostyrene (PFS) onto PVDF. The swelling and wettability measurements were performed in acetone and acetonitrile. It was found that the solvent uptake of the irradiated PVDF was lower compared to pristine PVDF. These results showed that the electrolyte/separator affinity may be adjusted by using the proposed method.

2.2   Nonfluorinated polymers and their composites

Most commonly used nonfluorinated polymers for supercapacitors are polyimides, polyanilines (PANIs), poly-4-vinyl phenol (PVPs), polyarylene ethers (PAE) and their different combinations with organic and inorganic additives. There are also other polymers including polyarylene ether nitrile (PEN), polybenzoxazines, polyamides (PA), poly-4-vinylpyridine (PVP), polybipyridines (PBP), polyurethanes (PU) and carbohydrate polymers (CP) for supercapacitor applications. Polymer dielectrics are the preferred materials of choice for capacitive energy-storage applications because of their potential for high dielectric breakdown strengths, low dissipation factors and good dielectric stability over a wide range of frequencies and temperatures, despite having inherently lower dielectric constants relative to ceramic dielectrics. These are amenable for processing into films at a relatively lower cost. Nonfluorinated polymers are useful for different supercapacitor applications.

Air forces required thermally robust compact capacitors for operation in a variety of aerospace power conditioning applications. These applications typically use polycarbonate (PC) dielectric films in wound capacitors for operation within a temperature range of -55 °C to 125 °C. Deng et al. [24] fabricated a radio frequency microelectromechanical system (RFMEMS) switched capacitor that exhibited low pull-in voltage. They proposed planarization of polyimide surface, resulting in polyimide surface roughness less than 10 nm. They tested the device under hold-down conditions of over 90 s, the actuation voltage dropped to 11.6 V, and the capacitance ratio never changed.

Abdalla et al. [25] prepared nano-composites for improving the energy-storage efficiency of electrostatic capacitors. For this purpose, they studied the dielectric properties of four highly pure amorphous polymer films: Polymethyl methacrylate (PMMA), polystyrene, polyimide, and poly-4-vinylpyridine. Comparison of dielectric properties of these polymers revealed the higher breakdown performance of polyimide (PI) and PMMA. The adding of colloidal silica to PMMA and PI resulted in a net decrease in the dielectric properties compared to the pure polymer. They demonstrated the effects of polymer purity, nanoparticle size, and film morphology factors on the energy storage performance of the capacitor.

In et al. [26] proposed laser-based facile fabrication of flexible micro-supercapacitors without tedious photolithographic patterning of porous carbon and metal current collectors. They fabricated flexible micro-supercapacitors based on laser carbonization of polyimide sheets. Localized pulsed laser irradiation rapidly converted the pristine polyimide surface into an electrically conductive porous carbon structure under ambient conditions. Thus, the polyimide sheet acted as both a precursor for the carbonization and a flexible substrate. They examined effects of various laser parameters to enhance electrical properties and altering the morphology of the carbonized structures. They produced interdigitated electrode patterns directly on the polyimide sheets by programmed laser scanning. These workers introduced a solid-state polyvinyl alcohol-phosphoric acid gel electrolyte into the active electrode area to realize a flexible all-solid-state micro capacitor assembly. The specific capacitance of the supercapacitors up to ~800 $\mu F/cm^2$ was achieved at a voltage scan rate of 10 mV/s with good capacitance retention under mechanical bending. The expected electrical double layer behavior was realized from the cyclic voltammetry measurements.

Treufeld et al.[6] attached a set of 12 new polyimides (PIs) with one or three polar CN (carbon-nitrogen) dipoles directly to the aromatic diamine part and studied their electric energy storage properties using broadband dielectric spectroscopy (BDS) and electric displacement-electric field (D-E) loop measurements to assess their potential in aerospace applications as high temperature film capacitors. They found that adding of highly polar nitrile groups to the PI structure increased permittivity and thus improved electrical energy storage property, especially at high temperatures, and 3 CN dipoles were better than 1 CN dipole. From the BDS results on PIs having 3 nitrile groups, the enhancement in permittivity from permanent dipoles decreased with dianhydride in the order of pyromellitic dianhydride (PMDA)>4,4'-oxydiphthalic dianhydride (OPDA)>1,1,1,3,3,3-hexafluoropropane dianhydride (6FDA)>4,4'-benzophenonetetracarboxylic dianhydride (BTDA). Meanwhile, the increase in permittivity also decreased in the order of para-para, meta-para, and meta-meta linkage in the diamine, suggesting that the para-para linkage favored easier dipole rotation than the meta-meta linkage. From the D-E loop study, the PIs with a combination of PMDA dianhydride and a para-para linkage exhibited the highest discharged energy density and a reasonably low loss.

Rivadeneyra et al. [27] designed and fabricated low-cost and flexible humidity sensors, using an inkjet-printing process based on the principles of the capacitor and the ability of a polyimide to absorb humidity. They fabricated the sensor by printing silver interdigitated electrodes on a thin polyimide film of 75 μm thickness. This work presented a reliable, fast, simple and low-cost manufacturing process to make small humidity sensors with low thermal drift and high temporal stability. They integrated these sensors into inkjet-printed RFID tags for monitoring of environmental humidity in diverse applications.

Zhang et al. [28] reported the first continuous fabrication of inkjet-printed polyimide films, which were used as insulating layers for the production of capacitors. These workers prepared polyimide ink from its precursor poly (amic) acid and directly printed onto a hot substrate (at around 160°C) to initialize a rapid thermal imidization. They printed polyimide films with good surface morphologies between two conducting layers to fabricate capacitors by carefully adjusting the substrate temperature, droplet spacing, droplet velocity, and other printing parameters. The highest capacitance value of 2.82 ± 0.64 nF was in favor of the polyimide inkjet printing approach as an efficient way for producing dielectric components of microelectronic devices.

Dobrzynska and Gijs [29] developed a flexible-substrate-based three-axial force sensor that has a high potential for use in skin-like sensing applications. It was a finger-shaped electrode capacitor whose operation based on the measurement of a capacitance change induced upon applying a three-axial load. An overall flexibility of the sensor and elasticity of the capacitor's dielectric were obtained by integrating three polymers in the sensor's technology process, namely polyimide, parylene-C, and polydimethylsiloxane, combined with standard metallization processes. They also theoretically modeled the sensor's capacitance and its three-axial force sensitivity. Moreover, they [30] also prepared a flexible force sensor using polyimide as a flexible substrate and as an elastic dielectric between two levels of finger-shaped aluminum electrodes was used to develop a technology for the realization of polyimide micro-features with gentle slopes to facilitate subsequent metallization processes. Smooth polyimide slopes were obtained by combining lithographic resist-reflow techniques with dry etching procedures and typical force sensitivity of 1-2 fF/N was realized using an impedance analyzer.

Lin et al. [31] proposed a novel method to measure the distribution of occlusive force on the crown using a flexible force sensor array. The basic principle of the developed flexible force sensor array was based on the strong piezoelectrical property of multilayer ceramic capacitors (MLCC) with high mechanical strength and low-cost ideal for large forces measuring applications. They fabricated a 3x3 MLCC force sensor array in a polyimide-based flexible circuit film (FCCL) and used to measure the corresponding force distribution in a simulated horse teeth mold. They indicated that the MLCC force sensor array has a potential to be an ideal component for bite force measurement.

Film capacitors that perform well at temperatures exceeding 150°C and have energy densities in excess of 1 J/cm^3 have been found useful for many applications in automotive, geophysical exploration, aerospace, and the military. Faradox Energy Storage, Inc. has produced film capacitors using amorphous oxides (OxFilm®) as the dielectric material. The capacitors were made by depositing thin films of an oxide dielectric on both sides of a double metalized polyimide substrate to form dielectric-coated electrodes. Faradox has fabricated 2″ long, 1/4″ diameter 2 uF, 100 VDC capacitors using silicon dioxide as the dielectric material. Jamison and Balliette [32] presented test data demonstrating that OxFilm capacitors have relatively stable properties over a wide temperature range. It was suggested that these capacitors can be cycled between 0°C and 200°C without degrading their performance.

Lee et al. [33] proposed methods for vertically aligned liquid crystal displays to obtain multi-domains with different threshold voltages without additional transistors or capacitors. They submitted an ultraviolet light through a photomask and polyimide partially decomposed by spatially assorted surface anchoring energy or inducing the adsorption of reactive monomers.

Industrial technology research institute's (ITRI) developed a plastic substrate technology of coating a polyimide solution directly onto the glass substrate with a de-bonding layer (DBL). The DBL provides a weak interface between it and the plastic film, which allows the film to be easily released after the thin-film transistor (TFT) process. Two key innovations, the flexible substrate material, and the de-bonding technology make the technology a success. The flexible display can be easily released from the glass carrier without any damage. Chen and Ho [34] first integrated a two-transistor, one-capacitor circuit in the backplane, and then the color OLED was deposited on the TFT backplane to form an active-matrix organic light emitting device (AMOLED) display.

Mativenga et al. [35] implemented circuits with high-performance amorphous-indium gallium-zinc oxide (TFTs) on polyimide/polyethylene-terephthalate plastic substrates. AC driving of pull-down TFTs gave the gate driver an improved lifetime of over ten years.

Dumitru et al. [36] used polyanionic proton conductor, named poly (4-styrene sulfonic acid) (PSSH), to gate an organic thin-film transistor (OFET) based on p-type poly (2, 5-bis (3-tetradecylthiophen-2-yl)thienol [3,2-b]thiophene) (pBTTT-C14) organic semiconductor. Upon applying negative gate bias, large electric double layer capacitors (EDLCs) are formed at gate-PSSH and pBTTT-PSSH interfaces due to the proton migration in the polyelectrolyte. These workers evaluated different device configurations and investigated their electrical performance along with their implementation on flexible Kapton.

Thin polymer films in the presence of high electric fields undergo partial discharge and have characteristic acoustic emissions. It is hoped that these acoustic signals can aid in anticipating the failure of the films, thereby providing a tripwire to reduce the electric voltage in high-energy capacitor applications before a failure actually occurs and the capacitor is permanently damaged. Hanley and Tucholski [37] compared the acoustic emission signals receive from Kapton films of different thicknesses (7-55 μm). A laser Doppler vibrometer with a frequency response from 0-22 kHz was used to study surface vibrations of a gold-coated polymer sample as voltage was raised at a controlled rate of 500 V/s from 0 V to material failure. The results of these tests demonstrated the relationship between characteristic frequencies of the acoustic emission and the polymer thickness.

Zampetti et al. [38] presented the design and fabrication of a humidity sensor on ultra-thin (8 μm) flexible polyimide substrate. The ultra-thin flexible substrate preserved when a read-out electronic interface was integrated by using polycrystalline silicon thin film transistor technology. The sensor device was a capacitor where a thin layer of [bis(benzo-cyclobutene)] was used as a dielectric sensitive material between two metal electrodes. The electrode layout has been designed with the aid of numerical simulations in order to optimize the sensor performances. The fabricated sensor showed sensitivity to a relative humidity of 0.38% (RH) and a linearity R^2 (0.996) in the range of 10-90 RH%.

Tiwari et al. [39] fabricated organic-organic nanoscale composite thin-film (NCTF) by using a sol-gel method where the precursor solution has been achieved with organic additives using solution deposition of 1-bromoadamantane and triblock copolymer (Pluronic P123, BASF, EO_{20}-PO_{70}-EO_{20}). The sol-gel process was used to make metal-insulator-metal capacitor (MIM) comprising of a nanoscale (10 nm-thick) thin-film on a flexible polyimide (PI) substrate at room temperature. Scanning electron microscope and atomic force microscope revealed that the deposited NCTFs were crack-free, uniform, highly resistant to moisture absorption, and well adhered on the Au-Cr/PI. The electrical properties of 1-bromoadamantane-P123 NCTF were characterized by dielectric constant, capacitance, and leakage current measurements. The 1-bromoadamantane-P123 NCTF on the PI substrate showed low leakage current density (5.5Å - 10-11cm^{-2}) and good capacitance (2.4 F at 1 MHz). In addition, the calculated dielectric constant of 1-bromoadamantane-P123 NCTF was 1.9, making it a suitable candidate for use in future flexible electronic devices as a stable intermetal dielectric. The improved electrical insulating properties of 1-bromoadamantane-P123 NCTF were due to the optimized dipole moments of the van der Waals interactions.

Nanocomposites can provide high capacitance densities, ranging from 5 nf/inch^2 to 25 nF/inch, depending on composition, particle size and film thickness. Das et al. [40] examined nanocomposites or materials in the area of printable and flexible technology. These workers used several substrates including polyimide and LCP (liquid crystal polymer) in flexible technology for laminate chip carrier and printed wiring board (PWB). Variety of printable nanomaterials for electronic packaging including nanocapacitors and resistors were developed. They fabricated a variety of printable discrete resistors with different sheet resistances, ranging from 1 ohm to 120 Mohm, processed on large panels (19.5 × 24 inches). It was suggested that low resistivity nanocomposites, with volume resistivity in the range of 10^{-4} ohm-cm to 10^{-6} ohm-cm, depending on composition, particle size, and loading can be used as conductive joints for high frequency and high density interconnect applications.

Ahmad et al. [41] demonstrated flexible catheter probes for magnetic resonance imaging (MRI) of the bile. The probes with a cytology brush can accept a resonant RF detector and are designed for sticking in the duct via a non-magnetic endoscope during endoscopic retrograde cholangiopancreatography (ERCP). The narrow enough coil (<3 mm) can pass through the biopsy channel of the endoscope and is sufficiently flexible to turn upto 90° to enter the duct. Coils were fabricated as multi-turn electroplated conductors on a flexible base, and two designs formed on SU-8 (epoxy-based negative photoresist) and polyimide substrates were compared. It was observed that careful control of thermal load required to obtain useable mechanical properties from SU-8, and that polyimide/SU-8 composites offered improved mechanical reliability. It was possible to obtain sub-millimeter resolution in 1H MRI experiments at 1.5 T magnetic field strength using test phantoms and in-vitro liver tissue.

Choi et al. [42] prepared polyimides with a low dielectric constant and excellent adhesion from a diamine containing phosphine oxide and fluorine groups, bis(3,3'-aminophenyl-2,3,5,6-tetrafluoro-4-trifluoromethyl phenyl phosphine oxide (mDA7FPPO), and rigid-rod type dianhydride containing fluorine groups, such as 3,6-di(3',5'-bis(trifluoromethyl)-phenyl)pyromellitic dianhydride (12FPMDA). The polyimides were synthesized via the known two-step process, (preparation of poly (amic-acid) followed by solution imidization) and characterized by FT-IR, NMR, DSC, TGA, and TMA. In addition, the solubility, intrinsic viscosity, dielectric constant and adhesive properties of the polyimides were also evaluated. For comparison, 3,6-di(4'-trifluoromethylphenyl) pyromellitic dianhydride (6FPMDA) and 3,6-diphenyl pyromellitic dianhydride (DPPMDA) were utilized The prepared polyimides exhibited high Tg (276-314 °C), excellent thermal stability (>500 °C in air), good adhesive property (104.7-126.3 g/mm), good solubility, and very low dielectric constant (2.34-2.89).

Kavetskiy et al. [43] calculated beta particle surface fluxes for tritium, Ni-63, Pm-147, and Sr-90 sources. The high current density was experimentally achieved from Pm-147 oxide in the silica-titana glass. A 96 GBq (2.6 Ci) Pm-147 4π-source with flux efficiency greater than 50% was used for constructing a direct charge capacitor with a polyimide coated collector and vacuum as electrical insulation. The capacitor connected to high resistance loads produced up to 35 kV. Overall conversion efficiency was over 10%.

Balde et al. [44] developed low-cost passive sensors on objects of various forms to suit technological processes of microelectronics on the flexible substrate for reaching excellent resolutions without any damage. Passive components were deposited on flexible polymer support such as PET (Polyethylene terephthalate) or Kapton® (Polyimide). The use of paper as an economic and ecological substrate has generated much interest. However, the brittleness of this material due to moisture and temperature has been a major technical challenge for its potential application in electronics. Shao et al. [45] presented the design and implementation of a chipless radio frequency identification (RFID) tag on a flexible substrate. These workers realized the tag onto flexible polyimide substrate using a toner-transferring process and also constructed a detection system to measure the proposed tag. The measurement results showed that the tag can provide excellent readability of more than 20 cm reading range. It was concluded that this tag can provide a meaningful approach toward the realization of ultralow-cost radio frequency identification (RFID) tags attached to low-value items.

Meena et al. [46] developed a new organic (1-bromoadamantane) ultrathin film as a gate dielectric, which was deposited by a sol-gel spin-coating process on a flexible polyimide substrate at room temperature. They demonstrated the practical properties of the film in the metal-insulator-metal (MIM) capacitor such as dielectric constant as well as bending result of leakage current density and breakdown voltage to be better related to such fundamental adhesion nature over a flexible substrate.

Good surface quality of plastic substrates is essential to reduce pixel defects during roll-to-roll fabrication of flexible display active matrix backplanes. Standard polyimide substrates have a high density of "bumps" from fillers and belt marks and other defects from dust and surface scratching. Some of these defects could be the source of shunts in dielectrics. Almanza-Workman et al. [47] developed a proprietary UV curable planarization material that can be coated by roll-to-roll processes and engineered it to have low shrinkage, excellent adhesion to polyimide, high dry etch resistance, and great chemical and thermal stability.

Ma et al.[48] established a relationship between chemical functionalities and dielectric properties using polyimides as potential polymer materials with high dielectric constants of up to 7.8, low dissipation factors (<1%) and high energy density (15 J/cm3). Watanabe et al. [49] used ionic liquid (IL)-based polymer electrolytes comprising of block copolymers and polyimides to obtain easily processable ionic polymer actuators with high performance and durability. Zhou et al. [50] prepared the parallel-plate capacitive humidity sensor with a polyimide (PI) film and obtained some promising results in improving response time.

Chen et al. [51] designed and fabricated a fully integrated wireless inductance-capacitance (LC) coupling microsensor by microelectromechanical systems (MEMS) technology. These workers produced sensing loop by connecting a deformable parallel-plated capacitor and a planar spiral inductor with a Ni (80) Fe (20) core. Polyimide and PMMA were used to isolate and package the devices with a typical dimension of the sensors of 5 × 5 mm 2 × 0.77 mm. They fabricated different electroplated inductive coils (30, 40, and 60 turns) to connect with a 4 × 4 mm^2 plate capacitor in series and setup LC sensing module for measuring liquid-level induced frequency responses. Experimental results showed that frequency response decreased as liquid level increased and sensitivity was about 7.01 kHz/cm with a deviation less than 2%. The developed planar spiral inductor with high permeability magnetic core were supposed to provide a wide range of frequency variation in LC sensing applications.

Venkat et al. [52] focused their research on the generation and dielectric evaluation of metalized, thin free-standing films derived from high-temperature polymer structures such as fluorinated polybenzoxazoles, post-functionalized fluorinated polyimides and fluorenyl polyesters incorporating diamond-like hydrocarbon units. The studies were mainly concentrated on variable temperature dielectric measurements of film capacitance and dissipation factor as well as the effects of thermal cycling, (up to a maximum temperature of 350 °C) on film dielectric performance. Initial investigations clearly pointed to the dielectric stability of these films for high-temperature power conditioning applications, as indicated by their relatively low-temperature coefficient of capacitance (TCC) (~2%) over the entire range of temperatures. They found some of the films to exhibit good dielectric breakdown strengths (up to 470 V/μm) and a film dissipation factor of the order of <0.003 (0.3%) at the frequency of interest (10 kHz) for the intended applications. The measured relative dielectric permittivities of these high-temperature polymer films were in the range of 2.9-3.5

Kim et al. [53] synthesized and fabricated polyaniline nanoparticles (PANI NPs) as charging elements for organic memory devices. The PANI NPs charging layer was self-assembled by epoxy-amine bonds between 3-glycidylpropyl trimethoxysilane functionalized dielectrics and PANI NPs. A memory window of 5.8 V (δV FB) represented by capacitance-voltage hysteresis for metal-pentacene-insulator-silicon capacitor was reported. In addition, the program/erase operations controlled by gate bias (-/+90 V) in the PANI NPs embedded pentacene thin film transistor device with polyvinylalcohol dielectric on flexible polyimide substrate were demonstrated.

Yang et al. [54] prepared a dielectric film with ultrahigh thermal stability on crosslinked polyarylene ether nitrile (PEN by solution-casting of polyarylene ether nitrile terminated phthalonitrile (PEN-Ph) combined with post self-crosslinking at high temperature. The film showed a 5% decomposition temperature over 520 °C and a glass transition temperature (Tg) around 386 °C. In addition to exhibiting stable dielectric constant and low dielectric loss in the frequency range of 100-200 kHz and temperature range of 25-300 °C. The temperature coefficient of dielectric constant was less than 0.001 °C-1 even at 400 °C. By cycling heating and cooling up to ten times or heating at 300 °C for 12 h, the film showed good reversibility and robustness of the dielectric properties. The crosslinked PEN film will be a potential candidate as high-performance film capacitor electronic devices materials used at high temperature.

2.2.1 Polyimides with inorganics

There are two broad categories of dielectric materials: paraelectric (class 1) and ferroelectrics (class 2). $BaTiO_3$, Pb_xZr_{1-x}, TiO_3 and Ba_xSr_{1-x}, TiO_3 are some examples for ferroelectrics which can exhibit dielectric constants up to three orders of magnitude higher than those of paraelectric materials such as SiO_2, Al_2O_3, Ta_2O_5, TiO_2 and BCB (Table 1)[55].

In order to increase dielectric properties of thin films for electrochemical devices, there are some paraelectric and ferroelectric composite applications been utilized with HPP type polymers. For this purpose most commonly the used polymer is polyimide but polyarylene ethers, PVP, PEN, epoxide and also polybenzoxasine type polymers have some applications as well. Some HPP polymers with paraelectric and ferroelectric substances are summarized in the following section.

Table 1. Dielectric constants for paraelectric and ferroelectric materials [55].

Name	Composition	Dielectric constant
BCB (Benzocyclobutene)	organic	2.7
Polyimide	organic	3.6
Silicon dioxide	SiO_2	3.7
Diamond-like carbon	$sp^2 + sp^3$ C	4-5
Magnesium fluoride	MgF	5
Silicon nitride	Si_3N_4	7
Aluminum nitride	AlN	9
Aluminum oxide	Al_2O_3	9
Amorphous tantalum oxide	Ta_2O_5	23
Silicon carbide	SiC	40
Hexagonal tantalum oxide	Ta_2O_5	50
Titanium oxide	TiO_2	50
Barium strontium titanate (BST)	$BaSrTiO_3$	Up to 1000
Lead zirconate titanate (PZT)	$PbZr_xTi_{1-x}O_3$	Up to 2000
BPZT	$Ba_{0.8}Pb_{0.2}(Zr_{0.12}Ti_{0.88})O_3$	Up to 3300
Barium titanate	$BaTiO_3$	Up to 5000

2.2.2 Polyimide (PI)/Al_2O_2 applications

Meena et al. [56] developed an Ag@Al_2O_3/polyimide composite film of a high dielectric constant for embedded capacitor applications. For this purpose, synthesized nano-sized Ag particles through redox reaction of silver nitrate in N, N-dimethylformamide were treated with aluminum isopropoxide (AIP) to construct Al_2O_3 thin layer on the surface of Ag@ Al_2O_3. These nano-particles were later on doped in polyimide by in-situ polymerization to prepare the Ag@ Al_2O_3/polyimide composite films. The obtained results indicated that a typical core-shell structure was formed and the sizes were all in nanometer level for Ag@Al_2O_3 particles. The Ag@ Al_2O_3 nanoparticles were well dispersed in the PI matrix and a slight decrease in mechanical properties of composite films appeared after Ag@ Al_2O_3 particles' addition. Moreover, the dielectric constant (k) increased gradually with Ag@Al_2O_3 content. Alias et al. [57] prepared polyimide (PI)/Al_2O_3 composite films by incorporating different micron-sized α- Al_2O_3 contents into PI derived from pyromellitic dianhydride and 4,4'- oxydianiline via ultrasonication for use in the fabrication of capacitors. It was observed that the dielectric constant, dielectric loss and thermal stability of PI/Al_2O_3 were increased with the increase in the added content of α- Al_2O_3.

2.2.3 Polyimide (PI)/SiO_2applications

There are many organic and inorganic based dielectric materials (Table 1) but deposition/etching processes and plate configurations for integrating the devices are different and it is also difficult to provide the entire range without either the highest or lowest-valued components occupying too large of an area, exhibiting excessively high series resistance, or being too small to fabricate with acceptable tolerance. Therefore, it becomes often necessary to mix integrated capacitor technologies on a single substrate. Benzocyclobutene (BCB), polyimide and SiO_2 provide a small enough capacitance and sufficient tolerance for the bottom end of the range while anodized metals or ferroelectric powders in epoxy thin films can cover decoupling, termination and some energy storage into the range of hundreds of nF. Ferroelectric thin films provide much higher dielectric constants but are currently difficult to integrate, especially for small values, and have less temperature, frequency, and bias stability than paraelectric. Once it becomes possible to form ferroelectric thin films at temperatures low enough for organic substrates these films will be very useful for energy storage. Yan et al [58] fabricated MIS trench capacitors with a diameter of ~6 μm and depth of ~54 μm successfully with polyimide insulator step coverage better than 30% and evaluated C-V characteristics and leakage current properties of the MIS trench capacitor under thermal treatment. The experimental results showed that the minimum capacitance density was around 4.82 nF/cm^2, and the leakage current density after 30 cycles of thermal shock tests became stable and it was around 30 nA/cm^2 under a bias voltage of 20 V.

Grabowski et al. [59] demonstrated experimentally the dominant factors in broad structure-performance relationships and they compared the dielectric properties of four high-purity amorphous polymer films (polymethyl methacrylate, polystyrene, polyimide, and poly-4-vinylpyridine) incorporating uniformly dispersed silica colloids (up to 45% v/v). The findings indicated that adding colloidal silica to higher breakdown strength amorphous polymers (polymethyl methacrylate and polyimide) caused a reduction in dielectric strength as compared to the neat polymer. Alternatively, low breakdown strength amorphous polymers (poly-4-vinylpyridine and especially polystyrene) with comparable silica dispersion showed similar or even improved breakdown strength for 7.5-15% v/v silica. At ~15% v/v or greater silica content, all the polymer NC films exhibited breakdown at similar electric fields, implying that these loading failure becomes independent of the polymer matrix and dominated by the silica content.

Huang et al. [60] employed the sol-gel method to synthesize an inorganic-organic photosensitive composite material, silica/polyimide. Silica was covalently bonded to polyimide by the introduction of 1,4-aminophenol as a coupling agent. They used a photosensitive cross-linking agent, 2-(dimethylamino) ethyl methacrylate to engender UV sensitivity to the composite. Using field emission scanning electron microscope imaging at very high magnifications, the composite film exhibited a porous cross-sectional structure. It was observed that pores were interconnected to form numerous continuous channels within the matrix and silica particles attached to the pore wall of the polyimide host. Depending on the preparation conditions, the size of the silica particles could vary from 20 nm to a few μm. The incorporation of silica improved the thermal and dimensional stabilities of the polyimide. Also, because the silica/polyimide composite had a porous interior structure, it had a rather low dielectric constant with the lowest achievable value being 1.82.

2.2.4 Polyimide (PI)/barium titanate (BaTiO$_3$) applications

The rapid growth in electronic equipment has created a demand for advanced devices that are flexible, thin, and light in weight. This demand has led to the development of flexible and stretchable electronic devices as a core technology. The dielectric materials used in embedded capacitors and energy-storage devices highly desire high dielectric permittivity, good mechanical properties, and excellent thermal stabilities. This technique is also suitable to miniaturize electronic devices and enhance their performances and reliabilities. Therefore, it is of great importance to study high dielectric materials that can be used in embedded environments.

Xu et al. [61] fabricated polyimide (PI)/barium titanate (BaTiO$_3$) nanocomposites hybrid nanofibers from electrospinning process. The results showed that BaTiO$_3$ fillers were uniformly dispersed up to 50 vol% in PI matrix. The dielectric permittivity of the composite (50 vol% BaTiO$_3$) was 29.66 at 1 kHz and room temperature with a dielectric loss of 0.009. The dielectric permittivity showed marginal variation with temperature (up to 150°C) and frequency (100 Hz-100 kHz). It was concluded that PI/BaTiO$_3$ nanocomposites with high thermal stability and good mechanical properties will be a promising candidate for uses in embedded capacitors, especially under high-temperature circumstances. Liu et al. [62] developed novel flexible three-phase high-κ polyimide (PI) composites by the incorporation of covalently bonded BaTiO$_3$ at graphene oxide

(BaTiO$_3$@GO) hybrids in which the BaTiO$_3$@GO hybrids were prepared by the reaction of amino-modified BaTiO$_3$ particles and GO. The obtained composite showed the dielectric constant as 285 (100 Hz), which was eighty times higher than that of a pure PI film and lower dielectric loss (0.25). It was proposed that flexible, high-κ BaTiO$_3$@RGO/PI composites have promising potential applications for capacitors and electronic devices in the fields of the micro-electron.

To produce wearable computers, it is necessary to fabricate functional membranes that contain passive devices such as capacitors and resistors on resin sheets at low temperatures. These sheets can serve as mounting boards for various electronic devices by improving the technique for room-temperature aerosol deposition (ASD) of a ceramic material (post-LTCC technology). Imanaka et al. [63] prepared advanced devices that were flexible, thin and light in weight. The workers established a technology for forming a dielectric inorganic BaTiO$_3$ film with an excellent degree of crystallinity and favorable electric properties for use in the production of flexible and stretchable electronic devices on a polyimide sheet. Their method was suitable to form a homogeneous nanoparticle structure inside a film as well as to produce a capacitor film with a dielectric constant of 200 on a polyimide sheet at room temperature. They also produced barium titanate film by the deposition on a polyimide sheet and obtained a closely packed film with a high degree of sealing by means of structure control to establish large-particle anchors, measuring about 100 nm in diameter, at the interface between the ceramic film and the polyimide sheet [64]. The maximum dielectric constant of the film of barium titanate formed on the polyimide sheet by using the improved aerosol-deposition was 350.They produced a capacitor film with a three-dimensional capillary structure inside an apparent dielectric constant of 40,000 (room temperature) by using this method. To achieve rapid charging, it is necessary to improve the speed of the chemical reaction between the cathode material and the electrolyte. Their process, however, does not require any unnecessary additive, such as a binder, unlike conventional tape casting method, and forms a film consisting of nanoparticles of a high crystallinity, and is therefore effective for rapid charging.

Xie et al. [65] used barium titanate (BaTiO$_3$) particles with average size of 92 nm as inorganic nano-fillers and polyimide (PI) as a matrix to fabricate novel BaTiO$_3$/PI nanocomposite dielectric films. The film was patterned, sputtered and etched to fabricate a prototype embedded capacitor. The prototype embedded capacitor has dielectric permittivity higher than 15 at low frequency and electric breakdown strength above 58 MV/m. Furthermore, the etching and sputtering have little impact on the film's dielectric properties.

Kang et al. [66] demonstrated the feasibility of inkjet-printed passive components. All passive components such as resistor, capacitor, and inductor were inkjet printed on a polyimide (PI) substrate with various functional inks. For the insulator layer, poly-4-vinylphenol (PVP) and cross-linking agent (poly(melamine-co-formaldehyde)) were dissolved in ethanol. A mixture of poly(3,4-ethylene dioxythiophene) doped with polystyrene sulfonated acid (PEDOT:PSS) and ethylene glycol was used to print a resistor. Barium titanate ($BaTiO_3$) and soft ferrite (Ni-Zn) powders were added to the synthesized insulator solution to improve its dielectric and magnetic characteristics, respectively. A resistor-capacitor (RC) circuit based on the results of the printed passive components was also fabricated. The measured responses of the printed RC circuit were in good agreement with estimated results. Wu et al. [67] prepared barium titanate ($BaTiO_3$, BT) fibers via electrospinning using a sol-gel precursor, followed by a calcination process. Polyimide (PI) nanocomposite films were fabricated from the electrospun BT fibers using an in-situ dispersion polymerization method. The results demonstrated that the dielectric permittivity of the PI nanocomposite films with 30 vol% BT fibers was improved up to ~27 at 102 Hz, and the corresponding dielectric loss was relatively low (~0.015). The dielectric permittivity of the PI/BT-fiber composite films exhibited a slight dependence on temperature, while it was highly dependent on the calcination temperature of the electrospun BT fibers. This study opened a new path to optimize the dielectric properties of thermosetting polymer composite films with high energy storage density.

2.2.5 Polyimide (PI)/Cupper (Cu) based applications

Peng et al. [68] prepared polyimides containing bipyridine units using a newly synthesized diamine monomer, (5,5′-bis [(4-amino) phenoxy]-2,2′-bipyridine (BPBPA)). The dielectric constant up to 7.2, the dielectric loss < 0.04, and the energy density about 2.77 J/cm^3 for these as-synthesized polyimides were realized. Furthermore, the polyimides exhibited high glass transition temperature (Tg) of 275–320 °C and tensile strengths of 175–221 MPa. These promising polyimides have potential applications in high-temperature flexible polymer film capacitor operated at high temperature. These authors also synthesized novel polyimide-copper complexes (PICuCs) of high dielectric constant (up to 133) by polymerization, complexation, and imidization of bipyridine-containing diamine monomer. The PICuCs showed high dielectric constant because of the enhanced electronic depolarization. Moreover, the PICuCs illustrated better mechanical and thermal properties compared to neat PI to qualify as promising materials for polymer film capacitors. Gupta et al. [69] investigated the charge storage and retention characteristics of nanoparticle-laden thin polyimide film for application in non-volatile memory devices. They formed well-dispersed and uniform sized metallic copper nanoparticles (CuNPs) as embedded entities confined within the polyimide film that was cast from solution. The capacitance-voltage measurements showed the embedded CuNPs functioned as a floating gate in the metal-insulator-semiconductor-type capacitor and exhibited a large hysteresis window of 1.52 V. The C-t measurements conducted after application of charging bias of 5 V showed that the charge was retained beyond 20,000 s. The proposed technique holds promise for developing low-cost processes for memory devices that employ relatively inexpensive materials, and yet demonstrate very good performance.

The recent demonstrations of manufacturable multilevel Cu metallization have heightened interest to integrate Cu and low-κ dielectrics for future integrated circuits. For reliable integration of both materials, Cu may need to be encapsulated by barrier materials. Wong et al. [70] advocated the use of electrical testing techniques to evaluate the Cu+ drift behavior of low-κ polymer dielectrics. Specifically, they employed bias-temperature stress and capacitance-voltage measurements as their high sensitivities were well-suited for examining charge instabilities in dielectrics. Charge instabilities other than Cu+ drift were possible. For example, when low-K polymers came into direct contact with either a metal or Si, interface-related instabilities attributed to electron/hole injection were observed. To overcome these issues, they developed Cu+ drift evaluation a planar Cu/oxide/polymer/oxide/Si capacitor test structure. According to their study, Cu^+ ions were found to drift readily into poly(arylene ether) and fluorinated polyimide, but much more slowly into benzocyclobutene. A thin nitride cap layer prevented the penetration.

The industry is strongly interested in integrating low-κ dielectrics with Damascene copper. Integration of copper wiring with silicon dioxide (oxide) requires barrier encapsulation since copper drifts readily in the oxide. Loke et al. [71] evaluated and compared the copper drift properties in six low-κ organic polymer dielectrics: parylene-F; benzocyclobutene; fluorinated polyimide; an aromatic hydrocarbon; and two varieties of poly(arylene ether).Copper/oxide/polymer/oxide/silicon capacitors subjected to bias-temperature stress to accelerate penetration of copper from the gate electrode into the polymer. The study showed that copper ions drifted readily into fluorinated polyimide and poly(arylene ether), more slowly into parylene-F, and even more slowly into benzocyclobutene. It was concluded that copper drift in these polymers was possibly retarded by increased crosslinking and enhanced by the polarity of the polymer.

2.2.6 Polyimide/ SnO₂applications

Agarwal et al. [72] presented a passive capacitor-based ethylene sensor using SnO_2 nanoparticles for the detection of ethylene gas using the nanoscale particle size (10 to

15nm) and film thickness (1300 nm) of the sensing dielectric layer in the capacitor. The SnO_2-sensing layer was deposited using room temperature dip coating process on flexible polyimide substrates with copper as the top and bottom plates of the capacitor. The capacitive sensor fabricated with SnO_2 nanoparticles as the dielectric showed a total decrease in capacitance of 5pF when ethylene gas concentration was increased from 0 to 100 ppm. A 7pF decrease in capacitance was achieved by introducing a 10nm layer of platinum (Pt) and palladium (Pd) alloys deposited on the SnO_2 layer. This also improved the response time by 40, recovery time by 28, and selectivity of the sensor to ethylene mixed in a CO_2 gas environment by 66.

Wu et al. [73] embedded tin dioxide quantum dots (SnO_2 QDs) in a polyimide (PI) layer as a carrier transport in a volatile memory device. Current-voltage (I-V) curves showed that the Ag/PI/SnO_2 QDs/PI/indium-tin-oxide (ITO) memory device had the ability to write, read, and refresh the electric states under various bias voltages. The capacitance-voltage (C-V) curve for Ag/PI/SnO_2 QDs/PI/ p-Si capacitor exhibited a counter-clockwise hysteresis, indicative of the existence of sites occupied by carriers. The origin of the volatile memory effect was attributed to holes trapping in the shallow traps formed between QD and PI matrix, which determined the carrier transport characteristics in the hybrid memory device.

2.2.7 Polyimide/Zr based applications

PI has the advantage of being used to very high temperatures. Bestaoui-Spurr et al. [74] fabricated energy density capacitors using polyimide (PI) films containing a layered material, zirconium orthophosphate, $ZrO(H_2PO_4)_2xH_2O$ (α-ZrP) for better dielectric properties. They examined the effects of water or other impurities on all three types of material. For example, the relative permittivities of all the composites were decreased strongly when water was removed from the materials. They observed impurity or water-associated loss peaks in all three types of materials. The effect of water was to decrease the dielectric strength of the composites. However, for both the wet and dry materials, the dielectric strength exhibited maximum values at a loading of about 5wt% α-ZrP. Polyimide filled with zirconium orthophosphate, $ZrO(H_2PO_4)_2 \cdot xH_2O$ (ZrP), layered materials have been used as dielectrics [75]. The dielectric constant and also a loss for the composites were found to increase with increasing loading of the layered compound, and water also contributed to both the dielectric constant and dielectric loss. The dielectric strength was maximum at a loading of about 5 wt-%. The effect of water was to decrease the dielectric strength.

Zou et al. [76] investigated wide temperature dielectric properties and corona resistance of Upilex-S polyimide (PI) films filled with Zirconium dioxide (ZrO_2)

nanoparticles.ZrO_2/PI nanodielectrics exhibited the stable dielectric properties and high energy density as well as charge-discharge efficiency below 300 °C. Corona resistance testing showed that even a small amount of nanofillers can improve the lifetime of PI significantly. These favorable performance features made polyimide nanocomposites attractive for high energy density capacitor applications at higher temperatures.

Chu et al.[77] demonstrated a new flexible metal-insulator-metal capacitor using 9.5-nm-thick ZrO_2 film on a plastic polyimide substrate based on a simple and low-cost sol-gel precursor spin-coating process. The as-deposited ZrO_2 film under suitable treatment of oxygen (O_2) plasma and subsequent annealing at 250 °C exhibited superior low leakage current density (9.0×10^{-9}) A/cm^2 at an applied voltage of 5 V and maximum capacitance density (13.3 fF/μm^2 at 1 MHz). The film was completely oxidized when O_2 plasma employed at relatively low temperature and power (30 W), hence enhancing the electrical performance of the capacitor indicating that the O_2 plasma reaction was a most effective process for the complete oxidation of the sol-gel precursor at the relatively low processing temperature.

2.2.8 Polyimide/TiC, TiO_2 applications

Weng et al. [78] prepared series of polyimide/titanium carbide (PI/TiC) composites with different TiC contents using the ultrasonic dispersion and in-situ polymerization method. The potential use of a PI/TiC composite film in an embedded capacitor was demonstrated. The PI/TiC composites exhibited appropriate mechanical properties and moderate electric breakdown strength. Dielectric investigation evidenced that the dielectric constant and the dielectric loss of these composites increased with the increase of the volume fraction of TiC particles. The composite with 20 vol% TiC particles showed a highest dielectric constant of 37 while retaining an appropriate dielectric loss of 0.026, as compared with the dielectric constant (3-4) of neat polyimide resin. In addition, the dielectric properties of the composites displayed good stability within a wide range of frequency.

Ling et al. [79] prepared several of nano-titanium carbide doped polyimide (nano-TiC/PI) composite films via in-situ polymerization. The TiC particles dispersed homogeneously in PI matrix when its' content was kept below 15 vol. %. The addition of TiC particles effectively improved the thermal stability of composite films. In comparison to a pure polyimide film, the dielectric constant of the composite film consisting of 5 vol. % TiC was increased by the order of thirteen times while there was a little increase in the loss tangent. Moreover, the composite films exhibited good dielectric stability over a wide range of frequency. The development of these novel polyimide composite films with favorable dielectric properties for embedded capacitor applications is interesting from the technical point of view.

Xia et al. [80] prepared nano-titanium carbide doped polyimide (nano-TiC/PI) composite films via in-situ polymerization. Before the addition, the TiC nanoparticles were modified by oleic acid (OA) to improve their dispersivity inside PI matrix. Compared with the pure-TiC/PI composite films, the dielectric loss increased slightly. Results of dielectric breakdown test indicated that the average breakdown field strength was improved over 67% (57KV/mm) than pure-TiCIPI composites with 5vol. %. (34 KV/mm). Therefore, the development of these polyimide composite films with favorable dielectric properties was supposed useful for embedded-capacitor applications.

Chang et al. [81] prepared double shell hollow spheres by encapsulating the polymeric hollow spheres with TiO_2 shells without the template removal process. They incorporated TiO_2 encapsulated hollow spheres with the polyamic acid solutions were imidized to prepare the porous polyimide films. The final porous texture of the film depended on the as-doped arrangement and the properties of the hollow spheres. The TiO_2 shell exhibited good thermal stability and chemical resistance to prevent the collapse of hollow spheres during the polyamic acid solution/hollow particle mixing and the imidization process. The effects of preparation conditions on the morphology of the encapsulated hollow spheres were investigated. The dielectric constant of the porous polyimide films (pore diameter around 0.6 µm) decreased to 2.8.

2.2.9 Polyimide/carbon nanotubes and carbon nanofibers

He et al. [82] prepared a series of polyimide composites with various mass fractions of multi-walled carbon nanotubes (MWNTs) by in-situ polymerization. To increase the chemical compatibility of carbon nanotubes with the polyimide matrix, MWNTs were treated with an acid mixture and sulfoxide chloride in turn. The composite films exhibited good thermomechanical properties. The storage modulus increased significantly by increasing MWNT content and decreasing the enhancement of temperature. The films' glass transition temperature increased with enhancing of MWNTs frictions. The dielectric constants of the composites decreased with increasing frequency and increased sharply with the addition of MWNTs, a is a favorable feature for practical use in anti-static materials and embedded capacitors. Zhang, B. et al [83] embedded a three-phase composite with multiwall carbon nanotube (MWNT) and $BaTiO_3$ particles into polyimide (PI) substrates by in-situ polymerization for high-technology fields that require new high-dielectric-permittivity materials. The $BaTiO_3$-MWNT/PI composite with low dielectric loss displayed relatively high dielectric constant and good flexibility due to the low concentrations of MWNT and $BaTiO_3$. The $BaTiO_3$ was found beneficial for the formation of a conductive network of MWNT in PI matrix. They successfully developed new polyimide nanocomposite films as flexible capacitor packaging materials with a low-frequency dielectric constant of 16.8 and the dielectric loss of 0.057.

He, G. et al. [84] synthesized fluorinated polyimide (PI) by two-step reaction between 4,4'-(hexafluoroisopropylidene) diphthalic anhydride and 2,2'-bis(trifluoromethyl)-4,4'diaminobiphenyl. PI composites, with various mass fractions of multi-walled carbon nanotubes (MWNTs), were prepared either by in-situ polymerization or blending process. To increase the chemical compatibility of carbon nanotubes with the PI matrix, MWNTs were treated with an acid mixture and sulfoxide chloride by turns. Results showed that the modified MWNTs are dispersed homogeneously in the matrix. The thermal stability of the nanocomposites was slightly lower than that of the pure PI. The storage modulus and glass transition temperature of the composite films were higher than that of PI matrix. The dielectric constants of the composites increased sharply, supporting its favorable practical uses in anti-static materials and embedded capacitors.

Imaizumi et al. [85] presented printable high-performance polymer actuators comprising an ionic liquid (IL), soluble polyimide, and ubiquitous carbon materials. The developed polymer electrolytes comprising of a soluble sulfonated polyimide (SPI) and IL, 1-ethyl-3-methylimidazolium bis(trifluoromethanesulfonyl)amide ([C$_2$mim][NTf2]) exhibited acceptable ionic conductivity up to 1×10^{-3} S cm^{-1} and favorable mechanical properties (elastic modulus $>1 \times 10^7$ Pa). The authors developed polymer actuators based on SPI/[C$_2$mim][NTf2] electrolytes using inexpensive activated carbon (AC) together with highly electron-conducting carbon such as acetylene black (AB), vapor grew carbon fiber (VGCF), and Ketjen black (KB). The resulting polymer actuators have a trilaminar electric double-layer capacitor structure, consisting of a polymer electrolyte layer sandwiched between carbon electrode layers. Displacement, response speed, and durability of the actuators depended on the combination of carbons. Especially the actuators with mixed AC/KB carbon electrodes exhibited relatively large displacement and high-speed response, retained about 80% of the initial displacement even after more than 5000 cycles. There was a correlation between the generated force of the actuators and the elastic modulus of SPI/[C$_2$mim] [NTf$_2$] electrolytes. The displacement of the actuators was proportional to the accumulated electric charge in the electrodes, regardless of carbon materials, and agreed well with the previously proposed displacement model.

Wang et al. [86] activated the polyimide-based carbon nanofiber non-woven fabrics at 750°C with H$_2$O$_2$ in a nitrogen atmosphere to increase their specific surface areas and capacitances. The higher activating degree of the nanofibers resulted in the higher specific surface area and a better capacitance performance of carbon nanofiber electrodes.

The capacitance of 174.2 F/g was obtained after activation for 6 h when 1 mol/L H_2SO_4 was used as the electrolyte.

Le et al.[87] prepared self-supported and binder-free micro-porous polyimide (PI)-based carbon fibers by polymer blend electrospinning technology followed by thermal treatment without activation and evaluated electrochemically for supercapacitor applications. Polyvinyl pyrrolidone (PVP) was successfully used pore forming template by controlling the crosslinking between PVP and PI precursor via imidization process. The results indicated that the specific capacitance was not only attributed to optimized pore structures and surface chemistry but also attributed to the wettability of the electrolyte. The improved rate performance related to the reduced ion transportation distance derived from the nanofibers.

2.2.10   Polyimides/other inorganic ingredients

Mo et al. [88] prepared $Bi_{3.95}Er_{0.05}Ti_3O_{12}$ (BErT) based thin films on flexible polyimide (PI) substrates at room temperature by pulsed laser deposition. The BErT thin films obtained under low oxygen pressures were dense, uniform, and crack-free with an amorphous structure. The BErT thin films are supposed to find interesting applications in flexible optoelectronic devices and embedded capacitors.

Cao and Zeng [89] proposed a novel approach of laser micro pen integrated direct writing for fabrication of the thick film electrostatic-controlled gap-tuning capacitors. The authors adopted commercial gold paste (Au) as the material of structural electrodes and polyimide colloid as the material of sacrificial layer and fabricated the thick film gap-tuning capacitor arrays on the fused silica glass substrate. The electrical analysis showed the maximum tuning range (MTR) of the thick film gap-tuning capacitor as 30.35% with the practical pull-in voltage of 39 V, and the Q value as 50 under the frequency of 1 GHz.

Wang et al. [90] blended CP2 polyimide (prepared from 6FDA and 1,3-bis(3-aminophenoxy)benzene) with (1-50 wt.%) detonation nanodiamonds (DND) - functionalized) with 4-(2,4,6-trimethylphenoxy)benzoic acid The blends as thin films were successfully used in high-energy-density capacitors because of their stable dielectric properties over a wide temperature range (-55 to 300°C) at frequencies up to 100 kHz. Both dielectric storage and loss increased substantially with the increase of DND content. Surface functionalization (with the above benzoic acid derivative) significantly reduced the dielectric loss, while the use of acetone-washed DNDs had no effect on the dielectric loss. DND was also blended with CP2 via in-situ polymerization and it offered little effect on the dielectric properties.

Cvetkovic et al.[91] demonstrated the feasibility of fabrication method combining classic and stencil lithography for producing simple organic rectifying circuits for use in radio frequency range. They fabricated devices on a flexible polyimide foil by two aligned levels of stencil lithography using Atomic layered deposition (ALD) deposited high-κ hafnium-oxide, (HfO_2 -experimental κ ~ 19).

Diahamand and Locatelli [92] identified space-charge-limited currents in thin polyimide film capacitor structures as the main conduction process in the very high-temperature range from 320°C to 400°C before the breakdown. Total trap density was estimated as $1.5 \times 1017 cm^{-3}$. and finally, the mobility temperature dependence of free charges between 1.6×10^{-6} and 2.3×10^{-6} cm2 V^{-1} s $^{-1}$ in the range from 340°C to 400°C was reported,

Villasenor [93] developed new generation biochemical sensors for earlier detection of medical conditions as the mapping of human genomes. The sensors based on ion-sensitive field-effect transistors (IsFETs) were useful to obtain qualitative as well as quantitative results. In these sensors, by using metal-insulator-metal (MIM) structures were integrated into series to the gate of submicron MOSFET devices to obtain highly sensitive and ultra-low-power pH sensors. One MIM capacitor enabled external polarization of the MOSFET device, while the second MIM capacitor connected to a sensing plate whose surface was either a thick polyimide layer or the last metallization level. The IsFETs were then used to create immunological-sensitive FETs (ImFETs), which can be used to detect not only pH but also antigens and antibodies of specific pathogens that cause a wide array of infectious diseases that could be markers used to detect chronic degenerative diseases.

Meena et al. [94] fabricated metal-insulator-metal (MIM) capacitors using 10-nm-thick zirconium-silicate ($ZrSi_xO_y$) and hafnium-silicate ($HfSi_mO_n$) thin dielectric films on the flexible polyimide substrate by a sol-gel process. The sol-gel films were oxidized at low temperature (~250 °C)by employing oxygen plasma to enhance the electrical performance. The oxygen plasma has been accepted as a most effective process at low temperature to surface oxidation of a dielectric film for the flexible organic device. The results showed the satisfactory electrical characteristics for the corresponding films with low leakage current densities ~10^{-9} Acm^{-2} at 5V and maximum-capacitance densities 12.10 ($ZrSi_xO_y$) and 14.32 fF/μm2 ($HfSi_mO_n$), at 1MHz. These combinatorial thin oxide films based MIM capacitors have been assumed to be very suitable for future flexible devices. J.S.et al. [95] fabricated fully flexible MIM capacitors on 25 μm thin polyimide (PI) substrates via the surface sol-gel process using 10-nm-thick zirconium-silicate ($ZrSi_xO_y$) and hafnium-silicate ($HfSi_mO_n$) films as gate dielectrics. Both films treated with oxygen (O_2) plasma followed by annealing (ca. 250 12°C) showed amorphous phase. These films in sandwich-like MIM configuration on the PI substrates exhibited the low leakage current densities of 7.1×10^{-9} and 8.4×10^{-9} A/cm^2 at the applied electric field of 10 MV/cm and maximum capacitance densities of 7.5 and 5.3 fF/μm 2 at 1 MHz, respectively. In addition, these films in MIM capacitors showed the estimated dielectric constants of 8.2 and 6.0, respectively. Prior to use of flexible MIM capacitors the reliability test by applying day-dependent leakage current density measurements were carried out to 30 days. The films of silicate-surfactant mesostructured materials have special interest to be used as gate dielectrics in future for flexible metal-oxide-semiconductor devices.

Yamada et al. [96] examined core resin film, electrode configuration and processing technology of a SrTiO$_3$ (STO) thin-film capacitor suitable for fabrication at reduced costs. The conclusions were drawn following results (1) Fabrication of an STO thin-film capacitor on a polyimide or glass-epoxy film was possible. (2) A Ru/STO/Cr-Cu three-layer-film configuration was highly promising. (3) Batch processing of the pattern formation of the 300-nm-thick STO thin-film capacitor by all chemical-etching techniques was also possible. In addition, it was confirmed that the STO thin-film capacitor has a flat-frequency response up to 10 GHz, a relative dielectric constant of 17 and a capacitor density of 500pF/mm^2.

2.2.11   Other Nonfluorinated Polymers and their inorganic composites

There is a new class of thermosetting polymers being developed as an alternative to the traditional phenolics, epoxies, bismaleimides, and polyimides. These polymers also provide features that are not found in the conventional materials, such as excellent dimensional stability, low water absorption, high char yield, good electrical properties, and wide molecular design flexibility. Moreover, benzoxazine monomers can be polymerized without the use of strong acid or alkali as catalysts. Manuspiya and Ishida [97] found polybenzoxazines as promising materials for the fabrication embedded capacitors or printed circuit board. Dielectric constants of polybenzoxazine as reported by various studies were noticed to be influenced by the molecular structure of benzoxazine monomers. The dielectric properties of the polybenzoxazine composites were influenced by several factors such as the molecular structure of the benzoxazine, ceramic volume fraction, the distribution of ceramic powders, and the types of connectivities between ceramic fillers.

Ceramics with a perovskite structure such as lead zirconate titanate (PZT), barium titanate (BaTiO$_3$), and BST (Ba:Sr)TiO$_3$ have been generally selected as dielectric boosters. However, it is necessary to ensure the proper surface modification of ceramic filler in polybenzoxazine composites; which not only increases the dielectric constant dramatically but also improves the distribution and lowers the dielectric loss of the polybenzoxazine composite.

Tang et al. [98] prepared polyarylene ether nitrile (PEN)/barium titanate (BT) nanocomposite films by a continuous ultrasonic dispersion fabrication process and optimization of the process through the investigations. Scanning electron microscopy (SEM) showed that BT nanoparticles with diameters less than 100 nm were well dispersed in the polymer matrix. The PEN/BT nanocomposite films exhibited excellent dielectric properties, without sacrificing tensile strength or thermal stability, when compared with those of pristine polymer. Furthermore, they observed that the nanocomposite films have good flexibility and could be curled as easily as pure PEN films.

Huang et al. [99] developed randomly oriented multiwalled carbon nanotube (MWCNT)/polyarylene ether nitrile (PEN) composite films by a solution casting method. The as-prepared films were subjected monoaxial hot-stretching in an oven to enhance their orientations and crystallinities. Results showed that the hot-stretching process enhanced the mechanical and thermal properties significantly. The electrical conducting pathways during the monoaxial hot-stretching of MWCNT/PEN composite films were examined. Since the amount of MWCNT filler was close to the percolation threshold, ca. 6 wt%, the dielectric properties, electrical conductivity, breakdown strength and energy density were found to be very sensitive to the stretch ratio. The electrical conductivity of composites with a 50% stretch ratio increased from 5.2×10^{-5} S cm^{-1} to 1.6×10^{-4} S cm^{-1} (1 kHz). Besides, the dielectric constant of composites with a 50% stretch ratio increased significantly from 378.0 to 1298.1 (100 Hz). Most importantly, the composites with 50% stretch ratio compensate some dielectric constant with breakdown strength, and finally, the energy density of the composites with a 50% stretch ratio increases by about 40%, from 2.51 to 3.50 J cm^{-3}, and has huge potential to be used as organic film capacitors. Novel nanocomposite films have also been obtained from polyarylene ether nitrile (PEN), barium titanate ($BaTiO_3$), and multiwalled carbon nanotubes (MWCNTs) by the solution casting method combined with continuous ultrasonic dispersion technology [100]. The micromorphology, as well as the thermal, mechanical, and dielectric properties of the nanocomposite films, were thoroughly investigated. All of the nanocomposite films exhibited excellent thermal stability and mechanical strength with initial decomposition temperatures up to 492 °C and the tensile strength 82 MPa. Besides, the nanocomposite films exhibiting excellent flexibility, these films can be curled easily into cylinders with several layers. Furthermore, the film with 1.5 wt. % MWCNTs and 20 wt.% $BaTiO_3$ had the best comprehensive dielectric properties, with potential for application in organic film capacitors.

Mardare et al. [101] evaporated thermally ultrapure aluminum onto various plastics (polyethylene 2.6-naphthalate, PEN; polyethylene terephthalate, PET; polyimide, PI and glass for comparison) and potentiostatically anodizing citric buffer. They monitored anodization procedure coulometrically and each alumina film formed was characterized by impedance spectroscopy. The resulting anodic alumina films were amorphous (proven by X-ray diffraction, XRD) and acted as a dielectric material in a solid state capacitor with Au top electrode. The authors evaluated capacitors characteristics using IV curves and frequency domain measurements. The performance of the capacitors demonstrated low leakage currents as well as dielectric losses. The oxide formation factors and capacities for each substrate coulometrically were determined. The ratio between apparent formation factor and the projected area has been taken as a useful parameter to determine the surface roughness. This surface roughness together with the high purity aluminum films and the anodization compression was responsible for the unexpected high mechanical stability of the composite material.

2.3 Fluorinated and nonfluorinated polymers and their composites

Zampetti et al. [102] presented the design and fabrication of a fully flexible sensorial system, composed of three different sensor units implemented on an ultrathin polyimide substrate of 8 μm thick. Each unit contained a capacitive chemical sensor integrated with readout electronics. The sensors were parallel plate capacitors with the top electrode properly patterned to allow analytes diffusion into the dielectric that acted as chemical interactive material. Three different polymers, poly(tetrafluoroethene) (PTFE), poly(methyl 2-methylpropenoate) (PMMP) and benzocyclobutene (BCB), were used as dielectrics. A ring oscillator circuit, implemented with polysilicon thin film transistors (PS-nTFT) was used to convert the capacitance variations into frequency shifts. The electronic tests showed oscillating frequencies of about 211±2 kHz and negligible frequency shifts under different bending radius conditions. Furthermore, the system response to some alcohols concentrations (Methanol, ethanol, 1-butanol, and 1-propanol) was reported and data analysis proved that the system was capable to discriminate methanol from ethanol.

Lin et al. [103] used (Polyaniline) nanomaterials doped with Nafion network these materials delivered the best cycle performance, maintaining 70.7% of its initial capacitance after 1000 cycles. Subramaniam et al. [104] illustrated the electrochemical performance of solid-state EDLCs constructed using composites of perfluorosulfonic acid polymer (Nafion) with micro porous polytetrafluoroethylene (PTFE) membrane and with cellulose acetate (CA) as an electrolyte with carbon as electrodes (surface area: 260 m^2 g^{-1}). Scanning electron microscopy was used to study the morphology of the composite electrolyte while micro-Raman and IR measurements were performed to determine the integrity of the composite. The performance of the EDLC with perfluorosulfonic acid and PTFE composite electrolyte was comparable to the performance of the EDLC with the pure perfluorosulfonic acid polymer electrolyte. These authors obtained a specific capacitance of 16 F g^{-1} for this EDLC with a maximum working potential of 2.0 V. An increase in the equivalent series resistance value from 0.08 Ω for the Nafion EDLC to 4.1 Ω for the Nafion/PTFE composite EDLC was noticed. However, the performance of the EDLC with the composite of Nafion/CA was poor due to a substantial increase in Electron spin resonance (ESR).

Maita et al. [105] presented an ultra-flexible tactile sensor, in a piezo-electric oxide-semiconductor FET configuration. The sensor was composed of a poly[vinylidenefluoride-co-trifluoroethylene] capacitor with an embedded readout circuitry that was based on nMOS polysilicon electronics and integrated directly on polyimide. The ultra-flexible device was designed according to an extended gate configuration. The sensor exhibited enhanced piezoelectric properties.

The composite insulation consisting of PTFE film layers with interstitial silicone gel, emerging as a new alternative type of insulation structure has been very successfully used for over 15 years in dry type capacitance-graded core insulation of high voltage equipment, such as wall bushings, current transformers, and cable terminals. This composite insulation structure was found to have excellent partial discharge performance with the unique characteristic of self-attenuating when in operation, sometimes down to non-detectable levels. Numerous observations were described from early field-testing, laboratory testing and dissection analysis. The consistency of the phenomenon, even after 15 years of operation was confirmed. Simulation experiments were able to reproduce the characteristic and demonstrate that the level of voltage applied, the structure design and the presence of silicone gel govern the self-attenuation. Hu, S. et al. [106] proposed a mechanism showing that the initial partial discharge helped to raise the breakdown voltage of the capacitor elements, thus reducing the onset of further discharges.

Most of the electric insulation materials used in power generation and energy storage have been polymeric dielectric materials. Some examples to cite include, winding insulation in power transformers, electric starter/generators, electric actuators, high-frequency resonators, and power converters; dielectric films in power capacitors or energy storage devices; encapsulation in electronic components and devices. These polymeric dielectric materials have limited continuous use because of their electrical properties deteriorating at high temperatures. The temperature dependence and thermal aging effect on their dielectric breakdown have been of prime concerns for analyzing their failure mechanism to prevent the catastrophic breakdown of the power system in aviation and military platforms. Yin et al.[107] have presented and discussed, the thermal aging effect on breakdown strength, mechanical properties and physical features of various polymeric insulation, such as polyimide (PI), polyetherimide (PEI), polyetheretherketones (PEEK), and perfluoroalkoxy (PFA) films.

Pecora et al. [108], presented a flexible pyroelectric sensor composed of a PVDF-TrFE capacitor realized on a ultra-thin polyimide film (5 μm thick), integrated with a n-channel low-temperature polysilicon thin film transistor also fabricated on ultra-thin polyimide (8 μm thick). Exploiting a multi-foil approach, the pyroelectric capacitors and the transistors were attached one over the other reaching a final thickness of about 15 μm. The bottom contact of the sensor capacitance was connected to the gate of the transistor by a silver ink, while, for bias and load resistances, external elements were used. The active sensor area was defined by a circular capacitor with a diameter of about 2 mm. In order to enhance PVDF-TrFE pyroelectric properties, an external stepwise voltage was applied to the structure up to values of 160 V at a temperature of about 80 °C. The devices were then tested, at different working frequencies (up to 800 Hz) under specific infrared radiation provided by He-Ne laser, with a wavelength of 632 nm and maximum power of 5 mW. An output signal of tens of millivolt was observed at 10 Hz, exploiting the pre-amplification of polysilicon thin film transistor.

Triboelectric nanogenerators, new promising energy harvesting technology has a simple working mechanism, produces high output power and has the ability of harvesting energy from a wide range of sources under various conditions. Rodrigues et al.[109] proposed a rotary triboelectric nanogenerator (TENG) using polytetrafluoroethylene (PTFE) and Nylon 6.6 as triboelectric materials for harvesting energy from water flows. The optimization of the developed setup for the effective generation of electricity showed that the best configuration was a triboelectric structure with four Nylon and one PTFE plate for the maximum used water flow of 44 L/min. With this configuration, the rotary TENG delivered a mean voltage value of ~102.2 V, a short-circuit current density ~120 mA/m^2 and a maximum power density of ~6.1 W/m^2. Finally, with this device, it was possible to light up more than 50 serial-connected light emitting diodes, fully charge a 1.0 μF capacitor with 15.2 V in just 65 s and feed a commercial temperature and humidity sensor. The developed rotary TENG can be applied to autonomous sensors that monitor water supply systems using the energy produced by water movement in plumbing.

Jabbarnia et al. [110] combined polyvinylidene fluoride and polyvinylpyrrolidone polymers with carbon black nanoparticles (50 nm) and electrospun them to fabricate nanofibrous membranes for supercapacitor separators. Different weight percentages (0, 0.25, 0.5, 1, 2, and 4 wt %) of carbon black nanoparticles were dispersed in N, N-dimethylacetamide, and acetone prior to the electrospinning processes at a variable voltage, pump speed, and tip-to-collector distances. The morphology, thermal, mechanical, hydrophobic, and electrochemical characterization of the nanofibrous membrane were analyzed using different techniques, such as scanning electron microscopy, differential scanning calorimetry, capacitance bridge, thermogravimetric analysis, dynamic mechanical analyzer, and water contact angle. Effects of annealing and UV irradiation exposures on the nanofibrous membranes were also investigated. The physical properties of the nanocomposite separators were significantly improved by carbon black inclusions in the polymeric structures, which may be useful for the applications of supercapacitor separators and other energy storage devices.

Guha et al. [111] demonstrated low-operating voltage, high mobility, and stable organic field-effect transistors (OFETs) using both non-polar and polar polymeric dielectrics such as poly(4-vinyl phenol) (PVP), poly methyl methacrylate (PMMA), and polyvinylidene fluoride-trifluoroethylene (PVDF-TrFe), dissolved in solvents of varying dipole moment. They used solvents with high dipole moment for dissolving the polymer dielectric and increased charge carrier mobilities by orders of magnitude in pentacene OFETs compared to low dipole moment solvents. These pentacene-based OFETs exhibited outstanding high stability under bias stress and in the air with negligible shifts in the threshold voltage.

Syahidah et al. [112] fabricated a symmetrical electrical double-layer capacitor (EDLC) using gel polymer electrolyte (GPE) combined with carbon-based electrodes. They fabricated three cells, A, B and C using different compositions of active materials (activated or porous carbon), binder (PVdF-HFP) and conductivity enhancer (super-P). The physicochemical properties of the GPEs were characterized by electrochemical impedance spectroscopy (EIS), scanning electron microscopy (SEM), linear sweep voltammetry (LSV) and cyclic voltammetry (CV). The ionic conductivity at the ambient temperature of the GPE was $2.16 \times 10\text{-}4$ S cm^{-1} at 7.5 wt. % of Mg $(CF_3SO_3)_2$ with a ~2.6 V electro-chemical stability window. At the 1000th cycle, the specific capacitances of cells A, B and C were, 89 F g^{-1} 63 and 49 F g^{-1} respectively. Cell A showed excellent long-term cyclic stability (less than a 5% decrease in specific capacitance after 1000 cycles). The best operating voltage for cell A was 1.6 V with the specific capacitance 106 F g^{-1} after 500 cycles.

Yu et al. [113], reported nanocomposites of increased dielectric permittivity, enhanced electric breakdown strength and high-energy density based on surface-modified $BaTiO_3$ (BT) nanoparticles filled poly(vinylidene fluoride) polymer. Polyvinylpyrrolidone (PVP) was used as the surface modification agent and homogeneous nanocomposite films prepared by solution casting processing. The dielectric permittivity of the nanocomposite with treated BT was higher than those with untreated BT and reached the maximum value of 77 (1 kHz) at BT concentration of 55 vol%. The electric breakdown strength of the nanocomposite was greatly enhanced to 336 MV/m at BT concentration of 10 vol% and the calculated energy density was 6.8 J/cm^3. The results indicated that using PVP as surface modification agent, the dielectric permittivity and electric breakdown strength of the ceramic-polymer nanocomposite can be greatly enhanced and high-energy density for energy storage can also be achieved.

Le et al. [114] prepared self-supported and binder-free micro-porous polyimide (PI)-based carbon fibers by polymer blend electrospinning technology and subsequent thermal treatment without activation. Polyvinyl pyrrolidone (PVP) was successfully used as a pore-forming template by controlling the crosslinking between PVP and PI precursor via the imidization process. The maximal specific capacitance of 215 F g^{-1} based on a symmetrical two-electrode supercapacitor was achieved at 0.2 A g^{-1}. It is noteworthy to mention that the energy density was 7.5 Wh kg^{-1} with a power density of 0.05 kW kg^{-1} at a current density of 0.2 A g-1, and 5.0 Wh kg^{-1} with a high power density of 7.5 kW kg^{-1} at 30.0 A g^{-1}. The maximum power density was 20.0 kW kg^{-1} with an energy density of 3.0 Wh kg^{-1}. The results indicated that the specific capacitance was not only attributed to optimized pore structures and surface chemistry but also attributed to the wettability of the electrolyte. The improved rate performance should be related to the reduced ion transportation distance derived from the nanofibers. Syahidah et al. [115], prepared a gel polymer electrolyte (PVP/PVdF-HF-Mg(CF$_3$SO$_3$)$_2$) system for a symmetrical solid-state electrical double layer capacitor (EDLC) using 1-butyl-2,3-dimethylimidazolium tetrafluoroborate ([bdmim]BF$_4$) ionic liquid in poly(vinyl pyrrolidone)/poly(vinylidene fluoride-co-hexafluoropropylene) (PVP/PVdF-HFP) blend. Ionic liquid-incorporated gel polymer electrolyte (GPE) containing 7.5 wt. % [bdmim] BF$_4$ exhibited the optimum room temperature ionic conductivity of $(2.92 \pm 0.02) \times 10^{-3}$ S cm^{-1} and was electrochemically stable up to +2.4 V. The solid-state displayed the highest energy and power density of 14 W h kg^{-1} and 21 W kg^{-1}, respectively, at an operating potential of 1.0 V, matching the GPE-IL useful for symmetrical solid-state EDLC applications.

2.4   Polyarylene ethers and their composites

There is a need for polymeric capacitors with improved energy storage density and thermal stability. Shaver et al. [116] investigated the effects of polymer molecular structure and the molecular symmetry on Tg, breakdown strength, and relative permittivity. Out of four amorphous poly(arylene ether ketone)s investigated, two of the polymers had symmetric bisphenols while the remaining two had asymmetric bisphenols. Two contained trifluoromethyl groups while the other two had methyl groups. The symmetric polymers had Tg's of approximately 160 °C while the asymmetric polymers showed higher Tg's near 180 °C. The symmetric polymers had breakdown strengths around 380 kV/mm at 150 °C whereas the asymmetric counterparts had breakdown strengths about 520 kV/mm even at 175 °C, with the fluorinated polymers performing slightly better in both cases. The non-fluorinated polymers had higher relative permittivities than the fluorinated materials, with the asymmetric polymers being better in both cases.

Sivaraman et al. [117] fabricated an all-solid supercapacitor using polyaniline (PANI) and sulfonated poly(ether ether ketone) (SPEEK). They also fabricated a composite electrode from chemically-synthesized PANI, SPEEK, electronically conducting carbon black and polytetrafluoroethylene (PTFE). SPEEK acted as a separator as well as the electrolyte. The unit cell consisted of two electrodes made from a p dopable PANI composite electrode separated by a SPEEK membrane with a thickness of 50 μm. As the electrodes of the capacitor were made from same p dopable PANI material, the cell was type I (p-p) capacitor. The cell characteristics were investigated using cyclic voltammetry, charging-discharging, and impedance techniques. The authors found unit cell capacitance as 0.6 F, which corresponds to 27 F per g of the active polymer material. The fabrication of [118] all solid supercapacitor based on polyaniline (PANI) and crosslinked sulfonatedpoly[ether ether ketone] (XSPEEK) has also reported [118]. The crosslinker used for sulfonated poly[ether ether ketone] (SPEEK) was 1,4-bis(hydroxymethyl) benzene. The XSPEEK was used as both solid electrolyte and separator membrane. Supercapacitors using various PANI/XSPEEK weight ratios were fabricated and characterized by cyclic voltammetry and galvanostatic charge-discharge studies. The supercapacitor with PANI/XSPEEK weight ratio 1:0.5, exhibited a specific capacitance of 480 F g^{-1}, the highest reported value for a supercapacitor based on a proton conducting solid polymer electrolyte and PANI. The detailed electrochemical impedance spectroscopy analysis showed that the complex capacitance of the supercapacitor depended on the XSPEEK content. The time constant (t0), derived from the imaginary part of complex capacitance decreased with increase in the XSPEEK content in the supercapacitor. Cycle life characteristics of the supercapacitor showed a decrease in specific capacitance during initial cycles but get stabilized afterward.

Na et al. [119] used a novel strategy to prepare a high-performance micro-porous polymer electrolyte (MPE) from a poly(arylene ether ketone) (PAEK)/poly(ethylene glycol) grafted poly(arylene ether ketone) (PAEK-g-PEG) polymer composite membrane matrix incorporating chitosan-based $LiClO_4$ gel electrolyte. The authors investigated morphology, porosity, and thermal and mechanical properties of the PAEK/PAEK-g-PEG polymer blend membrane matrix. They implemented a simple but effective method involving the addition of chitosan to improve the liquid retention capacity of the MPE.

The effects of chitosan concentration on the liquid uptake and leakage behaviors as well as the ionic conductivity of the electrolyte were evaluated. The synthesized MPE exhibited an ionic conductivity as high as 8×10^{-3} S cm^{-1} at room temperature. More importantly, these workers reported the fabrication of a novel solid-state electric double layer capacitor (S-EDLC) using the synthesized MPE and activated carbon electrodes. Alternating current (AC) impedance measurement results showed the good interfacial compatibility of the MPE and the activated carbon electrodes. Meanwhile, the as-prepared S-EDLC showed a specific capacitance of 118.63 F g^{-1}, with an energy density of 7.87 W h kg^{-1} and power density being 95.97 W kg^{-1} at a current density of 1 A g^{-1}. Furthermore, the S-EDLC exhibited greater cell-cycling stability after 5000 charge/discharge cycles than did the EDLC, indicating that this novel MPE should be suitable for use in EDLCs and other energy storage systems.

Mukaigawa et al. [120] conducted measurement of Cu drifts in methylsilsesquiazane-methyl silsesquioxane dielectric films in the presence of an electric field using bias-temperature stress (BTS) and capacitor-voltage (CV) analyses as well as time-dependent dielectric breakdown (TDDB) stress. They estimated a number of Cu ions in the dielectric films making use of the flat band voltage shift ΔVFB from the BTS. Comparing the flat band voltage measured by CV analysis with the leakage current integrated over time, it was concluded that the main content of the leakage current during BTS was the ionic current that can be attributed to the drift of Cu and mobile ions. The Cu ions were responsible for the leakage current during TDDB stress to increase. The drift rate of Cu in methyl silsesquioxane was lower than the reported values in polyarylene ether (PAE) and fluorinated polyimide (FPI) but was higher than that in case of plasma enhanced chemical vapor deposition (PEC VD)-SiON.

Wu et al. [121] investigated the integration of very thin sputtered Ta and reactively sputtered TaN barriers with Cu and a low-dielectric-constant (low-κ) layer of poly(arylene ether) (PAE-2). It was found that Cu readily penetrated into PAE-2 in order to degrade its dielectric strength in metal-insulator-semiconductor capacitors of Cu/PAE-2/Si structure at temperatures as low as 200 °C. Very thin Ta and TaN films of 25 nm thickness sandwiched between Cu and the low-κ dielectric served as effective barriers during a 30 min thermal annealing at temperatures up to 400 and 450 °C, respectively. The authors proposed a failure mechanism of outgassing induced gaseous stress of PAE-2 under the Ta film to explain its premature barrier degradation. The TaN barrier did not suffer from this gaseous stress problem because of its stronger adhesion to PAE-2 than that of Ta to PAE-2, leading to a better long-term reliability.

2.5 Polyvinylpyrrolidone (PVP) based applications

Constantinescu et al. [122] prepared laser-induced transfer for the printing of multilayered micro capacitors using the dielectric film made of PVP and Ag electrodes. The thermal behavior of the polymer with respect to the laser processing and also the structure and electrical properties of the capacitors Ag/polyvinyl phenol (PVP) multilayered pixels printed by a laser-induced forward transfer (LIFT) technique for thin film micro capacitor applications were discussed. By selecting adequate printing parameters (e.g. donor thickness, laser fluence, background pressure), authors explained the fabrication of functional microcapacitors. At ~350 μm in lateral size and 300 nm thickness of the dielectric film, the pixels have capacities in the picofarad range.

Zhao et al. [123] fabricated lead-free ferroelectric $Na_{0.5}Bi_{0.5}TiO_3$ (NBT) thick films on $LaNiO_3/Si$ (100) substrates by using a polyvinylpyrrolidone (PVP)-modified sol-gel technique. Dielectric properties, energy-storage performance, and leakage current characteristics were investigated in detail. It was noticed that at 100 kHz, the capacitance density of the NBT thick films was 295 nF/cm^2. The maximum recoverable energy-storage density and efficiency of the sample were 12.4 J/cm^3 and 43% at 1200 kV/cm, respectively. The low leakage current density of about 1×10^{-5} A/cm^2 was observed at 700 kV/cm at room temperature. These results indicated that the lead-free ferroelectric NBT thick films might be the promising candidates for high energy-storage capacitors application

Wu et al. [124] prepared an all-solid-state supercapacitor with polyaniline (PANI) as the anode, WO_3 as the cathode and Polyvinylpyrrolidone (PVP)-$LiClO_4$ as the solid electrolyte. The results showed good electrochromic and electrochemical performance of supercapacitor, which can thus result in rapid and reversible color changes under different operating voltages (−1 V ~ 1 V). The changing of color during the discharge process indicated the changes of energy level. This research proposed a new path for the development of intelligent all-solid-state supercapacitor in the future.

Huo et al. [125] prepared a series of quaternary ammonium functionalized poly(arylene ether sulfone)/poly(vinylpyrrolidone) (PAES-Q/PVP) composite membranes for separators and composite polymer electrolytes (CPEs) of electrical double-layer capacitors (EDLCs) by solution blending. They fabricated EDLCs using composite membranes and activated carbon electrodes. The authors concluded that the optimized solid-state EDLC with PAES-Q/PVP composite membrane containing 40 wt% PVP entrapping 6 M KOH served as CPE exhibited a specific capacitance of 140.85 F g^{-1} and an energy density of 4.81 W h kg^{-1} at a current density of 0.1 A g^{-1}. Electrochemical cycle performance demonstrated indicated almost 100% cycling retention and near 97% coulombic efficiency over 5000 charge-discharge cycles at 0.5 A g^{-1}

During the initial development of wearable computing devices, the conductive fibers of Al thin film on cylindrical poly(ethylene terephthalate) (PET) monofilament were fabricated by thermal evaporation. Their electrical current-voltage characteristics curves were excellent for incorporation into wearable devices such as fiber-based cylindrical capacitors or thin film transistors. Their surfaces were modified by UV exposure and dip coating of acryl or PVP to investigate the surface effect. The conductive fiber with PVP coating showed the best conductivities because the rough surface of the PET substrate transformed into a smooth surface. The conductivities of PET fiber with and without PVP were 6.81 × 103 Ω^{-1}cm^{-1} and 5.62 × 103 Ω^{-1}cm^{-1}, respectively. Liu and Kim et al. [126] studied the deposition of Al thin film on the cylindrical PET as well as on PET fiber using SEM, AFM, conductivities and thickness measurements. Hillocks were observed on the surface of conductive PET fibers that developed during Al thermal evaporation because of severe compressive strain and plastic deformation induced by large differences in thermal expansion between PET substrate and Al thin film. The hillocks grew longitudinally, not transversely.

Yang et al. [127] controlled the evaporation behavior and the resulting morphology of inkjet-printed dielectric layers using a mixed-solvent system to fabricate uniform poly-4-vinylphenol (PVP) dielectric layers without any pinholes. The mixed-solvent system consisted of two different organic solvents: 1-hexanol and ethanol. The effects of inkjet-printing variables such as overlap condition, substrate temperature, and different printing sequences (continuous and interlacing printing methods) on the inkjet-printed dielectric layer were investigated. On the basis of their studies, the authors concluded that all-inkjet-printed capacitors without electrical short-circuiting can be successfully fabricated using the optimized PVP solution (VFE = 0.6) and the mixed-solvent system is playing an important role in the fabrication of high-quality inkjet-printed dielectric layers in various printed electronics applications.

McKerricher et al. [128] fabricated and characterized fully inkjet printed multilayer capacitors and inductors using poly 4-vinylphenol (PVP) ink as the dielectric layer and silver nanoparticle ink as the conductor. The authors tailored dissolving method to make radio frequency (RF) structures in a multilayer inkjet printing process. The vias realized in a 350-nm PVP film and exhibited resistance better than 0.1 Ω. Metal-insulator-metal (MIM) capacitors with densities of 50 pF/mm^{-2} demonstrated values ranging from 16 to 50 pF. The 16-pF capacitor showed a self-resonant frequency over 1.5 GHz. The successful practice of inductors and capacitors in an all inkjet printed multilayer process with vias was noted to be an important step towards future applications for large area and flexible RF systems.

Capacitor's energy density can be simply improved by using an operative voltage extension. An organic gel usage has a more stabilized potential window than an aqueous gel. Based on these concepts, Ramasamy et al [129]. examined a two electrode symmetric electric double layer capacitor (EDLC) based on commercially available activated carbon (YECA-Fuzhou Yihuan Carbon) electrodes at 22 °C using a gel electrolyte from polyvinyl pyrrolidine (PVP) in ethylene glycol (EG)-water with sodium sulfate as a salt. Infrared spectroscopy confirmed the electrolyte interactions. Cyclic Voltammetry (CV) results showed the capacitor's reversibility and symmetric EDL. The linear sweep voltammetry (LSV) was used to determine a 2 V applicability of the gel. The galvanostatic cycling (CD) studies revealed an improvement in the low rate aqueous gel based cell and its voltage limit. The authors observed ~ 21 F.g^{-1}, 0.7 kW.kg^{-1} and 11 Wh.kg -1 values for a typical specific capacitance, real power and energy density of the cell respectively with 10% replacement of water by glycol which was accepted to be significantly higher than the aqua gel based cell (~ 22 F.g^{-1}, 0.55 kW.kg^{-1} and 6 Wh.kg^{-1}).

Yu et al [130] prepared a gel polymer electrolyte (PVA-PVP-H$_2$SO$_4$-MB) using methylene blue (MB) redox mediator into polyvinyl alcohol/polyvinyl pyrrolidone (PVA/PVP) blend for a quasi-solid-state supercapacitor. The authors determined electrochemical properties of the supercapacitor with this gel polymer electrolyte by cyclic voltammetry, galvanostatic charge-discharge, electrochemical impedance spectroscopy, and self-discharge measurements. Addition of MB mediator increased the ionic conductivity of gel polymer electrolyte by 56% up to 36.3 mS·cm^{-1}, and the series resistance reduced, because of the more efficient ionic conduction and higher charge transfer rate, respectively. The electrode specific capacitance of the supercapacitor with PVA-PVP-H 2SO$_4$-MB electrolyte was 328 F·g^{-1}, increasing by 164% compared to that of the MB-undoped system at the same current density of 1 A·g^{-1}. At the same time, the energy density of the supercapacitor increased from 3.2 to 10.3 Wh·kg-1. The quasi-solid-state supercapacitor showed excellent cyclability over 2000 charge/discharge cycles.

Sellam et al. [131] used nonaqueous proton conducting polymer electrolyte of ionic liquid 1-ethyl 3-methyl imidazolium hydrogen sulfate (EMIHSO$_4$) immobilized in the blend of poly (vinyl alcohol) (PVA) and poly (vinyl pyrrolidone) (PVP) polyelectrolyte with structural, thermal and electrochemical properties including high ionic conductivity (6.2×10^{-2} S cm^{-1} at 20 °C) to separate poly(3,4-ethylene dioxythiophene)-poly(styrene sulfonate) (PEDOT-PSS) and PEDOT-PSS/hydrous ruthenium oxide based electrodes.

They demonstrated its excellent suitability in supercapacitor fabrication with the introduced electrodes.

Choi et al. [132] studied the non-vacuum electrohydrodynamic atomization (EHDA) of solution processable poly 4-vinylphenol (PVP) ink for the fabrication of dielectric thin films. The authors achieved, the optimum flow rate/applied potential by using an operating envelope in EHDA and optimized parameters to generate an electrohydrodynamic jet, which subsequently is disintegrated into droplets to deposit a uniform thin film of PVP on indium-tin-oxide (ITO)-coated polyethylene terephthalate (PET) substrate with an average thickness of ~100 nm at constant substrate at a speed of 0. 3 mm/s. The capacitor was analyzed for the current/voltage (I-V) and the capacitance/voltage (C-V) characteristics metal-insulator-semiconductor (MIS) i. e., ITO/PVP/Poly (3, 4-ethylene dioxythiophene) poly (styrenesulfonate) (PEDOT: PSS).

Rodríguez et al. [133] used dip coated PVP (polyvinylpyrrolidone) and $LiClO_4$ films as solid ionic conductors in transparent energy storage systems. The authors studied the dependence of the ionic conductivity of $PVP/LiClO_4$ films on temperature, lithium salt concentration, and residual solvent content or humidity. These films were used in a symmetrical supercapacitor (PEDOT/PVP/PEDOT and analyzed for their behaviors. The experimental differences in the performance of a lithium-ion-based energy storage devices using these films as solid electrolyte were attributed to the changes in the residual content of the solvent used during their preparation or the hygroscopic nature of the electrolyte.

Commodity and engineering polymers such as PE, PP, and PA, etc. have strong hydrophobic character and therefore, not suited for use as separators in supercapacitors due to their poor wetting properties for various electrolytes. In order to overcome this problem, radiation-induced oxidation of these polymers in air or grafting with hydrophilic monomers can be used. For example, treating polyamide and polypropylene non-woven fabrics with low energy plasma resulted in cleaner polymer surfaces with improved wettability for the electrolyte. These modifications showed a significant decrease of internal resistance of the supercapacitor, thus increasing the usable power of the device. The swelling ratio and ionic conductivity of the hydrogel electrolyte membranes increased on the addition of PVP. The specific capacitance of flexible super-capacitor reached to 111 F g^{-1} at 20 % mass fraction level of PVP in the electrolyte membrane. [1].

2.6   Carbohydrate polymers

Carbohydrate polymers such as cellulose are suitable candidates as membranes for use in supercapacitors but their poor mechanical strength has limited their use. However, radiation crosslinking has been found useful to improve the mechanical properties of cellulosic materials, leading to cellulose-based separators capable of forming hydrogels that can absorb a significant amount of liquid electrolyte within their network. Fei et al. [134] have successfully crosslinked carboxymethyl-cellulose (CMC) by radiation. The density of crosslinking was a function of the absorbed radiation dose and degree of substitution in the cellulosic compound. The swelling was reduced at very high crosslinking density. Wach et al. [135] examined the effects of radiation types, environment and dose rates on the degree of crosslinking of CMC. They found out that Electron beam (EB) irradiation exhibited higher gel fraction (up to 90%) compared with 60% obtained by irradiation with UV-rays. Polypropylene (PP) has been used in aqueous electrolyte-based supercapacitors. However, PP is a semicrystalline hydrophobic polymer with low wettability in the liquid electrolytes. One of the ways to increase the wettability of the polymer has been the incorporation of hydrophilic moieties at the surface. This was easily achieved by radiation-induced grafting of acrylic acid (AAc) on PP. Such separators were first developed for alkaline batteries. Stepniak et al. [136] investigated the dependence of capacity on the wettability properties of the electrode and the separator by grafting of AAc on both activated carbon(AC) and PP using the plasma-induced grafting technique. The plasma activated samples were treated with UV to polymerize the AAc. It was found that the specific capacitance increased with increase in the degree of grafting. However, excessive grafting decreased the surface area due to the formation of double layer forms leading to a decrease in the capacitance. Nevertheless, the contact angle measurement showed that PP was turned from hydrophobic into the hydrophilic structure, which helped to absorb more electrolytes and decreased the electrolytic area resistance. The resistance was also found to decrease from 120,000 to 40–50 mΩ/cm^2 depending on the grafting degree.

Sinar and Knopf [137] fabricated extensively used Interdigitated capacitors (IDC) for a variety of chemical and biological sensing applications. They synthesized an electrically conductive aqueous graphene ink stabilized in deionized water using the polymer of a nontoxic hydrophilic food grade cellulose derivative carboxymethylcellulose (CMC). The water-based graphene ink was then used to fabricate IDC sensors on mechanically flexible polyimide substrates. In commercial machine weaving processes, significant mechanical deformations occur, mostly due to high bending, and shear forces associated with the process. Bending radii in textiles can be as small as 165 μm during weaving, corresponding to a strain of about 15%. Furthermore, textiles sensors are exposed to harsh physical and chemical environments during operation. To avoid device failure, the sensor active area has to be fully encapsulated for protection during process operations. Ataman et al. [138] presented low-power gas and temperature sensing textile integration and post-weaving characterization of a robust platform. The platform, consisting of an interdigitated planar capacitor for gas sensing and a resistive temperature sensor, was fabricated on a 50 μm thick flexible Kapton E® film using a simple roll-to-roll compatible process, and particularly targeted disposable textile products, such as smart air filters, and medical garments. In order to demonstrate the versatility of the platform, the sensors were functionalized for humidity sensing by spray-coating of a 10 μm thick cellulose acetate butyrate (CAB) polymer layer. They resulted in that printing and functionalizing these IDC sensors on bendable substrates that will lead to new innovations in healthcare and medicine, food safety inspection, environmental monitoring, and public security.

For microelectronic applications, biodegradable and biocompatible nanomaterials such as microfibrillated cellulose (MFC) have attracted attention. Couderc et al. [139] developed an innovative electrostatically actuated mechanical switch device made of a microfibrillated cellulose sheet coated with a thin polyimide layer. The studied MFC sheets revealed a fibrous-like morphology composed of cellulose nanofibres leading to a high surface roughness. Moreover, the porous microstructure and the hydrophilic nature of the MFC sheet induced poor dielectric properties. These shortcomings made MFC sheets relatively unsuitable for electronic applications. In order to overcome these drawbacks, both sides of the MFC sheet were coated with a thin polyimide layer, which greatly improved the dielectric properties, moisture sensitivity, and sheet surface roughness. The coated sheet was then patterned in order to be used as a substrate for the fabrication of a micromechanical switch.

Chitnis et al. [140] presented a minimally invasive implantable pressure sensing transponder for continuous wireless monitoring of intraocular pressure (IOP). The transponder was designed to make the implantation surgery simple while still measuring the true IOP through direct hydraulic contact with the intraocular space. Furthermore, when IOP monitoring completed, the design allows physicians to easily retrieve the transponder. The device consists of three main components: 1) a hypodermic needle (30 gauge) that penetrates the sclera through pars plana and establishes direct access to the vitreous space of the eye; 2) a micromachined capacitive pressure sensor connected to the needle back-end; and 3) a flexible polyimide coil connected to the capacitor forming a parallel LC circuit whose resonant frequency is a function of IOP. Most parts of the sensor sit externally on the sclera and only the needle penetrates inside the vitreous space. In-vitro tests showed a sensitivity of 15 kHz/mmHg with approximately 1-mmHg resolution. One month in-vivo implants in rabbits confirmed biocompatibility and functionality of the device.

Hudspeth and Kaya [141] fabricated and characterized small, wearable devices that would use humidity and temperature measurements as metrics using novel collagen-based relative humidity sensors for health monitoring. They used a natural by-product of meat and leather industries, collagen as an interesting and inexpensive alternative to polyimide dielectric sensing materials. For being useful as a health monitoring tool, the device needs to respond quickly and predictably to humidity changes. Collagen has been considered to be a viable humidity sensing material for use in capacitive relative humidity (RH) sensors. Gelatin, a partially hydrolyzed form of collagen was used to allow for easier spin coating. Authors have successfully fabricated devices by depositing a collagen thin film (1.2 μm) via spin coating, followed by Au/Pd electrodes (60 nm) via sputter coating. They used a plastic mask made from a rapid prototyping machine during physical vapor deposition (PVD) to pattern electrodes. This simple method replaced more complicated photolithography processes. Interdigitated electrodes (rather than parallel plate electrodes) formed a 6 mm wide, planar capacitor structure that has little dependence on the dielectric thickness and not affected by dielectric swelling. Initial findings indicated that these devices very closely match the results of the commercial relative humidity sensor used for reference. The capacitance-humidity relationship was shown to be non-linear, with an average change of 3 fF for every 1% change in RH around 60% RH, and an average change of 7 fF for every 1% change in RH around 80% RH.

3.   Challenges and future directions

The state-of-the-art polymer of biaxially oriented polypropylene (BOPP) film dielectric has a maximal energy density of 5 J/cm^3 and a high breakdown field of 700 MV/m, but a limited dielectric constant (~2.2) and a reduced breakdown strength above 85 °C [6]. Great efforts have been put into exploring high-temperature polymer dielectrics to fulfill the demand of high-temperature applications. Although there are some differences in the mechanisms of electrochemical energy systems, most of these depend on the polymer electrolyte membranes/separators and have two main functions i) to separate the electrochemically active masses (electrodes) and ii) to intercede the electrochemical reactions occurring at the anode and cathode by conducting specific ions.

The chemical stability and durability (shelf lifetime) properties are very important for all electrochemical energy systems. The PIs with a combination of PMDA dianhydride and a para-para linkage exhibited the highest discharged energy density and a reasonably low loss [6]. Design and fabrication of a fully flexible sensorial system on an ultrathin polyimide application of a ring oscillator circuit, implemented with polysilicon thin film transistors (PS-nTFT) to convert the capacitance variations into frequency shifts with very good results [101]. Ultra-thin flexible polyimide substrates have been used for different purposes mainly as humidity sensors [38, 108] and sometimes developed on demand [106]. All-solid supercapacitor applications using polyarylene ether, PVP type polymers have promising results and presented to be an exploited new path for the development of intelligent all-solid-state supercapacitor in the future [119, 124].

Quaternary ammonium functionalized poly(arylene ether sulfone)/poly(vinylpyrrolidone) (PAES-Q/PVP) composite membranes for separators and composite polymer electrolytes (CPEs) of electrical double-layer capacitors (EDLCs) by solution blending gave very good results [125]. For gel type electrolytes, significantly higher values obtained by using organic gels than the aqua gels since they have a more stabilized potential window than aqueous gels.

Nanocomposites can provide high capacitance densities, ranging from 5 nf/inch^2 to 25 nF/inch, depending on composition, particle size, and film thickness as conductive joints for high frequency and high density interconnect applications [40].Organic-organic nanoscale composite thin-film (NCTF) dielectric by solution deposition with the sol-gel process used to make a metal-insulator-metal capacitor (MIM) [39].The low-cost flexible substrate is another research area and there are different applications for ultralow-cost RFID tags [45]. Flexible, high-κ BaTiO$_3$@RGO/PI composites have promising potential applications for capacitors and electronic devices in the fields of microelectronics [62]. Ti-based nanocomposite films with favorable dielectric properties is immeasurably potential for embedded-capacitor applications [80].Dielectric inorganic BaTiO$_3$ film with an excellent degree of crystallinity and favorable electric properties for use in the production of flexible and stretchable electronic devices on a polyimide sheet has been introduced as another promising technology [64]. The dielectric constants of the composites decreased with increasing frequency and increased sharply with the addition of MWNTs, which is favorable for practical use in anti-static materials and embedded capacitors. BaTiO$_3$-MWNT/PI composites [84] increased the dielectric constants sharply. BErT/PI based thin films gave encouraging results in flexible optoelectronic devices and embedded capacitors [88]. High temperature, flexible polymer film capacitor applications used for developing low-cost processes for memory devices that employ relatively inexpensive materials, and Cu based nanocomposite applications demonstrated very good performances [70]. Wide temperature dielectric properties and corona resistance were obtained using polyimide and ZrO$_2$ nanocomposites for high energy density capacitor applications at high temperature. Fabrication of metal-insulator-metal (MIM) capacitors using 10-nm-thick zirconium-silicate (ZrSi$_x$O$_y$) and hafnium-silicate (HfSi$_m$O$_n$) thin dielectric films on the flexible polyimide substrate by sol-gel process has created special interest to be used as gate dielectrics in future for flexible metal-oxide-semiconductor devices [94]. $Na_{0.5}Bi_{0.5}TiO_3$ (NBT) thick films on $LaNiO_3$/Si (100) substrates by using a polyvinylpyrrolidone (PVP)-modified sol-gel technique might be the promising candidates for high energy-storage capacitors application [123].

Polyimides, fluorinated polymers, and PVP type high-performance polymer (HPP) separator applications seem to continue with some modifications.

Crosslinking decreases the swelling property of the polymer while increasing mechanical properties at high temperatures. This can be achieved by using a) chemical cross-linking either by chemical treatment (either by sol-gel and/or thermal treatments using crosslinkers like polyamines, polyphenols or peroxides type) or b) irradiation (UV or high-energy radiation sources such as EB).

Divinylbenzene (DVB), 1,2-bis(p,p-vinylphenyl)ethane (BVPE), and triallyl cyanurate (TAC) type cross linkers were used effectively for radiation-induced polymerization to improve the stability of the membranes enormously. DVB/TACor DVB/BVPE type combinations were found to further improve the stability. Double crosslinking and grafting are the other important options to improve the properties. The most established battery separator was obtained by radiation-induced grafting of AAc onto PE and PP films, used in commercial silver oxide batteries, nickel-cadmium batteries, and alkaline button batteries applications. In particular, more research is needed to establish the correlations between the preparation conditions and the performance of the last batteries [1].

Investigating various preparation methods using electrolytes with important performance parameters of high conductivity and stability has been another important research area in the super-capacitors. HPP usage and radiation crosslinking or grafting showed strong potential for the preparation of solid polymer electrolytes for supercapacitor applications. However, studies in radiation crosslinking or grafting are limited and controlling the conductivity and stability has not been fully explored and more research should continue to prepare new solid polymer electrolytes [1]. In conventional radiation-induced grafting, the lack of molecular weight control of graft chains as well as their heterogeneity and chain architecture, have been the principal drawbacks. With new techniques like Atom-transfer radical-polymerization (ATRP) and Reversible Addition-Fragmentation Chain Transfer (RAFT), it is possible to control these parameters by predetermining molecular weight and very narrow distribution (D = 1.1–1.3). As a result, it is also easy to control physicochemical properties of conductivity, mechanical and transport properties of the membranes by controlling graft chain lengths with very narrow distributed RAFT polymers [1].

Polymer membranes using various monomer/polymer combinations, structure, and composition of the membranes will help to tune not only the obtained polymer electrolyte membranes but also their properties using radiation-induced grafting method. Carboxylic acid groups containing radiation-grafted membranes have been commercialized for battery application after their performance and durability were tested and established for many years. However, for lithium batteries and vanadium redox flow batteries and supercapacitors, radiation-grafted membranes have some drawbacks. A balance between high conductivity, mechanical stability, safety and cost has to be always considered. Cost competitiveness can be achieved by reducing the absorbed dose applied in grafting and economizing the amount of monomer used in radiation-induced grafting [1].

Non-volatile memory capacitor, metal-ferroelectric-insulator-semiconductor memory element, field effect transistor type memory and high energy storage capacitor with ultrathin films have been developed using epitaxially grown fluorinated polymers with very promising properties [14.15]. Nano-sized fillers (SiO_2, Al_2O_3, Ta_2O_5, TiO_2 and BCB, different kinds of ceramics, TiC, TiO_2, AlO_2, $BaTiO_3$, Pb_xZr_{1-x}, TiO_3 and Ba_xSr_{1-x}, TiO_3, etc.) have been used to prepare flexible composite film and also packaged memory device applications following specific production processes [16,17,18].

Different supercapacitor fabrication methods have been used for different applications by using different types of membranes/separators. Flexible all solid-state micro capacitor assemblies have been developed using laser-based facile fabrication [26]. Embedded capacitor technique has been very useful for miniaturizing electronic devices and enhancing their performance and reliability. Therefore, it is of great importance to study high dielectric materials that can be used in the embedded environment. However, it increases the package manufacturing cost, thus, tradeoff studies are necessary on a given product design to understand if the improved performance and package factors required the additional cost for that specific product [56].

Printing technologies provide a simple procedure to build electronic circuits on low-cost flexible substrates. Inkjet-printing process and continuous ink-jet printing processes have been used for fabrication of inkjet-printed films [27, 28]. The mixed-solvent systems are expected to play an important role in the fabrication of high-quality inkjet-printed dielectric layers for their utilization in various printed electronics applications [127]. Inkjet printed multilayer capacitors and inductors using poly 4-vinylphenol (PVP) ink as the dielectric layer and silver nanoparticle ink as the conductor presented to be an important step towards the large area and flexible Radio frequency (FRF) systems [128]. Lithographic resist-reflow techniques with dry etching procedure is the another fabrication technique [31] for force type capacitors [29, 30]. Flexible-universal-plane solution for quality flexible displays and other non-display applications used active-matrix organic light-emitting-diode (AMOLED) displays in color OLED using plastic substrates [34]. Roll-to-roll fabrication of flexible film production is important and UV curable planarization improved good surface quality of plastic substrates to reduce pixel defects material since some of these defects could be the source of shunts in dielectrics [47].

Conclusions

Polymer dielectrics are the preferred materials (separators and/or membranes) of choice for capacitive energy-storage applications because of their potential for high dielectric breakdown strengths, low dissipation factors and good dielectric stability over a wide range of frequencies and temperatures, despite having inherently lower dielectric constants relative to ceramic dielectrics. There is an enormous amount of research on this subject to achieve cost-effective, highly conductive, and durable materials to improve commercialization of electrochemical systems of supercapacitors and batteries.

Separators and/or membranes are mainly obtained by following three attractive methods: 1) use of high-performance polymers (HPP) 2) use of radiation-induced polymerization products and 3) use of their organic and inorganic based composites. Thermoplastic and thermoset membranes have been significantly improved by applying several strategies, including the selection of starting polymeric materials (mainly fluorinated polymers and polyimides) and their different combinations with organic and inorganic ingredients in thermoplastic and thermoset forms, adopting methods of controlled grafting or crosslinking and tuning the crosslinking density. In particular, blending, composite applications and crosslinking effectively reduced the extent of degradation and enhanced durability under real operating conditions. Molecular weight and structures of the polymers, nature of materials, blending and their amounts in the compositions, operational processes and application areas have been important parameters and required more research. Polymer purity, nanoparticle size, film morphology factors, granulation type and electroplating experimental conditions have been found to influence enormously the dielectric strength and the energy storage performance of the capacitors [25].

Carbohydrate polymers has another important potential in different microelectronic applications like continuous wireless monitoring systems, pH, humidity sensors, small wearable devices, chemical and biological sensing applications etc. [138,139,140,141].

Long-term in-situ testing of the membranes under actual battery or supercapacitor conditions is necessary to evaluate the real potential of the new materials.

References

[1] M.M. Nasef, SA. Gürsel, D. Karabelli, O. Guven, Radiation-grafted materials for energy conversion and energy storage applications, Prog. Polym. Sci. 63 (2016) 1–41. https://doi.org/10.1016/j.progpolymsci.2016.05.002

[2] https://en.wikipedia.org/wiki/Supercapacitor

[3] R. Kotz, M. Carlen, Principles and applications of electrochemical capacitors, Electrochim. Acta 45 (2000) 2483–98. https://doi.org/10.1016/S0013-4686(00)00354-6

[4] ACD Leon, Q. Chen, NB. Palaganas, JO. Palaganas, J. Manapat, RC. Advincula, High performance polymer nanocomposites for additive manufacturing applications, React. Funct. Polym. 103 (2016) 141–155. https://doi.org/10.1016/j.reactfunctpolym.2016.04.010

[5] X. Peng, W. Xu, L. Chen, Y. Ding, T. Xiong, S. Chen, H. Hou, Development of high dielectric polyimides containing bipyridine units for polymer film capacitor, React. Funct. Polym. 106 (2016) 93-98. https://doi.org/10.1016/j.reactfunctpolym.2016.07.017

[6] I. Treufeld, D.H. Wang, B.A. Kurish, L.-S. Tan, L. Zhu, Enhancing electrical energy storage using polar polyimides with nitrile groups directly attached to the main chain, J. Mater. Chem. A 2 (2014) 20683-20696. https://doi.org/10.1039/C4TA03260H

[7] Editor-in-chief, Lide R. D. (1913–1995). Strength of chemical bonds. Boca Raton: CRC Press In CRC Handbook of Chemistry and Physics 75th Edition (Special Student Edition), 9-51–9-73.

[8] H. Sheng, M. Wei, A.D. Aloia, G. Wu, Heteroatom polymer-derived 3d high-surface-area and mesoporous graphene sheet-like carbon for supercapacitors, ACS Appl. Mater. Interf. 8 (44) (2016) 30212-30224. https://doi.org/10.1021/acsami.6b10099

[9] Z. Zhu, S. Tang, J. Yuan, X. Qin, Y. Deng, R. Qu, G.M. Haarberg, Effects of various binders on supercapacitor performances, Int. J. Electrochem. Sci. 11 (2016) 8270-8279. https://doi.org/10.20964/2016.10.04

[10] J.M. Rosas, R.Ruiz-Rosas, R. Berenguer, D. Cazorla-Amorós, E. Morallón, H. Nishihara, T. Kyotani, J. Rodríguez-Mirasol, T. Cordero, Easy fabrication of superporous zeolite templated carbon electrodes by electrospraying on rigid and

flexible substrates, J. Mater. Chem. A 4 (2016) 4610-4618.
https://doi.org/10.1039/C6TA00241B

[11]   X. Meng, D. Deng, , Bio-inspired synthesis of α-Ni(OH)2 nanobristles on various
       substrates and their applications, J. Mater. Chem. A 4 (2016) 6919-6925.
       https://doi.org/10.1039/C5TA09329E

[12   P. Staiti, F. Lufrano, Nafion® and Fumapem® polymer electrolytes for the
       development of advanced solid-state supercapacitors, Electrochim. Acta 206
       (2016) 432-439. https://doi.org/10.1016/j.electacta.2015.11.103

[13]   T. Hibino, K. Kobayashi, M. Nagao, S. Kawasaki, High-temperature supercapacitor with
       a proton-conducting metal pyrophosphate electrolyte, Sci. Rep. 5 (2015) 7903.
       https://doi.org/10.1038/srep07903

[14]   Y.J. Park, S.J. Kang, Y. Shin, R.H. Kim, I. Bae, C. Park, Non-volatile memory
       characteristics of epitaxially grown PVDF-TrFE thin films and their printed
       micropattern application, Curr. Appl. Phys. 11 (2011) 30-34.
       https://doi.org/10.1016/j.cap.2010.11.119

[15]   Y.J. Park, S.J. Kang, B. Lotz, M. Brinkmann, A. Thierry, K.J. Kim, C. Park,
       Ordered ferroelectric PVDF-TrFE thin films by high throughput epitaxy for
       nonvolatile polymer memory, Macromolecules 41 (2008) 8648-8654.
       https://doi.org/10.1021/ma801495k

[16]   K.P. Murali, S. Rajesh, O. Prakash, A.R. Kulkarni, R. Ratheesh, Preparation and
       properties of silica filled PTFE flexible laminates for microwave circuit
       applications, Compos. Part A Appl. Sci. Manuf. 40 (2009) 1179-1185.
       https://doi.org/10.1016/j.compositesa.2009.05.007

[17]   F. Xiang, H. Wang, M.L. Zhang, X. Yao, Frequency-temperature compensation
       mechanism for bismuth based dielectric/PTFE microwave composites, J.
       Electroceram. 21 (2008) 457-460. https://doi.org/10.1007/s10832-007-9208-1

[18]   R. Nowak, P. Hart, D. Alcoe, Large thin organic PTFE substrates for multichip
       applications Proceedings - Electronic Components and Technology Conference 2
       (2005)1359-1363.

[19]   M. Nowogrodzki, Improved electroplated PTFE for high performance capacitor
       applications, EIC 1971 - Proceedings of the 10th Electrical Insulation Conference,
       7460805 (2016) 174-176.

[20]   Y. Su, G. Xie, T. Xie, H. Zhang, Z. Ye, Q. Jing, H. Tai, X. Du, Y. Jiang, Wind energy harvesting and self-powered flow rate sensor enabled by contact electrification, J. Phys. D. 49 (2016) 215601. https://doi.org/10.1088/0022-3727/49/21/215601

[21]   M.-K. Kim, M.-S. Kim, S.-E. Jo, Y.-J. Kim, Triboelectric-thermoelectric hybrid nanogenerator for harvesting frictional energy, Smart Mater. Struct. 25 (2016) 125007. https://doi.org/10.1088/0964-1726/25/12/125007

[22]   P. Sivaraman, S.K. Rath, V.R. Hande, A.P. Thakur, M. Patri, A.B. Samui, All-solid-supercapacitor based on polyaniline and sulfonated polymers, Synt. Met. 156 (2006) 1057–1064. https://doi.org/10.1016/j.synthmet.2006.06.017

[23]   L. Dumas, E. Fleury, D. Portinha, Wettability adjustment of PVDF surfaces by combining radiation-induced grafting of (2,3,4,5,6) pentafluorstyrene and subsequent chemoselective "click type" reaction. Polymer 55 (2014) 2628–34. https://doi.org/10.1016/j.polymer.2014.04.002

[24]   Z. Deng, H. Wei, S. Fan, J. Gan, Design and analysis a novel RF MEMS switched capacitor for low pull-in voltage application, Microsyst. Technol. 22 (2016) 2141-2149. https://doi.org/10.1007/s00542-015-2604-6

[25]   S. Abdalla, F. Al-Marzouki, A. Obaid, S. Gamal, Effect of addition of colloidal silica to films of polyimide, polyvinylpyridine, polystyrene, and polymethylmethacrylate nano-composites, Materials 9 (2016) 104. https://doi.org/10.3390/ma9020104

[26]   J.B. In, B. Hsia, J.H. Yoo, S. Hyun, C. Carraro, R. Maboudian, C.P. Grigoropoulos, Facile fabrication of flexible all solid-state micro-supercapacitor by direct laser writing of porous carbon in polyimide, Carbon 83 (2015) 144-151. https://doi.org/10.1016/j.carbon.2014.11.017

[27]   A. Rivadeneyra, J. Fernández-Salmerón, M. Agudo, J. A López-Villanueva, L.Fermín Capitan-Vallvey, A.J. Palma, Design and characterization of a low thermal drift capacitive humidity sensor by inkjet-printing, Sens. Actut. B Chem. 195 (2014) 123-131.

[28]   F. Zhang, C. Tuck, R. Hague, Y. He, E. Saleh, Y. Li, C. Sturgess, R. Wildman, Inkjet printing of polyimide insulators for the 3D printing of dielectric materials for microelectronic applications, J. Appl. Polym. Sci. 133 (2016) 43361. https://doi.org/10.1002/app.43361

[29] J.A. Dobrzynska, M.A.M. Gijs, Polymer-based flexible capacitive sensor for three-axial force measurements, J. Micromech. Microeng. 23 (2013) 015009. https://doi.org/10.1088/0960-1317/23/1/015009

[30] J.A. Dobrzynska, M.A.M. Gijs, Capacitive flexible force sensor, Procedia Eng. 5 (2010) 404-407. https://doi.org/10.1016/j.proeng.2010.09.132

[31] K.-R. Lin, T.-H. Liu, S.-W. Lin, C.-H. Chang, C.-H. Lin, occlusive bite force measurement utilizing flexible force sensor array fabricated with low-cost multilayer ceramic capacitors (MLCC), TRANSDUCERS 2009-15th International Conference on Solid-State Sensors, Actuators and Microsystems, 5285593, pp. 2238-2241

[32] K.D. Jamison, B. Balliette, High temperature performance of oxide film capacitors, IMAPS International Conference on High Temperature Electronics Network, HiTEN 2011, pp.21-26

[33] J-H. Lee, H. Jin, J-W. Kim, K-H. Kim, B.W. Park, T-H. Yoon, H. Kim, K-C.Shin, H.S. Kim, Formation of liquid crystal multi-domains with different threshold voltages by varying the surface anchoring energy, J. Appl. Phys. 112 (2012) 054107. https://doi.org/10.1063/1.4747909

[34] J. Chen, J.-C. Ho, A flexible universal plane for displays, Inf. Disp. 27 (2011) 6-9.

[35] M. Mativenga, M.H. Choi, J.W. Choi, J. Jang, Transparent flexible circuits based on amorphous-indium-gallium-zincoxide thin-film transistors, IEEE Electr. Device L. 32 (2011) 170-172. https://doi.org/10.1109/LED.2010.2093504

[36] L. Dumitru, K. Manoli, M. Magliulo, L. Torsi, Comparison between different architectures of an electrolyte-gated organic thin-film transistor fabricated on flexible Kapton substrates, Proceedings of the 2013 5th IEEE International Workshop on Advances in Sensors and Interfaces, IWASI 2013 ,6576074. https://doi.org/10.1109/IWASI.2013.6576074

[37] D.P. Hanley, E.J. Tucholski, Acoustic signatures of partial electric discharges in different thicknesses of Kapton, Proceedings of Meetings on Acoustics, 9, 065004,2010. https://doi.org/10.1121/1.3486243

[38] E. Zampetti, S. Pantalei, A. Pecora, A. Valletta, L. Maiolo, A. Minotti, A. Macagnano, G. Fortunato, A. Bearzotti, Design and optimization of an ultra-thin flexible capacitive humidity sensor, Sens. Actuat. B Chem. 143 (2009) 302-307.

[39] J.N. Tiwari, J.S. Meena, C.-S. Wu, N. Tiwari, M.C. Chu, F.-C. Chang, F.-H. Ko, Thin-film composite materials as a dielectric layer for flexible metal-insulator-metal capacitors, Chem. Sus. Chem. 3 (2010) 1051-1056. https://doi.org/10.1002/cssc.201000118

[40] R.N. Das, F.D. Egitto, B. Wilson, M.D. Poliks, V.R. Markovich, Polymer nanocomposites, printable and flexible technology for electronic packaging, International Symposium on Microelectronics, IMAPS 2009, 995-1000

[41] M.M. Ahmad, R.R.A Syms, I.R. Young, B. Mathew, W. Casperz, S.D. Taylor-Robinson, C.A. Wadsworth and W.M.W Gedroyc, Catheter-based flexible microcoil RF detectors for internal magnetic resonance imaging, J. Micromech. Microeng, 19 (2009) 074011. https://doi.org/10.1088/0960-1317/19/7/074011

[42] J.K. Choi, K.Y. Paek, T.-H. Yoon, Adhesive and dielectric properties of novel polyimides with bis(3,3′-aminophenyl)-2,3,5,6-tetrafluoro-4-trifluoromethyl phenyl phosphine oxide (mDA7FPPO), Eur. Polym. J. 45 (2009) 1652-1658. https://doi.org/10.1016/j.eurpolymj.2009.03.009

[43] A. Kavetskiy, G. Yakubova, Q. Lin, D. Chan, S.M. Yousaf, K. Bower, J.D. Robertson, A. Garnov, D. Meier, Promethium-147 capacitor, Appl. Radiat. Isot. 67 (2009) 1057-1062. https://doi.org/10.1016/j.apradiso.2009.02.084

[44] M. Balde, F. Jacquemoud-Collet, B. Charlot, P. Combette, B. Sorli, Microelectronic technology on paper substrate, DTIP 2012 - Symposium on Design, Test, Integration and Packaging of MEMS/MOEMS, 6235335, pp. 140-143

[45] B. Shao, Q. Chen, R. Liu, L.-R. Zheng, Design of fully printable and configurable chipless RFID tag on flexible substrate, Microw. Opt. Technol. Lett. 54 (2012) 226-230. https://doi.org/10.1002/mop.26499

[46] J.S. Meena, M.-C. Chu, R. Singh, H-P D. Shieh, P.-T.Liu, F.-H. Ko, Controlled deposition of new organic ultrathin film as a gate dielectric layer for advanced flexible capacitor devices, J. Mater. Sci. Mater. Electron. 24 (2013) 1807-1812. https://doi.org/10.1007/s10854-012-1016-y

[47] A.M. A.-Workman, A. Jeans, S. Braymen, R. E. Elder, R. A. Garcia, A.de la F. Vornbrock, J. Hauschildt, E. Holland, W. Jackson, M. Jam, F. Jeffrey, K. Junge, H-J. Kim, O. Kwon, D. Larson, J. Maltabes, P. Mei, C. Perlov, M. Smith, D. Stieler, C.P. Taussig, S. Trovinger, Zhao, Planarization coating for polyimide

substrates used in roll-to-roll fabrication of active matrix backplanes for flexible displays, HP Laboratories Technical Report 23 (2012).

[48]   R. Ma, AF. Baldwin, C. Wang, I. Offenbach, M. Cakmak, R. Ramprasad, G.A Sotzing, Rationally designed polyimides for high-energy density capacitor applications, ACS Appl. Mater. Interfaces 6 (2014) 10445-10451. https://doi.org/10.1021/am502002v

[49]   M. Watanabe, S. Imaizumi, T. Yasuda, H. Kokubo, Ion gels for ionic polymer actuators, (Chapter) in: Asaka, Kinji (et al.) (Eds.), Soft Actuators: Materials, Modeling, Applications, and Future Perspectives, Springer, Tokyo, 2014, pp.141-156.

[50]   W-H. Zhou, L.-C. Wang, L.-B. Wang, numerical study of the structural parameter effects on the dynamic characteristics of a polyimide film micro-capacitive humidity sensor, IEEE Sens. J. 16 (2016) 5979-5986. https://doi.org/10.1109/JSEN.2016.2579644

[51]   Y.H. Chen, H.C. Chang, C.C. Lai, I.N. Chang, Fabrication and application of a wireless inductance-capacitance coupling microsensor with electroplated high permeability material NiFe, J. Phys. Conf. Ser. 266 (2011) 012066. https://doi.org/10.1088/1742-6596/266/1/012066

[52]   N. Venkat, T.D. Dang, Z. Bai, V.K. McNier, J. N. DeCerbo, B.-H. Tsao, J.T. Stricker, High temperature polymer film dielectrics for aerospace power conditioning capacitor applications, Mater. Sci. Eng. B Solid State Mater. Adv. Technol. 68 (2010) 16-21.

[53]   Y-H. Kim, M. Kim, S. Oh, H. Jung, Y. Kim, T-S Yoon, Y.-S. Kim, H.H. Lee, Organic memory device with polyaniline nanoparticles embedded as charging elements, Appl. Phys. Lett. 100 (2012) 163301. https://doi.org/10.1063/1.4704571

[54]   R. Yang, R. Wei, K. Li, Tong, K. Jia, X. Liu, Crosslinked polyarylene ether nitrile film as flexible dielectric materials with ultrahigh thermal stability, Sci. Rep. 6 (2016) 36434. https://doi.org/10.1038/srep36434

[55]   R. Ulrich, L. Schaper, Materials options for dielectrics in integrated capacitors, Proceedings of the International Symposium and Exhibition on Advanced Packaging Materials Processes, Properties and Interfaces (2000) 38-43. https://doi.org/10.1109/ISAPM.2000.869240

[56]   L.Z. Liu, X.H. Gao, L. Weng, H. Shi, C. Wang, Preparation of high-dielectric-constant $Ag@Al_2O_3$/polyimide composite films for embedded capacitor

applications, Proceedings of the IEEE International Conference on Properties and Applications of Dielectric Materials, (2012) 6318901.

[57] A. Alias, Z. Ahmad, A.B. Ismail, Preparation of polyimide/Al2O3 composite films as improved solid dielectrics, Mater. Sci. Eng. B 176 (2011) 799-804. https://doi.org/10.1016/j.mseb.2011.04.001

[58] Y.Y. Yan, M. Xiong, B. Liu, Y.T. Ding, Z.M. Chen, Low capacitance and highly reliable blind through-silicon-vias (TSVs) with vacuum-assisted spin coating of polyimide dielectric liners, Sci. China Technol. Sci. 59 (2016) 1581-1590. https://doi.org/10.1007/s11431-016-0266-6

[59] C.A. Grabowski, S.P. Fillery, N.M. Westing, C. Chi, J.S. Meth, M.F. Durstock, R.A. Vaia, Dielectric breakdown in silica-amorphous polymer nanocomposite films: The role of the polymer matrix, ACS Appl. Mater. Interf. 5 (2013) 5486-5492. https://doi.org/10.1021/am4005623

[60] S-H. Huang, T-M. Don, W-C. Lai, C-C. Chen, L-P. Cheng, Porous structure and thermal stability of photosensitive silica/polyimide composites prepared by sol-gel process, J. Appl. Polym. Sci. 114 (2009) 2019-2029. https://doi.org/10.1002/app.30790

[61] W. Xu, Y. Ding, S. Jiang, W. Ye, X. Liao, H. Hou, High permittivity nanocomposites fabricated from electrospun polyimide/$BaTiO_3$ hybrid nanofibers, Polym. Compos. 37 (2016) 794-801. https://doi.org/10.1002/pc.23236

[62] L. Liu, Y. Zhang, W. Tong, L. Ding, P.K. Chu, P. Li, Polyimide composites composed of covalently bonded $BaTiO_3$@GO hybrids with high dielectric constant and low dielectric loss, RSC Adv. 6 (2016) 86817-86823. https://doi.org/10.1002/pc.23236

[63] Y. Imanaka, H. Amada, F. Kumasaka, Dielectric and insulating properties of embedded capacitor for flexible electronics prepared by aerosol-type nanoparticle deposition, Jpn. J. Appl. Phys. 52 (2013) 05DA02. https://doi.org/10.7567/JJAP.52.05DA02

[64] Y. Imanaka, H. Amada, F. Kumasaka, Microelectronics packaging application using post-LTCC technology, 2011 IMAPS/ACerS 7th International Conference and Exhibition on Ceramic Interconnect and Ceramic Microsystems Technologies, CICMT 2011.

[65] D. Xie, X. Wu, T.-L. Ren, L.-T. Liu, Z.-M. Dang, Fabrication and characterization of embedded capacitors based on novel nanocomposite as dielectric materials,

Nami. Jishu yu Jingmi Gongcheng/Nanotechnology and Precision engineering, 8 (2010) 460-464.

[66] B.J. Kang, C.K. Lee, J.H. Oh, All-inkjet-printed electrical components and circuit fabrication on a plastic substrate. Microelectron Eng, 97 (2012) 251-254. https://doi.org/10.1016/j.mee.2012.03.032

[67] Y.-H. Wu, J.-W. Zha, Z.-Q. Yao, F. Sun, R.K.Y. Li, Z.-M. Dang, Thermally stable polyimide nanocomposite films from electrospun $BaTiO_3$ fibers for high-density energy storage capacitors, RSC Adv. 5 (2015) 44749-4475. https://doi.org/10.1039/C5RA06684K

[68] X. Peng, W. Xu, L. Chen, Y. Ding, S. Chen, X. Wang, H. Hou, , Polyimide complexes with high dielectric performance: Toward polymer film capacitor applications, J. Mater. Chem. C 4 (2016) 6452-6456. https://doi.org/10.1039/C6TC01304J

[69] R.K. Gupta, D.Y. Kusuma, P.S. Lee, M.P. Srinivasan, Copper nanoparticles embedded in a polyimide film for non-volatile memory applications, Mater. Lett. 68 (2012) 287-289. https://doi.org/10.1016/j.matlet.2011.10.099

[70] S.S. Wong, A.L.S. Loke, J. T. Wetzel, P. H. Townsend, R.N. Vrtis, M. P. Zussman, Electrical reliability of Cu and low-κ dielectric integration, Materials Research Society Symposium - Proceedings, 511 (1998) 317-327. https://doi.org/10.1557/PROC-511-317

[71] A.L.S. Loke, S.S. Wong, N.A. Talwalkar, J.T. Wetzel, P.H. Townsend, T. Tanabe, R.N. Vrtis, M.P. Zussman, D. Kumar, Evaluation of copper penetration in low-κ polymer dielectrics by bias-temperature stress, Materials Research Society Symposium - Proceedings, 565 (1999) 173-187. https://doi.org/10.1557/PROC-565-173

[72] M. Agarwal, M.D. Balachandran, S. Shrestha, K. Varahramyan, SnO2 nanoparticle-based passive capacitive sensor for ethylene detection, J. Nanomater. (2012) 145406.

[73] C. Wu, F. Li, T. Guo, T.W. Kim, Carrier transport in volatile memory device with SnO_2 quantum dots embedded in a polyimide layer, Jpn. J. Appl. Phys. 50 (2011) 095003.

[74] N.B. Spurr, C.A. Edmondson, M.C. Wintersgill, J.J. Fontanella, T. Adams, Effect of nanoparticles on the dielectric properties of polyimide, Smart Mater. Struct. 20 (2011) 094001. https://doi.org/10.1088/0964-1726/20/9/094001

[75] N. Bestaoui-Spurr, T. Adams, C. Rhodes, C.A. Edmondson, J.J. Fontanella, M.C. Wintersgill, Polymer nanocomposites for high energy storage capacitors, ASME 2010 Conference on Smart Materials, Adaptive Structures and Intelligent Systems, SMASIS 2010,2, pp. 257-263

[76] C. Zou, D. Kushner, S. Zhang, Wide temperature polyimide/ZrO2 nanodielectric capacitor film with excellent electrical performance, Appl. Phys. Lett. 98 (2011) 082905. https://doi.org/10.1063/1.3559623

[77] M-C. Chu, J.S. Meena, C-C. Cheng, H-C. You, F-C. Chang, F.-H. Ko, Plasma-enhanced flexible metal-insulator-metal capacitor using high-k ZrO_2 film as gate dielectric with improved reliability, Microelectron Reliab. 50 (2010) 1098-1102. https://doi.org/10.1016/j.microrel.2010.05.004

[78] L. Weng, Q.-S. Xia, L.-W. Yan, L.-Z. Liu, Z. Sun, In situ preparation of polyimide/titanium carbide composites with enhanced dielectric constant, Polym. Compos. 37 (2016)125-130. https://doi.org/10.1002/pc.23162

[79] W. Ling, Q. Xia, L. Yan, C. Wang, M. Cao, L. Liu, Preparation, morphology and dielectric properties of nano-TiC/polyimide composite films, Polym. Polym. Compos. 22 (2014) 123-128.

[80] Q-S. Xia, W. Ling, L.-W. Yan, L-Z. Liu, M.-C. Cao, J.-W. Liu, Preparation and dielectric properties of nano-TiC/polyimide composite films as embedded-capacitor application, 8th International Forum on Strategic Technology 2013, IFOST 2013 – Proceedings, 1, 6616976, pp. 230-232, 2013.

[81] C-J. Chang, M.-H. Tsai, G.-S. Chen, M.-S. Wu, T.-W. Hung, Preparation and properties of porous polyimide films with TiO_2/polymer double shell hollow spheres, Thin Solid Films 517 (2009) 4966-4969. https://doi.org/10.1016/j.tsf.2009.03.201

[82] G. He, T. Zhong, L. Long, J. Wen, H. Li, Preparation and properties of the polyimide thin films reinforced by acylchloride-functionalized multiple-walled carbon nanotubes, J. Compos. Mater. 47 (2013) 3041-3051. https://doi.org/10.1177/0021998312461822

[83] B. Zhang, W. Yin, Y. Lu, L. Wan, The preparation and properties of BaTiO3-carbon nanotube/polyimide three-phase composites by in-situ polymerization for flexible package circuit, ICEPT-HDP 2012 Proceedings - 2012 13th International Conference on Electronic Packaging Technology and High Density Packaging, 6474616 (2012) 272-274.

[84] G. He, J. Zhou, K. Tan, H. Li, Preparation, morphology and properties of acyl chloride grafted multiwall carbon nanotubes/fluorinated polyimide composites, Compos. Sci. Technol. 71 (2011) 1914-1920. https://doi.org/10.1016/j.compscitech.2011.09.006

[85] S. Imaizumi, Y. Ohtsuki, T. Yasuda, H. Kokubo, M. Watanabe, Printable polymer actuators from ionic liquid, soluble polyimide, and ubiquitous carbon materials, ACS Appl. Mater. Interface, 5 (2013) 6307-6315. https://doi.org/10.1021/am401351q

[86] S.-Q. Wang, Z.-H. Yang, H.-B. Li, X.-Y. Li, Q.-H. Guo, H2O2-activated polyimide-based carbon nanofiber non-woven fabrics and their capacitances, Nami Jishu yu Jingmi Gongcheng/Nanotechnology and Precision Engineering, 7 (2009) 195-200.

[87] T. Le, Y. Yang, Z. Huang, F. Kang, Preparation of microporous carbon nanofibers from polyimide by using polyvinyl pyrrolidone as template and their capacitive performance, J. Power Sources 78 (2015) 683-692. https://doi.org/10.1016/j.jpowsour.2014.12.055

[88] Z. Mo, G. Wu, D. Bao, Room-temperature preparation and dielectric properties of amorphous $Bi_{3.95}Er_{0.05}Ti_3O_{12}$ thin films on flexible polyimide substrates via pulsed laser deposition method, Appl. Surf. Sci. 258 (2012) 5323-5327. https://doi.org/10.1016/j.apsusc.2012.01.140

[89] Y. Cao, X. Zeng, Laser micropen integrated direct writing for fabrication of thick film gap-tuning capacitor, Microelectron. Eng. 114 (2014) 7-11.

[90] D.H. Wang, S.P. Fillery, M.F. Durstock, L.M. Dai, R.A. Vaia, L.S. Tan. Nanodiamond/polyimide high temperature dielectric films for energy storage applications, Adv. Mat. Res. (2013) 785-786. https://doi.org/10.4028/www.scientific.net/AMR.641-642.785

[91] N.V. Cvetkovic, K. Sidler, V. Savu, J. Brugger, D. Tsamados, A.M. Ionescu, Organic half-wave rectifier fabricated by stencil lithography on flexible substrate, Microelectron. Eng. 100 (2012) 47-50. https://doi.org/10.1016/j.mee.2012.07.110

[92] S. Diaham, M.-L. Locatelli, Space-charge-limited currents in polyimide films, Appl. Phys. Lett. 101 (2012) 242905-4. https://doi.org/10.1063/1.4771602

[93] J.F. Villasenor, Breaking paradigms in biochemical sensing, Electronic Products, 54 (2012).

[94] J.S. Meena, M.-C. Chu, F.-H. Ko, Flexible MIM capacitors using zirconium-silicate and hafnium-silicate as gate-dielectric films, INEC 2010- 2010 3rd International Nanoelectronics Conference, Proceedings 5425076, pp. 992-993.

[95] J.S. Meena, M.-C. Chu, C.-S. Wu, S. Ravipati, F.-H. Ko, Environmentally stable flexible metal-insulator-metal capacitors using zirconium-silicate and hafnium-silicate thin film composite materials as gate dielectrics, J. Nanosci. Nanotechnol. 11 (2011) 6858-6867. https://doi.org/10.1166/jnn.2011.4247

[96] H. Yamada, H. Okabe, K. Yamashita, Chemical wet etching process technology for fabricating SrTiO3 thin film capacitors on core resin films, JIEP, 12 (2009) 511-518. https://doi.org/10.5104/jiep.12.511

[97] H. Manuspiya, H. Ishida, Polybenzoxazine-based composites for increased dielectric constant, (Chapter 36) in: Ishida, H.; Agag, T (Eds.), H. Handbook of Benzoxazine Resins, Elsevier, Amsterdam, 2011, pp. 621-639.

[98] H. Tang, J. Zhong, J. Yang, Z. Ma, X. Liu, Flexible polyarylene ether nitrile/BaTiO$_3$ nanocomposites with high energy density for film capacitor applications, J. Electron. Mater. 40 (2011)141-148. https://doi.org/10.1007/s11664-010-1417-8

[99] X. Huang, K. Wang, K. Jia, X. Liu, Polymer-based composites with improved energy density and dielectric constants by monoaxial hot-stretching for organic film capacitor applications, RSC Adv. 5 (2016) 51975-51982. https://doi.org/10.1039/C5RA05029D

[100] X. Huang, Z. Pu, L. Tong, R. Zhao, X. Liu, Novel PEN/BaTiO3/MWCNT multicomponent nanocomposite film with high thermal stability for capacitor applications, J. Electron. Mater. 42 (2013) 726-733. https://doi.org/10.1007/s11664-012-2391-0

[101] A.I. Mardare, M. Kaltenbrunner, N.S. Sariciftci, S. Bauer, A.W. Hassel, Ultra-thin anodic alumina capacitor films for plastic electronics, Physica Status Solidi 209 (2012) 813-818. https://doi.org/10.1002/pssa.201100785

[102] E. Zampetti, L. Maiolo, A. Pecora, F. Maita, S. Pantalei, A. Minotti, A. Valletta, M. Cuscunà, A. Macagnano, G. Fortunato, A. Bearzotti, Flexible sensorial system based on capacitive chemical sensors integrated with readout circuits fully fabricated on ultra-thin substrate, Sens. Actuat. B Chem. 155 (2011) 768-774.

[103] Y.-C. Lin, X.-X. Zhong, H.-X. Huang, H-Q. Wang, Q-P Feng, Q-Yu Li, Q.-P. Feng, Q.-Y. Li, Preparation and application of polyaniline doped with different sulfonic acids for supercapacitor, Wuli Huaxue Xuebao/Acta Physico - Chimica Sinica 32 (2016) 474-480.

[104] C.K. Subramaniam, C.S. Ramya, K. Ramya, Performance of EDLCs using Nafion and Nafion composites as electrolyte, J. Appl. Electrochem. 41 (2011) 197-206. https://doi.org/10.1007/s10800-010-0224-5

[105] F. Maita, L. Maiolo, A. Minotti, A. Pecora, D. Ricci, G. Metto, G. Scundurra, G. Giusi, C.Ciofi, G. Fortunato, Ultraflexible Tactile piezoelectric sensor based on low-temperature polycrystalline silicon thin-film transistor technology, IEEE Sens. J. 15 (2015) 3819-3826. https://doi.org/10.1109/JSEN.2015.2399531

[106] S. Hu, E. Euvrard, R. Wang, Study on partial discharge self-attenuation characteristics of a new dry type composite insulation for HV current transformers and bushings, Transmission and Distribution Exposition Conference: 2008 IEEE PES Powering Toward the Future, PIMS 2008, 4517255

[107] W. Yin, P. Irwin, D. Schweickart, Dielectric breakdown of polymeric insulations aged at high temperatures,Proceedings of the 2008 IEEE International Power Modulators and High Voltage Conference, PMHVC, 4743713, pp. 537-542. https://doi.org/10.1109/IPMC.2008.4743713

[108] A. Pecora, L. Maiolo, F. Maita, A. Minotti, Flexible PVDF-TrFE pyroelectric sensor driven by polysilicon thin film transistor fabricated on ultra-thin polyimide substrate, Sens. Actuat. A Phys. 185 (2012) 39-43.

[109] C.R.S. Rodrigues, C.A.S. Alves, J. Puga, A.M. Pereira, J.O. Ventura, Triboelectric driven turbine to generate electricity from the motion of water, Nano Energy 30 (2016) 379-386. https://doi.org/10.1016/j.nanoen.2016.09.038

[110] A. Jabbarnia, W.S. Khan, A. Ghazinezami, R. Asmatulu, Investigating the thermal, mechanical, and electrochemical properties of PVdF/PVP nano fibrous membranes for supercapacitor applications, J. Appl. Polym. Sci. 133 (2016) 43707. https://doi.org/10.1002/app.43707

[111] S. Guha, G. Knotts, N.B. Ukah, Enhanced performance of all organic field-effect transistors and capacitors through choice of solvent, Technical Proceedings of the 2014 NSTI Nanotechnology Conference and Expo, NSTI-Nanotech 2014 , 3, pp. 45-48

[112] S.N. Syahidah, S.R. Majid, Ionic liquid-based polymer gel electrolytes for symmetrical solid-state electrical double layer capacitor operated at different operating voltages, Electrochim. Acta 175 (2015) 184-19. https://doi.org/10.1016/j.electacta.2015.02.215

[113] K. Yu, Y. Niu, Y. Zhou, Y. Bai, H. Wang, Nanocomposites of surface-modified BaTiO3 nanoparticles filled ferroelectric polymer with enhanced energy density, J. Am. Ceram. 96 (2013) 2519-2524. https://doi.org/10.1111/jace.12338

[114] T. Le, Y. Yang, Z. Huang, F. Kang, Preparation of microporous carbon nanofibers from polyimide by using polyvinyl pyrrolidone as template and their capacitive performance, J. Power Sources 278 (2015) 683-692. https://doi.org/10.1016/j.jpowsour.2014.12.055

[115] S.N. Syahidah, S.R. Majid, Ionic liquid-based polymer gel electrolytes for symmetrical solid-state electrical double layer capacitor operated at different operating voltages, Electrochim. Acta 175 (2015) 184-190. https://doi.org/10.1016/j.electacta.2015.02.215

[116] A.T. Shaver, K. Yin, H. Borjigin, W. Zhang, S.R. Choudhury, E. Baer, S.J. Mecham, J.S. Riffle, J.E. McGrath, Fluorinated poly(arylene ether ketone)s for high temperature dielectrics, Polymer 83 (2016) 199-204. https://doi.org/10.1016/S0378-7753(03)00606-2

[117] P. Sivaraman, V.R. Hande, V.S. Mishra, Ch. Srinivasa Rao, A.B. Samui, All-solid supercapacitor based on polyaniline and sulfonated poly(ether ether ketone), J. Power Sources 124 (2003) 351-354. https://doi.org/10.1016/S0378-7753(03)00606-2

[118] P. Sivaraman, R.K. Kushwaha, K. Shashidhara, V.R. Hande, A.P. Thakur, A.B. Samui, M.M. Khandpekar, All solid supercapacitor based on polyaniline and crosslinked sulfonated poly[ether ether ketone], Electrochim. Acta 55 (2010) 2451-2456. https://doi.org/10.1016/j.electacta.2009.12.009

[119] R. Na, G. Huo, S. Zhang, P. Huo, Y. Du, J. Luan, K. Zhu, G. Wang, A novel poly(ethylene glycol)-grafted poly(arylene ether ketone) blend micro-porous polymer electrolyte for solid-state electric double layer capacitors formed by incorporating a chitosan-based LiClO4 gel electrolyte, J. Mater. Chem. 4 (2016) 18116-18127. https://doi.org/10.1039/C6TA07846J

[120] S. Mukaigawa, T. Aoki, Y. Shimizu, T. Kikkawa, Measurement of copper drift in methylsilses quiazane-methylsilsesquioxane dielectric films, Jpn. J. Appl. Phys.

Part 1: Regular Papers and Short Notes and Review Papers 39 (4 B) (2000) 2189-2193.

[121] Z.-C. Wu, C.-C. Wang, R.-G. Wu, Y-L. Liu, P-S. Chen, Z-M. Zhu, M-C. Chen, J-F. Chen, C.-I. Chang, L.-J. Chen, Electrical reliability issues of integrating thin Ta and TaN barriers with Cu and low-κ dielectric, J. Electrochem. Soc. 146 (1999) 4290-4297. https://doi.org/10.1149/1.1392629

[122] C. Constantinescu, L. Rapp, P. Rotaru, P. Delaporte, A.P. Alloncle, Polyvinylphenol (PVP) micro capacitors printed by laser-induced forward transfer (LIFT): Multilayered pixel design and thermal analysis investigations J. Appl. Phy 49 (2016) 155301.

[123] Y. Zhao, X. Hao, M. Li, Dielectric properties and energy-storage performance of (Na 0.5Bi0.5)TiO3 thick films, J. Alloys Compd. 601 (2014) 112-115. https://doi.org/10.1016/j.jallcom.2014.02.137

[124] X. Wu, Q. Wang, W. Zhang, Y. Wang, W. Chen, Preparation of all-solid-state supercapacitor integrated with energy level indicating functionality, Synth. Met. 220 (2016) 494-501. https://doi.org/10.1016/j.jallcom.2014.02.137

[125] P. Huo, Y. Liu, R. Na, X. Zhang, S. Zhang, G. Wang, Quaternary ammonium functionalized poly(arylene ether sulfone)/poly(vinylpyrrolidone) composite membranes for electrical double-layer capacitors with activated carbon electrodes, J. Memb. Sci. 505 (2016) 148-156. https://doi.org/10.1016/j.memsci.2016.01.025

[126] Y. Liu, E. Kim, J.I. Han, Fabrication of thermally evaporated Al thin film on cylindrical PET monofilament for wearable computing devices, Electron. Mater. Lett. 12 (2016) 186-196. https://doi.org/10.1007/s13391-015-5239-y

[127] H.S. Yang, B.J. Kang, J.H. Oh, control of evaporation behavior of an inkjet-printed dielectric layer using a mixed-solvent system, J. Electron. Mater. 45 (2016) 755–763. https://doi.org/10.1007/s11664-015-4196-4

[128] G. McKerricher, J.G. Perez, A, Shamim, Fully inkjet printed RF inductors and capacitors using polymer dielectric and silver conductive ink with through vias, IEEE Trans. Electron Dev. 62 (2015) 1002-1009. https://doi.org/10.1109/TED.2015.2396004

[129] C. Ramasamy, J.P.Del Val, M. Anderson, An electrochemical cell study on polyvinylpyrrolidine aqueous gel with glycol addition for capacitor applications, Electrochim. Acta 135 (2014) 181-186. https://doi.org/10.1016/j.electacta.2014.04.169

[130] F. Yu, M. Huang, J. Wu, Z. Qiu, L.Fan, J.Lin, Y.Lin, A redox-mediator-doped gel polymer electrolyte applied in quasi-solid-state supercapacitors, J. Appl. Polym. Sci. 131 (2014) 39784. https://doi.org/10.1002/app.39784

[131] H.S.A. Sellam, High rate performance of flexible pseudocapacitors fabricated using ionic-liquid-based proton conducting polymer electrolyte with poly(3, 4-ethylenedioxythiophene):poly(styrene sulfonate) and its hydrous ruthenium oxide composite electrodes, ACS Appl. Mater. Interf. 5 (2013) 3875-3883. https://doi.org/10.1021/am4005557

[132] K.H. Choi, A. Ali, H.C. Kim, M.T. Hyun, Fabrication of dielectric poly(4-vinylphenol) thin films by using the electro hydrodynamic atomization technique, J. Korean Phys. Soc. 62 (2013) 269-274. https://doi.org/10.3938/jkps.62.269

[133] J. Rodríguez, E. Navarrete, E.A. Dalchiele, L. Sánchez, J.R. Ramos-Barrado, F. Martín, Polyvinylpyrrolidone-LiClO4 solid polymer electrolyte and its application in transparent thin film supercapacitors, J. Power Sources 237 (2013) 270-276. https://doi.org/10.1016/j.jpowsour.2013.03.043

[134] B. Fei, RA. Wach, H. Mitomo, F. Yoshii, T. Kume, Hydrogel of biodegradable cellulose derivatives: I. Radiation-induced crosslinking of CMC. J. Appl. Polym. Sci. 78 (2000) 278–2783. https://doi.org/10.1002/1097-4628(20001010)78:2%3C278::AID-APP60%3E3.0.CO;2-9

[135] RA. Wach, H. Mitomo, F. Yoshii, T. Kume, Hydrogel of biodegradable cellulose derivatives: II. Effect of some factors on radiation-induced crosslinking of CMC. J. Appl. Polym. Sci. 81 (2001) 3030–3037. https://doi.org/10.1002/app.1753

[136] I. Stepniak, A. Ciszewski Grafting effect on the wetting and electro-chemical performance of carbon cloth electrode and polypropylene separator in electric double layer capacitor. J. Power Sources 195 (2010) 5130–7. https://doi.org/10.1016/j.jpowsour.2010.02.032

[137] D. Sinar, G.K. Knopf, Printed graphene interdigitated capacitive sensors on flexible polyimide substrates, 14th IEEE International Conference on Nanotechnology, IEEE-NANO 2014.

[138] C. Ataman, T. Kinkeldei, G. Mattana, A. V. Quintero, F. Molina-Lopez, J. Courbat, K. Cherenack, D. Briand, G. Tröster, N.F.De Rooij, A robust platform for textile integrated gas sensors, Sens. Actuat. B Chem. 177 (2013) 1053-1061. https://doi.org/10.1016/j.snb.2012.11.099

[139] S. Couderc, O. Ducloux, B.J. Kim, T. Someya, A mechanical switch device made of a polyimide-coated microfibrillated cellulose sheet, J. Micromech. Microeng. 19 (2009) 055006.

[140] G. Chitnis, T. Maleki, B. Samuels, L.B. Cantor, B. Ziaie, A minimally invasive implantable wireless pressure sensor for continuous IOP monitoring, IEEE Trans. Biomed. Eng. 60 (2013) 250-256. https://doi.org/10.1109/TBME.2012.2205248

[141] M.A. Hudspeth, T.Kaya, Collagen as a humidity sensing dielectric material, Materials Research Society Symposium Proceedings,1427 (2012) 74-79. https://doi.org/10.1557/opl.2012.1415

Chapter 7

Hydrothermal Synthesis of Supercapacitors Electrode Materials

Christelle Pau Ping Wong[1], Chin Wei Lai[1], Joon Ching Juan[1]*

[1]Nanotechnology & Catalysis Research Centre (NANOCAT), Level 3, Block A, IPS Building, University of Malaya (UM), Kuala Lumpur 50603, Malaysia

*jcjuan@um.edu.my

Abstract

Nowadays, electrochemical capacitors have been considered as one of the important power sources to overcome the environmental problems such as depletion of fossil fuels and increase in energy demand. The electrochemical capacitors are considered promising energy storage devices because they store a large amount of energy in comparison to the conventional capacitors and deliver higher power than batteries. Electrochemical capacitor electrodes play a vital role in enhancing the electrochemical performance, especially their active materials (i.e., carbonaceous, transition metal oxide and conducting polymers). In general, carbonaceous materials exhibited large surface area and high conductivity and hence are widely used as electrode materials for electrical double-layer capacitor system. In contrast, transition metal oxides and conducting polymers have been adopted for fabrication of pseudocapacitor electrodes owing to their storage mechanisms. In order to fully utilize the advantages of electrical double-layer capacitors and pseudocapacitors, hybrid capacitors have attracted great attention in the field of electronic devices and electric vehicles. Thus, this chapter presents a critical review of recent developments of the hydrothermal synthesis technique for hybrid capacitors' electrodes in terms of their specific capacitance, voltage window, current density, cycling stability, and current density.

Keywords

Electrochemical Capacitors, Carbonaceous Materials, Electrical Double-layer Capacitor, Pseudocapacitor, Hybrid Capacitors, Hydrothermal Synthesis

Contents

1.   Introduction

Over the past few years, the increase in energy demand and energy consumption problem have led to a great deal of research towards developing and use of renewable energy such as hydropower, wind and solar energy. This renewable energy can reduce the depletion rate of fossil fuels and the effect of global warming. Nonetheless, such energy sources require the development of energy storage systems to store and maximum utilization of energies. Energy storage systems include secondary batteries, fuel cell, and electrochemical capacitors [1]. Among these, supercapacitors have emerged as an ideal energy storage device providing high power density, long cycling capability, and fast charge-discharge tendency as compared to batteries [2]. Because of these unique properties, supercapacitors are widely involved in high power demanding applications, especially hybrid electric vehicles (HEV). A supercapacitor is used to store the braking energy and release them over long times with a high cycle rate. This makes supercapacitors offer several advantages such as low fuel consumption and low CO_2 emissions.

There are two types of supercapacitors according to their storage mechanism, namely electrical double-layer capacitors (EDLCs) and pseudocapacitors [3]. EDLCs have high power density but suffer from low specific capacitance owing to the utilization of carbonaceous material, which stored energy by the electrostatic accumulation of charges in Helmholtz double layer at solution interface close to the surface of carbon instead of storing charges in the dielectric layer. In contrast, pseudocapacitors store energy through a conventional Faradaic route involving fast and reversible redox reactions between the electrolyte and electroactive materials on the surface of electrodes. Faradaic reactions occur due to the functional groups that present on the electrode surface and thus pseudocapacitor stores a higher amount of capacitance than EDLCs [4]. However, poor cycling stability is the major drawback of pseudocapacitor. A series of conducting polymers and transition metal oxides show pseudocapacitance properties. The electrochemical performance of supercapacitor has great relevance to the properties of adopted electrode materials and its synthesis methods.

Numerous synthesis methods including sol-gel, electrochemical deposition, arc discharge, chemical vapor deposition, chemical precipitation and hydrothermal have been employed to regulate the physical and chemical properties of supercapacitor electrode materials. Of the various synthesis methods, the hydrothermal technique has merits of inexpensive experimental setup, mild conditions, environmental friendly (use variety of precursors that less toxic and easier to handle), and large scale [5]. This chapter is aimed to provide an overview of investigations of electrochemical performance of supercapacitor electrode materials that have been synthesized through hydrothermal technique. Various examples are cited to prove that the hydrothermal method plays a vital role in preparing supercapacitor electrode.

2.   Supercapacitor

2.1   Types of supercapacitors

Depending on the charge storage mechanism, supercapacitors are categorized as electrochemical double-layer capacitors (EDLCs) and pseudocapacitors. The comparison of EDLCs and pseudocapacitors is shown in Table 1 [6]. It is noteworthy that both types of capacitors suffer from several disadvantages which restrict their usage in some applications. In order to overcome the limitations of EDLCs and pseudocapacitors, hybrid capacitors were developed as new energy storage devices, which are composed of double layer capacitance and pseudocapacitance materials into a single electrode, so-called symmetric hybrid (composite). Other than a symmetric hybrid, another type of supercapacitor is an asymmetric hybrid capacitor. It consists of carbonaceous material in one electrode and the Faradaic reaction occurs in another electrode. Unlike symmetric/asymmetric hybrid capacitors, battery-type hybrid capacitor stores charge through intercalation/deintercalation of Li ions similar to a battery mechanism.

Table 1. Comparison of EDLCs and pseudocapacitors.

Parameters	EDLCs	Pseudocapacitors
Storage mechanism	Non-Faradaic and electrostatic accumulation of charges in Helmholtz double layer at interface of electrode/electrolyte	Faradaic reaction and charge transfer take place at the interface of electrode/electrolyte.
Electrode material	Carbons: activated carbon (AC), carbon nanotube (CNT), carbon aerogel and graphene/reduced graphene oxide (rGO), etc.	(i) Transition metal oxides (RuO_2, MnO_2, WO_3, ZnO, NiO, Co_3O_4, Fe_3O_4, etc.) (ii) Conducting polymers (PANI, polypyrrole, polyacetylene and poly(3,4-ethylene dioxythiophene), etc.)
Merits	• Long cycle life	• High specific capacitance • High energy and power density
Demerits	• Low specific capacitance • Low energy and power density	• Poor cycling stability • Limited conductivity of metal oxides

2.2 Evaluation of supercapacitor performance

A series of tests and their equations are listed in Table 2 to calculate the capacitance, energy density and power density of the supercapacitor [7].

Table 2. Performance assessment of supercapacitor using different characterization techniques.

Characterization technique	Equations	Remarks
Cyclic voltammetry	$$C = \frac{\int i\,dt}{dV \times SR}$$	• Used in 3 electrode system • Redox peaks are notable and repeating deviations
Galvanostatic charge/discharge	$$C = \frac{I \times dt}{dV}$$	• Best determines from slope of discharge curves
Energy density	$$E = \frac{1}{2}CV^2$$	• Describes the amount of energy which can be stored • Assessment of practical performance
Power density	$$P_{max} = \frac{V^2}{4ESR} \; or \; \frac{E}{t}$$	• Describes the speed at which energy can be delivered • Assessment of practical performance

*C= capacitance; dV= change in potential; SR= scan rate; I= current; dt= discharge time; E= energy density; V= voltage; P= power density; ESR= equivalent series resistance.

2.3   Challenges to supercapacitors

Supercapacitors have many advantages in comparison to batteries and fuel cells. However, they still suffer from some challenges:

i. High cost: Cost is the major factor that influences the utilization of electrode material in every application in terms of raw material and adopted synthesis method. For example, ruthenium oxide (RuO_2), the most famous pseudocapacitor electrode material is not cost effective due to it being less abundant. Moreover, carbon materials with high specific surface area are also expensive (> US$ 100 per kg)[8].

ii. Low energy density: Majority of supercapacitors exhibit low energy density typically 1-10 Wh kg^{-1}, which is lower than a battery (10-100 Wh kg^{-1}) [9]. In order to meet the energy demands, several supercapacitors are required to connect in a series which led to an increase in internal resistance.

iii. High self-discharging rate: High self-discharge rate is another demerit of a supercapacitor with a range of 20 to 40% per day[10].

2.4   Applications of supercapacitors

Due to their relatively long cycle life and fast charge times, supercapacitors are widely used in various applications, such as hybrid electric vehicles (HEV), portable electronic device (i.e., mobile phones, camera, laptop, etc.) and industrial power backup. For example, the supercapacitor is coupled with a battery in hybrid electric vehicles to produce high power over a short time period during acceleration, braking, and cold starting. Companies like NEC, Maxwell, NESS, Kold Ban, PowerSystem Co., Panasonic, Elna, SAFT, ESMA, Econd, EPCOS, Chubu Electric Power, CAP-xx, ELIT, Korchip and AVX are responsible for manufacturing supercapacitors, owing to their potential for replacing batteries.

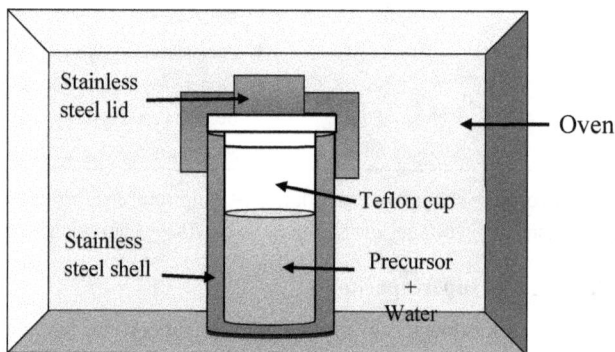

Fig. 1. Teflon-lined stainless steel autoclave.

3.   Hydrothermal Method

3.1   What is hydrothermal technique?

In general, hydrothermal is the most famous technique which is widely used for the synthesis of high purity nano-architectured materials (1D, 2D, and hierarchical structures) through the thermochemical conversion process. This reaction is carried out in a sealed system (usually Teflon-lined stainless steel autoclaves) at a temperature above the boiling point of the solvent used as shown in Figure 1. Water is used as an aqueous medium for gasification because the water molecules initiate the hydrolysis reaction and act as a source of H_2 [11]. Thus, the pressure is built up in the sealed container during the acceleration of temperature in the heating process. In addition, the hydrothermal method, with advantages of simple experimental setup, mild experimental conditions, low cost, production of high crystalline material and involvement of non-toxic chemicals, is considered to be promising for the synthesis of functional nanomaterials [12]. When organic or inorganic solvents are used as the media, the technique is known as solvothermal. As well known, hydrothermal methods have great potential in controlling the morphologies of materials as well as tuning their physical and chemical properties to meet the requirements of the application as supercapacitors. A number of parameters such as temperature, pH, reaction time and types of precursor/solvent, are the key factors that have to be considered during preparation of nanostructured materials of satisfactory morphology.

3.2   Carbonaceous materials

Up to date, carbonaceous materials have attracted extensive attention because of their unique properties such as high surface area and high electrical conductivity, which provide better electrochemical stability for supercapacitor than pseudocapacitance materials. Among the carbonaceous materials, activated carbon (AC) is the oldest and most widely used active materials for supercapacitor electrodes. It has also been used in manufacturing commercial supercapacitors for hybrid vehicles. The AC is normally synthesized through carbonization (600-900 °C) under inert atmosphere followed by physical or chemical activation of precursor (i.e., biomass, polymers, coal, petroleum and their derivatives) [13]. In physical activation, the carbon precursor is exposed to the oxidizing atmosphere (i.e., CO_2, steam, etc.) in the temperature range of 600-1200 °C while chemical activation involves the mixing of chemical (i.e., KOH, NaOH, or H_3PO_4) with carbon precursor before the carbonization treatment. The carbonization/activation treatment is then carried out in the temperature range of 400-900 °C, which is slightly lower than that required for the physical activation method. However, the requirements to employ high-temperature treatments have put a restriction on the adoption of this method for production of AC. In order to overcome these disadvantages, hydrothermal carbonization as a promising route has been utilized to synthesize carbon materials from carbon precursor without the use of toxic chemicals. Wei et al. [14] reported hydrothermally synthesized AC possesses large specific surface area up to 2387 m^2 g^{-1} and exhibited excellent specific capacitance value of 175 F g^{-1} at a current density of 20 A g^{-1}. The hydrochars were carbonized at 230-250 °C, which is significantly lower than the temperature required for conventional carbonization route. Wang et al. [14] prepared carbon nanosheets (CNS) from hemp through hydrothermal carbonization (180 °C) followed by KOH activation technique [15]. The resulted CNS with a surface area of 2287 m^2 g^{-1} demonstrated remarkable capacitance of 113 F g^{-1} at a current density of 100 A g^{-1} and temperature 20 °C. Porous graphene/activated carbon composite was successfully prepared via hydrothermal carbonization and subsequent two-step KOH activation by Zheng and his co-workers [16]. The resulted composite has a high specific surface area of 2106 m^2 g^{-1} and exhibited a specific capacitance of 210 F g^{-1} at a scan rate of 1 mV s^{-1}.

Other than AC, the hydrothermal technique has been widely employed to prepare rGO from GO. Zhou et al. [17] successfully synthesized rGO by a clean and simple hydrothermal dehydration approach. They pointed out that hydrothermal reduction not only removed the oxygen functional groups that attached on the graphene sheets but was also able to repair the sp^2 network. Xu et al. [18] reported that the hydrothermally synthesized graphene exhibited high electrical conductivity (5 mS cm^{-1}) as well as better specific capacitance of 175 F g^{-1} at a scan rate of 10 mV s^{-1}. Moreover, Manafi et al. [19] successfully synthesized multi-walled carbon nanotubes (MWCNTs) via the hydrothermal method without using catalysts.

3.3   Metal oxides/hydroxides

Among pseudocapacitance materials, transition metal oxides have been chosen as promising materials owing to their various oxidation states that enable multiple redox reactions to take place during the charge/discharge process. As a pseudocapacitive electrode material, RuO$_2$ has the advantages of high theoretical capacitance (1400-2000 F g^{-1}) [20], improved electrical conductivity and superior chemical as well as mechanical stability. These favorable features make RuO$_2$ as an attractive candidate for high electrochemical performance supercapacitor. However, its high cost and toxic nature have limited the use of RuO$_2$ for commercial purposes. As a result, numerous alternative electrode materials have emerged.

A range of active pseudocapacitive materials for use as electrode materials, namely MnO$_2$, Co$_3$O$_4$, NiO, V$_2$O$_5$, Fe$_2$O$_3$ are available. By controlling the reaction parameters (i.e., hydrothermal temperature, reaction time, concentration, etc.) in the hydrothermal route, different kinds of nano-architectures of metal oxides can be obtained as presented in Table 3. For example, Li et al. prepared MnO$_2$ with different nanostructures via the hydrothermal technique at 120 °C within 12 h in a solution containing MnSO$_4$•H$_2$O and (NH$_4$)$_2$S$_2$O$_8$ as precursors. The concentration of MnO$_2$ precursor was found to affect significantly the morphologies of MnO$_2$. At low concentration of MnO$_2$ precursor, MnO$_2$ nanorods could be obtained while the increase in the concentration of precursor favored the formation of MnO$_2$ hollow urchins. By adding polyvinylpyrrolidone (PVP) as surfactant into a mixed solution (MnSO$_4$•H$_2$O and (NH$_4$)$_2$S$_2$O$_8$), MnO$_2$ smooth balls were obtained [21]. Thus, this single step hydrothermal method is very suitable for synthesizing metal oxides of different nanostructures.

Table 3. *Hydrothermal synthesized metal oxides with different nanostructures and their electrochemical performances.*

Materials	Morphology	Specific Capacitance (F g^{-1})	Electrolyte	Voltage window (V)	Current density/ Scan rate	Ref.
RuO$_2$	Nanoparticles	210	0.5M H$_2$SO$_4$	0-1.0 (vs. Ag/AgCl)	2000 mV s^{-1}	[22]
MnO$_2$	Nanorods	270	6M KOH	0-0.5 (vs. Hg/HgO)	1 A g^{-1}	[21]
	Hollow urchins	215				
	smooth balls	242				
ε-MnO$_2$	Core-shell urchin	120	1M Na$_2$SO$_4$	0-0.8 (vs. SCE)	5 mV s^{-1}	[23]
α-MnO$_2$	Nanorods	106	1M Na$_2$SO$_4$	0-0.8 (vs. SCE)	4 A g^{-1}	[24]
	Nanowires	158				
MnO$_2$	Spherical	165	1M Na$_2$SO$_4$	0-1.0 (vs. SCE)	4 A g^{-1}	[25]
α-MnO$_2$	Urchin	151.5	1M Na$_2$SO$_4$	0-1.0 (vs. SCE)	1 A g^{-1}	[26]
α-MnO$_2$	Nanosilks	135	1M Na$_2$SO$_4$	0-1.0	0.15 A g^{-1}	[27]
Mn$_2$O$_3$	Nanocubics	191.1	0.5M Na$_2$SO$_4$	0-0.8 (vs. SCE)	0.1 A g^{-1}	[28]
Mn$_3$O$_4$	Nanooctahedrons	322	1M Na$_2$SO$_4$	-0.05-0.95(vs. Ag/AgCl)	5 mV s^{-1}	[29]
NiO	Nanoflakes	110.2	2M KOH	0-0.46 (vs. SCE)	2 A g^{-1}	[30]
NiO	Nanoflakes	400	2M KOH	0-0.55 (vs. Hg/HgO)	2 A g^{-1}	[31]
NiO	Nanowires	750	0.1 NaOH	0-0.8 (vs. SCE)	1 mV s^{-1}	[32]
NiO	Nanowall	270	1M KOH	0-0.55 (vs. SCE)	0.67 A g^{-1}	[33]
NiO	Nanosheet	567	2M KOH	-0.15-0.4(vs. Ag/AgCl)	1 A g^{-1}	[34]
NiO	Nanourchin	540.5	-	-0.1-0.5 (vs. SCE)	1 A g^{-1}	[35]
Co$_3$O$_4$	Nanostructures	1090	1M NaOH	-0.2-0.6 (vs. Hg/HgO)	10 mV s^{-1}	[36]
Co$_3$O$_4$	Nanorods	281	2M KOH	-0.25-0.55(vs. SCE)	5 mV s^{-1}	[37]
Co$_3$O$_4$	Microspheres	850	2M KOH	0-0.46 (vs. SCE)	1 A g^{-1}	[38]

217

Material	Morphology	Value	Electrolyte	Potential window	Current density	Reference
Co_3O_4	Nanoflakes	352	2M KOH	0.1-0.6 (vs. Hg/HgO)	2 A g^{-1}	[39]
Co_3O_4	Flakes	263	6M KOH	-0.4-0.55 (vs. Hg/HgO)	1 A g^{-1}	[40]
CeO_2	Nanoplates	602	3M KOH	0-0.35(vs. SCE)	0.5 A g^{-1}	[41]
V_2O_5	Nanorods	235	1M Na_2SO_4	-0.2-0.8 (vs. SCE)	1 A g^{-1}	[42]
	Nanoballs	161				
	Nanowires	177				
VO_2	Starfruit	216	0.5M K_2SO_4	-0.2-0.8 (vs. SCE)	1 A g^{-1}	[43]
MoO_2	Nanoparticles	620	1M H_2SO_4	-0.1-0.45 (vs. Ag/AgCl)	1 A g^{-1}	[44]
Fe_2O_3	Nanoshuttles	249	1M KOH	-1.0- -0.5 (vs. SCE)	0.5 A g^{-1}	[45]
Fe_2O_3	Nanoparticles	340.5	1M KOH	0-0.44 (vs. SCE)	1 A g^{-1}	[46]
Fe_3O_4	Cubic	118.2	1M Na_2SO_3	-1.0-0.1 (vs. SCE)	6 mA	[47]
WO_3	Nanofibers	797.05	2M H_2SO_4	-0.35-0.2 (vs. SCE)	0.5 A g^{-1}	[48]
WO_3	Nanopillars	421.8	0.5M Na_2SO_4	-0.5-0 (vs. SCE)	0.5 A g^{-1}	[49]

In general, MnO_2 can be constructed into different crystallographic forms, including α, β, γ, δ and λ forms, in which α-MnO_2 has been considered as the most promising candidate electrode for high performance supercapacitor owing to its unique tunnel structure of (2×2) and (1×1) that provides more pathways for ions diffusion while β-MnO_2 showed 1D channel (1×1) structure with tunnel size of 1.89 Å as illustrated in Figure 2 [50, 51]. This narrow tunnel size hindered the intercalation of cations, making a β-MnO_2 undesirable candidate for a supercapacitor electrode. However, Li et al. reported a remarkable observation that β-MnO_2 nanorods showed quite high specific capacitance of 270 F g^{-1} at a current density of 1 A g^{-1}, which could be ascribed to their nano-sized structures with diameters of 100 nm and 3-4 μm of length that shorten the ion diffusion path [21]. The results revealed that crystallographic structure was not the only factor that influenced the electrochemical performance of supercapacitors, but also their morphology and surface area. Wei and her co-workers prepared α-MnO_2 with different morphologies through a hydrothermal method using a $KMnO_4$ solution and different additives as starting materials. One-dimensional α-MnO_2 nanowires were obtained by adding urea while ammonium persulfate was used for MnO_2 nanorods. The as-prepared α-MnO_2 nanowires demonstrated much better electrochemical behaviors than MnO_2 nanorods, which attributed to the large surface area of nanowires as well as the availability of more active sites for electrochemical reaction [24].

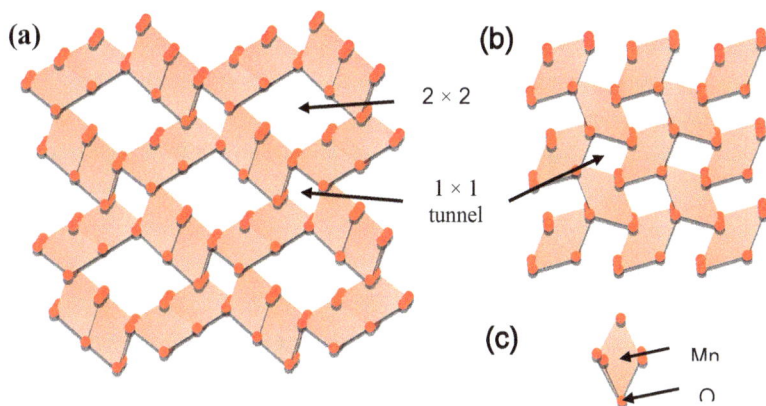

Fig. 2. The crystal structure of (a) α-MnO_2 and (b) β-MnO_2 along the [001] direction; (c) [MnO_6] octahedron [52].

It has also been shown that different preparation methods can strongly affect the electrochemical performance of NiO [31]. For example, mesoporous NiO nanoflakes synthesized via hydrothermal process, possess a higher specific surface area of 185 m^2 g^{-1} and better cycling performance than the common NiO nanoflakes (110 m^2 g^{-1}) prepared through chemical bath deposition (CBD).

Hydrothermal method not only improves the properties of materials by controlling their particle size and morphology but is also capable to directly deposit the desired material on top of a conductive substrate as well as allowes a close contact between active material and the current collector. Zhu et al. synthesized porous NiO nanowall arrays by hydrothermal method and a subsequent annealing process using nullaginite ($Ni_2(OH)_2CO_3$) as precursor [33]. The porous NiO nanowall arrays were directly built on flexible Fe-Co-Ni alloy without the use of any binder (i.e., PVDF or PTFE). As well known, some of the binders have environmental issues and a large amount of binder would reduce the electrode's conductivity, resulting in high equivalent series resistance (ESR) of supercapacitor [53]. According to the electrochemical test, the as-obtained NiO nanowall arrays exhibited specific capacitance as high as 270 F g^{-1} at a current density of 0.67 A g^{-1}. These textural features are favorable for electron transport between the active material and the current collector.

Other than NiO, hydrothermal methods also employed to synthesize Co_3O_4 by directly growing the cobalt-hydroxide-carbonate intermediate on the surface of substrates in a solution containing cobalt precursor and urea [36]. The intermediate was then annealed at high temperature (170-300 °C) to form nanostructured Co_3O_4. Experimental results showed that temperature played a crucial role in deciding the morphology of the Co_3O_4 nanostructures. At 70 °C, necklace-like Co_3O_4 nanorod arrays with 50 nm in diameter were formed while flower-like Co_3O_4 nanostructure with 22 nm in diameter was achieved at 90 °C, suggesting a bigger diameter of nanorods improved the accessibility of ions. However, the nanostructure of synthesized Co_3O_4 at 50 °C demonstrated the better electrochemical performance (1090 F g^{-1} at 10 mV s^{-1}) owing to high porous net-like microstructure with an average crystalline size of 8 nm. In such a case the diffusion path is shorter than the thicker structures that were obtained at a higher temperature.

Feng et al. developed an efficient one-step low-temperature hydrothermal method for the synthesis of Co_3O_4 microspheres by rapid oxidation of $[Co(NH_3)_6]^{2+}$ to $[Co(NH_3)_6]^{3+}$ followed by a recrystallization process [38]. Unlike ordinary preparation techniques involving high-temperature calcination step, Co_3O_4 microspheres synthesized at 100 °C hydrothermal temperature demonstrated excellent specific capacitance of 850 F g^{-1} at a current density of 1 A g^{-1}.

Furthermore, iron oxides are potential electrode materials for supercapacitors due to low cost and easy availability. The most common iron oxides are Fe_2O_3 and Fe_3O_4, which can be synthesized using the hydrothermal route. For example, Zhu et al. prepared hematite (Fe_2O_3) by employing a hydrothermal method using $FeCl_3\cdot 6H_2O$ in the presence of precipitation agent NaAc and surfactant of poly(vinylpyrrolidone) (PVP) [46]. The precursor concentration greatly affected the size and dispersion of hematite, showing that too low concentration of precursor was responsible for poor dispersion and small size of hematite. Nanoparticles of hematite with uniform size and good dispersity could be obtained by increasing the concentration of PVP to 1.00 g. The nanoparticles of 40 nm size exhibited higher specific capacitance (340.5 F g^{-1}) than 1 μm Fe_2O_3 (170.8 F g^{-1}) at current density of 1 A g^{-1}, indicating the smaller the size of hematite, the larger is the surface area for the adsorption/desorption of K^+ or H^+ from the electrolyte to take places.

3.3 Composites

Metal oxides are predicted to provide high specific capacitance owing to their storage mechanisms. Nonetheless, the experimental specific capacitances are usually lower than the theoretical value, probably due to their poor electrical conductivity and serious aggregation problem. In order to improve the capacitive behavior of metal oxides, introducing conductive additives such as activated carbon, graphene or rGO and carbon nanotubes into metal oxides to fabricate hybrid electrodes is believed to improve the electron transport. Reports based on carbon/metal oxides composites, such as CNT/Fe_3O_4, rGO/NiO, G/MnO_2, rGO/RuO_2, G/V_2O_5, GNS/Fe_2O_3, and rGO/Co_3O_4, have been published and shown that the addition of carbon could help to improve the charge transfer performance of metal oxide and its cycles life as summarized in Table 4.

Table 4. *Hydrothermal synthesized carbon/metal oxides composites and their electrochemical performances.*

Materials	Electrolyte	Specific capacitance (F g^{-1})	Voltage window (V)	Current density/ Scan rate	Cycling stability	Ref.
rGO/NiO	6M KOH	1016.6	0-0.5 (vs Ag/AgCl)	1 mV s^{-1}	Retained over 94.9 % of initial capacitance even after 5000 cycles	[54]
G/RuO$_2$	1M H$_2$SO$_4$	551	-0.2-0.8 (vs SCE)	1 A g^{-1}	Capacity retention of about 94.3 % after 2000 cycles	[55]
GNS/Fe$_2$O$_3$	6M KOH	320	-1.0 - -0.2 (vs SCE)	10 mA cm^{-2}	Capacity retention of about 97 % after 500 cycles	[56]
N-rGO/ Fe$_2$O$_3$	1M KOH	268.4	-1.0-0 (vs SCE)	2 A g^{-1}	Cyclic stability is decreased only 4.21 % after 2000 cycles	[57]
G/MnO$_2$	1M Na$_2$SO$_4$	315	0-1.0 (vs SCE)	0.2 A g^{-1}	Remained 87 % of capacitance after 2000 cycles	[58]
G/MnO$_2$	1M Na$_2$SO$_4$	218	0-1.0 (vs SCE)	5 mV s^{-1}	Retained about 94 % of capacitance after 1000 cycles	[59]
G/MnO$_2$	1M Na$_2$SO$_4$	560	0-0.5 (vs Ag/AgCl)	0.2 A g^{-1}	Remained 79 % of capacitance after 1000 cycles	[60]
G/MnO$_2$	1M Na$_2$SO$_4$	211.5	0-1.0 (vs Ag/AgCl)	2 mV s^{-1}	About 75 % capacitance retention after 1000 cycles	[61]
G/Mn$_3$O$_4$	1M Na$_2$SO$_4$	121	-0.3-0.7 (vs Ag/AgCl)	0.5 A g^{-1}	Almost no deterioration after 10000 cycles	[62]
G/Mn$_3$O$_4$	1M Na$_2$SO$_4$	171.5	-0.2-0.8 (vs SCE)	0.1 A g^{-1}	N/A	[63]
G/V$_2$O$_5$	1M Na$_2$SO$_4$	128.8	0-0.8	0.5 A g^{-1}	Retained 82 % of the initial	[64]

Material	Electrolyte	Specific capacitance	Potential window	Current density	Cycling stability	Ref.
rGO/RuO_2	1M H_2SO_4	471	-0.1-0.9 (vs Ag/AgCl)	0.5 A g^{-1}	Remained 92 % of capacity after 5000 cycles	[65]
rGO/Co_3O_4	2M KOH	263	0-0.8 (vs SCE)	0.2 A g^{-1}	Decay of capacitance of about 7.91 % after 1000 cycles	[66]
rGONS/Co_3O_4	1M KOH	445	0-0.6 (vs Hg/HgO)	0.5 A g^{-1}	N/A	[67]
G/NiO	6M KOH	429.7	0-0.35 (vs Hg/HgO)	0.2 A g^{-1}	Retained 86.1 % of the initial capacity after 2000 cycles	[68]
rGO/$Ni(OH)_2$	1M KOH	1667	-0.1-0.5 (vs SCE)	3.3 A g^{-1}	Maintained 92.5 % of capacitance after 1000 cycles	[69]
rGO/CNT/α-$Ni(OH)_2$	2M KOH	1320	0-0.45 (vs Ag/AgCl)	6 A g^{-1}	Decay of capacitance of about 7.8 % after 1000 cycles	[70]
CNT/Fe_3O_4	6M KOH	117.2	-1.0-0 (vs SCE)	10 mA cm^{-2}	Retained 91 % of capacitance after 500 cycles	[71]

G = Graphene; CNT = Carbon nanotubes; GNS = Graphene nanosheets; rGONS = Reduced graphene oxide nanosheets.

Yang et al. demonstrated that, under hydrothermal condition, GNS/Fe_2O_3 composites can be used to fabricate supercapacitor electrodes [56]. Fe_2O_3 nanorods were formed under the identical experimental conditions without adding GNS. GNS was prepared by the modified Hummers method and chemical exfoliation method using glucose as reducing agent. The capacitances of obtained composites, GNS and Fe_2O_3 were 320, 90 and 36 F g^{-1}, respectively, indicating that the GNS/Fe_2O_3 composite exhibited the highest specific capacitance. The enhanced specific capacitance of GNS/Fe_2O_3 composite may be attributed to the novel open architecture of Fe_2O_3 nanorods as well as the effective prevention of the restocking of GNS in order to increase the active sites for the electrochemical reaction.

Graphene-MnO_2 composites were produced on the surface of CVD grown 3D graphene foam using the hydrothermal method [60]. By changing the solution acidity, different structures of MnO_2 have been obtained as a (i) reticular structure in the absence of HCl; (ii) crumpled flower-like structure with 0.03M HCl; (iii) regular flower-like structure with 0.1M HCl and (iv) hollow MnO_2 nanotubes with 0.3M HCl. According to the charge/discharge curves, the measured specific capacitances were 188.4, 415, 560 and 284 F g^{-1} at a current density of 0.2 A g^{-1}, respectively. The composite with homogenous and regular flower-like MnO_2 offers the highest specific capacitance owing to its nanostructure, high conductive graphene and large surface area of 3D graphene, which facilitated ions transport into the electrode.

Lee et al. prepared G/Mn_3O_4 composite via a hydrothermal process using $KMnO_4$, graphene and ethylene glycol as starting materials. Prior hydrothermal step, graphene was formed by oxidation of graphite to graphite oxide through Hummers method followed by ultrasonication of graphite oxide to form GO. The resulted GO was then reduced to graphene via chemical exfoliation route using hydrazine hydrate as reducing agent [62]. The G/Mn_3O_4 composites exhibited higher specific capacitance than pure Mn_3O_4 nanorods, which were 121 and 25 F g^{-1} at a current density of 0.5 A g^{-1}, respectively. The increase in capacitance may be ascribed to the conductivity of graphene sheets.

In addition, the hydrothermal process is also known as a green process because of its closed system conditions and low consumption of chemicals as GO can be simultaneously reduced to rGO accompanied by the growth of metal oxides during the process. For example, Fan and his co-workers synthesized G/Mn_3O_4 composite by combining the reduction of GO and the growth of Mn_3O_4 via an *in-situ* one-pot hydrothermal process using hydrazine hydrate as multi-functional reagent [63]. Experimental results showed that G/Mn_3O_4 composite displayed a specific capacitance of

171 F g^{-1} at a current density of 0.1 A g^{-1}, which is two times higher than that of pure Mn$_3$O$_4$ nanoparticles (85 F g^{-1}).

Liu et al. [66] and Song et al. [67] employed the hydrothermal routes to synthesis rGO/Co$_3$O$_4$ and rGONS/Co$_3$O$_4$ composites using cobalt acetate (Co(Ac)$_2$) and CoCl$_2$•6H$_2$O as precursors, respectively. A specific capacitance of 445 F g^{-1} was obtained at a current density of 0.5 A g^{-1} in 1M KOH for the rGONS/Co$_3$O$_4$ composites prepared with 3 mmol of CoCl$_2$•6H$_2$O precursor. In contrast, a supercapacitor based on rGO/Co$_3$O$_4$ composite (1:2 of GO: Co(Ac)$_2$) has a good electrochemical performance with a specific capacitance of 263 F g^{-1} at a current density of 0.2 A g^{-1} than pure Co$_3$O$_4$ in 2M KOH electrolyte.

Moreover, graphene/NiO nanocomposite was prepared by a facile hydrothermal method using Ni(NO$_3$)$_2$•6H$_2$O, urea and graphite oxide as starting materials [68]. For comparison, pure NiO was synthesized using the same method without adding graphite oxide. The as-prepared graphene/NiO nanocomposite exhibited a higher specific capacitance of 429.7 F g^{-1} than pure NiO (125.3 F g^{-1}) at a current density of 0.2 A g^{-1}. The improvement in electrochemical performance of graphene/NiO nanocomposite has been attributed to (i) increased electronic conductivity of NiO which leads to high diffusive coefficient of ions in and out of the host material; (ii) the mesoporous structure of nanocomposite provide large surface area for electrolyte access and ensure sufficient redox reactions to take place; (ii) graphene serve as a matrix to maintain the NiO structure and thus resulting in excellent cycling performance.

Besides metal oxides, nickel hydroxide [Ni(OH)$_2$] has drawn tremendous attention as a promising electrode material for high-performance supercapacitors due to its low cost, high chemical/thermal stability and availability in various morphologies [72]. Nonetheless, Ni(OH)$_2$ also suffers from poor electrical conductivity (10^{-17} S cm^{-1}) like other metal oxides. Hence, integrating with high conductive carbon is one of the approaches to improve their supercapacitor applications. For example, Min et al. prepared rGO/Ni(OH)$_2$ composite through the hydrothermal route using GO and nickel foam (NF) as starting materials [69]. During the hydrothermal process, nickel metal was oxidized to form Ni(OH)$_2$ while GO was simultaneously reduced to rGO and it covers the surface of Ni(OH)$_2$ layer to generate a rGO/Ni(OH)$_2$/NF composite. The as-prepared composite electrode exhibited excellent capacitive performance with the high capacitance of 1667 F g^{-1} at a current density of 3.3 A g^{-1} in comparison with Ni(OH)$_2$/NF (409 F g^{-1}).

For further improvement, a ternary composite rGO/CNT/α-Ni(OH)$_2$ has also been synthesized by a one-step hydrothermal technique in which rGO/CNT served as a binder and conductive matrix for α-Ni(OH)$_2$ nanoparticles [70]. The composite with the optimum ratio of 20:2:2 for GO:CNT:Ni showed a specific capacitance as high as 1320 F g^{-1} at a current density of 6 A g^{-1}. Even at a high current density of 15 6 A g^{-1}, a specific capacitance of 1008 F g^{-1} can also be obtained after 1000 cycles. Results indicated that 3D rGO/CNT acted as a conductive matrix and allowed efficient charge transfer, resulting in high electrochemical capacitance.

Other than rGO, CNT has advantages of high surface area, excellent chemical/mechanical stability and superior conductivity which make them an ideal candidate for high electrochemical performance supercapacitor applications. For example, Guan prepared CNT/Fe_3O_4 nanocomposite through a simple hydrothermal method using $FeSO_4 \cdot 7H_2O$ as iron precursor [71]. The specific capacitances of CNT, Fe_3O_4 and CNT/Fe_3O_4 nanocomposite were 80.1, 36.1 and 117.2 F g^{-1}, respectively. The conductive CNTs prevent the agglomeration of Fe_3O_4 and offer more active sites for Faradaic reactions.

Conclusion

With increasing demands of energy and developing the utilization of renewable energy, supercapacitors emerged as ideal energy storage devices providing high power density, long cycling capability, and fast charge-discharge as compared to batteries. A supercapacitor is composed of electrode, electrolytes, and separator. Among these, the electrode plays a vital role in enhancing the electrochemical performance of supercapacitors, especially its active materials (i.e., carbonaceous, transition metal oxide and conducting polymers). In general, carbonaceous materials have advantages of large surface area and high conductivity and are widely used as electrode materials for electrical double-layer capacitors (EDLCs). In contrast, transition metal oxides and conducting polymers are adopted for fabrication pseudocapacitor electrodes owing to their storage mechanisms. In order to fully utilize the advantages of EDLCs and pseudocapacitors, hybrid capacitors have attracted great attention in the field of electronic devices and electric vehicles. However, supercapacitors still suffer from some demerits such as low energy density, high cost, and enhanced discharge rate. Taking these into consideration, synthesis of materials with suitable morphology and narrow particle size has been relatively important.

As mentioned above, there have been many reports on the synthesis of nanomaterials through hydrothermal technique. In comparison with other methods, the hydrothermal method has many advantages of large-scale production, avoid once of the use of toxic chemicals and generating high purity materials. The hydrothermal method can be used to synthesize nanomaterials of different structures and illustrating unique properties. Moreover, the hydrothermal method is also able to prepare materials for direct deposition on top of the conductive substrate. During the hydrothermal process, a number of parameters such as temperature, pH, reaction time and types of precursor/solvent, play the key role in controlling the morphologies of nanostructured materials. Like other techniques, the hydrothermal method also suffers from certain demerits. Being a closed system technique, it is difficult to understand the intermediate phase changes taking place in the solution. The hydrothermal method normally required quite a long time overall process development. The hydrothermal method is still far away from being fully understood and thus the hydrothermal method will continue to be exploited by many researchers to synthesize materials in coming years.

Acknowledgement

The authors thankfully acknowledge the financial support by MOSTI-Science Fund (03-01-03-SF1032), Grand Challenge Grant (GC002A-15SBS) and UMCIC Prototype (RU005F-2016) from University of Malaya.

References

[1]   Z. Yang, J. Zhang, M.C. Kintner-Meyer, X. Lu, D. Choi, J.P. Lemmon, J. Liu, Electrochemical energy storage for green grid, Chem. Rev. 111 (2011) 3577-3613. https://doi.org/10.1021/cr100290v

[2]   C. Gong, M. Huang, P. Zhou, Z. Sun, L. Fan, J. Lin, J. Wu, Mesoporous $Co_{0.85}Se$ nanosheets supported on Ni foam as a positive electrode material for asymmetric supercapacitor, Appl. Surf. Sci. 362 (2016) 469-476. https://doi.org/10.1016/j.apsusc.2015.11.194

[3]   C. Liu, Z. Yu, D. Neff, A. Zhamu, B.Z. Jang, Graphene-based supercapacitor with an ultrahigh energy density, Nano Lett. 10 (2010) 4863-4868. https://doi.org/10.1021/nl102661q

[4]   G. Yu, X. Xie, L. Pan, Z. Bao, Y. Cui, Hybrid nanostructured materials for high-performance electrochemical capacitors, Nano Energy 2 (2013) 213-234. https://doi.org/10.1016/j.nanoen.2012.10.006

[5]   Z. Liu, X. Huang, Z. Zhu, J. Dai, A simple mild hydrothermal route for the synthesis of nickel phosphide powders, Ceram. Int. 36 (2010) 1155-1158. https://doi.org/10.1016/j.ceramint.2009.12.015

[6]   M.M. Sk, C.Y. Yue, K. Ghosh, R.K. Jena, Review on advances in porous nanostructured nickel oxides and their composite electrodes for high-performance

supercapacitors, J. Power Sourc. 308 (2016) 121-140.
https://doi.org/10.1016/j.jpowsour.2016.01.056

[7]   S. Faraji, F.N. Ani, The development supercapacitor from activated carbon by
        electroless plating—A review, Renew. Sustain. Energ. Rev. 42 (2015) 823-834.
        https://doi.org/10.1016/j.jpowsour.2016.01.056

[8]   A. Burke, Ultracapacitors: why, how, and where is the technology, J. Power Sourc.
        91 (2000) 37-50. https://doi.org/10.1016/S0378-7753(00)00485-7

[9]   Y. Zhang, H. Feng, X. Wu, L. Wang, A. Zhang, T. Xia, H. Dong, X. Li, L. Zhang,
        Progress of electrochemical capacitor electrode materials: A review, Int. J.
        Hydrogen Energ. 34 (2009) 4889-4899.
        https://doi.org/10.1016/j.ijhydene.2009.04.005

[10]   H. Chen, T.N. Cong, W. Yang, C. Tan, Y. Li, Y. Ding, Progress in electrical
        energy storage system: A critical review, Progr. Nat. Sci. 19 (2009) 291-312.
        https://doi.org/10.1016/j.pnsc.2008.07.014

[11]   K. Tekin, S. Karagöz, S. Bektaş, A review of hydrothermal biomass processing,
        Renew. Sustain. Energ. Rev. 40 (2014) 673-687.
        https://doi.org/10.1016/j.rser.2014.07.216

[12]   A.A. Mirghni, M.J. Madito, T.M. Masikhwa, K.O. Oyedotun, A. Bello, N.
        Manyala, Hydrothermal synthesis of manganese phosphate/graphene foam
        composite for electrochemical supercapacitor applications, J. Colloid Interface Sci.
        494 (2017) 325-337. https://doi.org/10.1016/j.jcis.2017.01.098

[13]   L. Wei, G. Yushin, Nanostructured activated carbons from natural precursors for
        electrical double layer capacitors, Nano Energy 1 (2012) 552-565.
        https://doi.org/10.1016/j.nanoen.2012.05.002

[14]   L. Wei, M. Sevilla, A.B. Fuertes, R. Mokaya, G. Yushin, Hydrothermal
        carbonization of abundant renewable natural organic chemicals for high-
        performance supercapacitor electrodes, Adv. Energy Mater. 1 (2011) 356-361.
        https://doi.org/10.1002/aenm.201100019

[15]   H. Wang, Z. Xu, A. Kohandehghan, Z. Li, K. Cui, X. Tan, T.J. Stephenson, C.K.
        King'ondu, C.M. Holt, B.C. Olsen, Interconnected carbon nanosheets derived
        from hemp for ultrafast supercapacitors with high energy, ACS Nano 7 (2013)
        5131-5141. https://doi.org/10.1021/nn400731g

[16]   C. Zheng, X. Zhou, H. Cao, G. Wang, Z. Liu, Synthesis of porous
        graphene/activated carbon composite with high packing density and large specific

surface area for supercapacitor electrode material, J. Power Sourc. 258 (2014) 290-296. https://doi.org/10.1016/j.jpowsour.2014.01.056

[17]   Y. Zhou, Q. Bao, L.A.L. Tang, Y. Zhong, K.P. Loh, Hydrothermal dehydration for the "green" reduction of exfoliated graphene oxide to graphene and demonstration of tunable optical limiting properties, Chem. Mater. 21 (2009) 2950-2956. https://doi.org/10.1021/cm9006603

[18]   Y. Xu, K. Sheng, C. Li, G. Shi, Self-assembled graphene hydrogel via a one-step hydrothermal process, ACS Nano 4 (2010) 4324-4330. https://doi.org/10.1021/nn101187z

[19]   S. Manafi, M.B. Rahaei, Y. Elli, S. Joughehdoust, High-yield synthesis of multi-walled carbon nanotube by hydrothermal method, Can. J. Chem. Eng. 88 (2010) 283-286. https://doi.org/10.1002/cjce.20275

[20]   L. Chen, Y. Hou, J. Kang, A. Hirata, T. Fujita, M. Chen, Toward the theoretical capacitance of RuO_2 reinforced by highly conductive nanoporous gold, Adv. Energy Mater. 3 (2013) 851-856. https://doi.org/10.1002/aenm.201300024

[21]   N. Li, X. Zhu, C. Zhang, L. Lai, R. Jiang, J. Zhu, Controllable synthesis of different microstructured MnO_2 by a facile hydrothermal method for supercapacitors, J. Alloy. Comp. 692 (2017) 26-33. https://doi.org/10.1016/j.jallcom.2016.08.321

[22]   K. H. Chang, C. C. Hu, Hydrothermal synthesis of hydrous crystalline RuO_2 nanoparticles for supercapacitors, Electrochem. Solid State Lett. 7 (2004) A466-A469. https://doi.org/10.1149/1.1814593

[23]   P. Yu, X. Zhang, D. Wang, L. Wang, Y. Ma, Shape-controlled synthesis of 3D hierarchical MnO_2 nanostructures for electrochemical supercapacitors, Cryst. Growth Des. 9 (2008) 528-533. https://doi.org/10.1021/cg800834g

[24]   H. Wei, J. Wang, S. Yang, Y. Zhang, T. Li, S. Zhao, Facile hydrothermal synthesis of one-dimensional nanostructured α-MnO_2 for supercapacitors, Physica E: Low-dimensional Systems and Nanostructures 83 (2016) 41-46. https://doi.org/10.1016/j.physe.2016.04.008

[25]   J. L. Liu, L. Z. Fan, X. Qu, Low temperature hydrothermal synthesis of nano-sized manganese oxide for supercapacitors, Electrochim. Acta 66 (2012) 302-305. https://doi.org/10.1016/j.electacta.2012.01.095

[26]   S. Zhao, T. Liu, D. Shi, Y. Zhang, W. Zeng, T. Li, B. Miao, Hydrothermal synthesis of urchin-like MnO_2 nanostructures and its electrochemical character for

supercapacitor, Appl. Surf. Sci. 351 (2015) 862-868.
https://doi.org/10.1016/j.apsusc.2015.06.045

[27] Z. Li, Z. Liu, B. Li, D. Li, Q. Li, H. Wang, MnO_2 nanosilks self-assembled
      micropowders: Facile one-step hydrothermal synthesis and their application as
      supercapacitor electrodes, J. Taiwan Inst. Chem. Eng. 45 (2014) 2995-2999.
      https://doi.org/10.1016/j.jtice.2014.08.015

[28] W. Li, J. Shao, Q. Liu, X. Liu, X. Zhou, J. Hu, Facile synthesis of porous Mn_2O_3
      nanocubics for high-rate supercapacitors, Electrochim. Acta 157 (2015) 108-114.
      https://doi.org/10.1016/j.electacta.2015.01.056

[29] H. Jiang, T. Zhao, C. Yan, J. Ma, C. Li, Hydrothermal synthesis of novel Mn_3O_4
      nano-octahedrons with enhanced supercapacitors performances, Nanoscale 2
      (2010) 2195-2198. https://doi.org/10.1039/c0nr00257g

[30] Y. Z. Zheng, H. Y. Ding, M. L. Zhang, Preparation and electrochemical properties
      of nickel oxide as a supercapacitor electrode material, Mater. Res. Bull. 44 (2009)
      403-407. https://doi.org/10.1016/j.materresbull.2008.05.002

[31] X. Yan, X. Tong, J. Wang, C. Gong, M. Zhang, L. Liang, Synthesis of
      mesoporous NiO nanoflake array and its enhanced electrochemical performance
      for supercapacitor application, J. Alloy. Comp. 593 (2014) 184-189.
      https://doi.org/10.1016/j.jallcom.2014.01.036

[32] A. Paravannoor, R. Ranjusha, A. Asha, R. Vani, S. Kalluri, K. Subramanian, N.
      Sivakumar, T. Kim, S.V. Nair, A. Balakrishnan, Chemical and structural stability
      of porous thin film NiO nanowire based electrodes for supercapacitors, Chem.
      Eng. J. 220 (2013) 360-366. https://doi.org/10.1016/j.cej.2013.01.063

[33] J. Zhu, J. Jiang, J. Liu, R. Ding, H. Ding, Y. Feng, G. Wei, X. Huang, Direct
      synthesis of porous NiO nanowall arrays on conductive substrates for
      supercapacitor application, J. Solid State Chem. 184 (2011) 578-583.
      https://doi.org/10.1016/j.jssc.2011.01.019

[34] K.K. Purushothaman, I. Manohara Babu, B. Sethuraman, G. Muralidharan,
      Nanosheet-assembled NiO microstructures for high-performance supercapacitors,
      ACS Appl. Mater. Interfaces 5 (2013) 10767-10773.
      https://doi.org/10.1021/am402869p

[35] Y. Zhang, J. Wang, H. Wei, J. Hao, J. Mu, P. Cao, J. Wang, S. Zhao,
      Hydrothermal synthesis of hierarchical mesoporous NiO nanourchins and their

supercapacitor application, Mater. Lett. 162 (2016) 67-70.
https://doi.org/10.1016/j.matlet.2015.09.123

[36] H. Wang, L. Zhang, X. Tan, C.M. Holt, B. Zahiri, B.C. Olsen, D. Mitlin,
Supercapacitive properties of hydrothermally synthesized Co_3O_4 nanostructures, J.
Phys. Chem. C115 (2011) 17599-17605. https://doi.org/10.1021/jp2049684

[37] G. Wang, X. Shen, J. Horvat, B. Wang, H. Liu, D. Wexler, J. Yao, Hydrothermal
synthesis and optical, magnetic, and supercapacitance properties of nanoporous
cobalt oxide nanorods, J. Phys. Chem. C 113 (2009) 4357-4361.
https://doi.org/10.1021/jp8106149

[38] C. Feng, J. Zhang, Y. Deng, C. Zhong, L. Liu, W. Hu, One-pot fabrication of
Co_3O_4 microspheres via hydrothermal method at low temperature for high capacity
supercapacitor, Mater. Sci. Eng. B 199 (2015) 15-21.
https://doi.org/10.1016/j.mseb.2015.04.010

[39] B. Duan, Q. Cao, Hierarchically porous Co_3O_4 film prepared by hydrothermal
synthesis method based on colloidal crystal template for supercapacitor
application, Electrochim. Acta 64 (2012) 154-161.
https://doi.org/10.1016/j.electacta.2012.01.004

[40] L. Xie, K. Li, G. Sun, Z. Hu, C. Lv, J. Wang, C. Zhang, Preparation and
electrochemical performance of the layered cobalt oxide (Co_3O_4) as supercapacitor
electrode material, J. Solid State Electrochem. 17 (2013) 55-61.
https://doi.org/10.1007/s10008-012-1856-7

[41] N. Padmanathan, S. Selladurai, Shape controlled synthesis of CeO_2 nanostructures
for high performance supercapacitor electrodes, RSC Adv. 4 (2014) 6527-6534.
https://doi.org/10.1039/c3ra43339k

[42] J. Mu, J. Wang, J. Hao, P. Cao, S. Zhao, W. Zeng, B. Miao, S. Xu, Hydrothermal
synthesis and electrochemical properties of V_2O_5 nanomaterials with different
dimensions, Ceram. Int. 41 (2015) 12626-12632.
https://doi.org/10.1016/j.ceramint.2015.06.091

[43] J. Shao, X. Li, Q. Qu, H. Zheng, One-step hydrothermal synthesis of hexangular
starfruit-like vanadium oxide for high power aqueous supercapacitors, J. Power
Sourc. 219 (2012) 253-257. https://doi.org/10.1016/j.jpowsour.2012.07.045

[44] E. Zhou, C. Wang, Q. Zhao, Z. Li, M. Shao, X. Deng, X. Liu, X. Xu, Facile
synthesis of MoO_2 nanoparticles as high performance supercapacitor electrodes

and photocatalysts, Ceram. Int. 42 (2016) 2198-2203.
https://doi.org/10.1016/j.ceramint.2015.10.008

[45] X. Zheng, X. Yan, Y. Sun, Y. Yu, G. Zhang, Y. Shen, Q. Liang, Q. Liao, Y. Zhang, Temperature-dependent electrochemical capacitive performance of the α-Fe_2O_3 hollow nanoshuttles as supercapacitor electrodes, J. Colloid Interface Sci. 466 (2016) 291-296. https://doi.org/10.1016/j.jcis.2015.12.024

[46] M. Zhu, Y. Wang, D. Meng, X. Qin, G. Diao, Hydrothermal synthesis of hematite nanoparticles and their electrochemical properties, J. Phys. Chem. C 116 (2012) 16276-16285. https://doi.org/10.1021/jp304041m

[47] J. Chen, K. Huang, S. Liu, Hydrothermal preparation of octadecahedron Fe_3O_4 thin film for use in an electrochemical supercapacitor, Electrochim. Acta 55 (2009) 1-5. https://doi.org/10.1016/j.electacta.2009.04.017

[48] J. Xu, T. Ding, J. Wang, J. Zhang, S. Wang, C. Chen, Y. Fang, Z. Wu, K. Huo, J. Dai, Tungsten oxide nanofibers self-assembled mesoscopic microspheres as high-performance electrodes for supercapacitor, Electrochim. Acta 174 (2015) 728-734. https://doi.org/10.1016/j.electacta.2015.06.044

[49] M. Zhu, W. Meng, Y. Huang, Y. Huang, C. Zhi, Proton-insertion-enhanced pseudocapacitance based on the assembly structure of tungsten oxide, ACS Appl. Mater. Interfaces 6 (2014) 18901-18910. https://doi.org/10.1021/am504756u

[50] J. Zang, X. Li, In situ synthesis of ultrafine β-MnO_2/polypyrrole nanorod composites for high-performance supercapacitors, J. Mater. Chem. 21 (2011) 10965-10969. https://doi.org/10.1039/c1jm11491c

[51] D. Su, H.-J. Ahn, G. Wang, Hydrothermal synthesis of α-MnO_2 and β-MnO_2 nanorods as high capacity cathode materials for sodium ion batteries, J. Mater. Chem. A 1 (2013) 4845-4850. https://doi.org/10.1039/c3ta00031a

[52] Y. Dong, K. Li, P. Jiang, G. Wang, H. Miao, J. Zhang, C. Zhang, Simple hydrothermal preparation of α-, β-, and γ-MnO_2 and phase sensitivity in catalytic ozonation, RSC Adv. 4 (2014) 39167-39173. https://doi.org/10.1039/C4RA02654C

[53] L. Kouchachvili, N. Maffei, E. Entchev, Novel binding material for supercapacitor electrodes, J. Solid State Electrochem. 18 (2014) 2539-2547. https://doi.org/10.1007/s10008-014-2500-5

[54] P. Cao, L. Wang, Y. Xu, Y. Fu, X. Ma, Facile hydrothermal synthesis of mesoporous nickel oxide/reduced graphene oxide composites for high

performance electrochemical supercapacitor, Electrochim. Acta 157 (2015) 359-368. https://doi.org/10.1016/j.electacta.2014.12.107

[55] N. Lin, J. Tian, Z. Shan, K. Chen, W. Liao, Hydrothermal synthesis of hydrous ruthenium oxide/graphene sheets for high-performance supercapacitors, Electrochim. Acta 99 (2013) 219-224. https://doi.org/10.1016/j.electacta.2013.03.115

[56] W. Yang, Z. Gao, J. Wang, B. Wang, L. Liu, Hydrothermal synthesis of reduced graphene sheets/Fe$_2$O$_3$ nanorods composites and their enhanced electrochemical performance for supercapacitors, Solid State Sci. 20 (2013) 46-53. https://doi.org/10.1016/j.solidstatesciences.2013.03.011

[57] H. Liu, J. Zhang, D. Xu, L. Huang, S. Tan, W. Mai, Easy one-step hydrothermal synthesis of nitrogen-doped reduced graphene oxide/iron oxide hybrid as efficient supercapacitor material, J. Solid State Electrochem. 19 (2015) 135-144. https://doi.org/10.1007/s10008-014-2580-2

[58] Y. Liu, D. Yan, R. Zhuo, S. Li, Z. Wu, J. Wang, P. Ren, P. Yan, Z. Geng, Design, hydrothermal synthesis and electrochemical properties of porous birnessite-type manganese dioxide nanosheets on graphene as a hybrid material for supercapacitors, J. Power Sourc. 242 (2013) 78-85. https://doi.org/10.1016/j.jpowsour.2013.05.062

[59] S. Deng, D. Sun, C. Wu, H. Wang, J. Liu, Y. Sun, H. Yan, Synthesis and electrochemical properties of MnO$_2$ nanorods/graphene composites for supercapacitor applications, Electrochim. Acta 111 (2013) 707-712. https://doi.org/10.1016/j.electacta.2013.08.055

[60] X. Dong, X. Wang, J. Wang, H. Song, X. Li, L. Wang, M.B. Chan-Park, C.M. Li, P. Chen, Synthesis of a MnO$_2$–graphene foam hybrid with controlled MnO$_2$ particle shape and its use as a supercapacitor electrode, Carbon 50 (2012) 4865-4870. https://doi.org/10.1016/j.carbon.2012.06.014

[61] Z. Li, J. Wang, S. Liu, X. Liu, S. Yang, Synthesis of hydrothermally reduced graphene/MnO$_2$ composites and their electrochemical properties as supercapacitors, J. Power Sourc. 196 (2011) 8160-8165. https://doi.org/10.1016/j.jpowsour.2011.05.036

[62] J.W. Lee, A.S. Hall, J.-D. Kim, T.E. Mallouk, A facile and template-free hydrothermal synthesis of Mn$_3$O$_4$ nanorods on graphene sheets for supercapacitor electrodes with long cycle stability, Chem. Mater. 24 (2012) 1158-1164. https://doi.org/10.1021/cm203697w

[63]   Y. Fan, X. Zhang, Y. Liu, Q. Cai, J. Zhang, One-pot hydrothermal synthesis of
Mn$_3$O$_4$/graphene nanocomposite for supercapacitors, Mater. Lett. 95 (2013) 153-
156. https://doi.org/10.1016/j.matlet.2012.12.110

[64]   M. Lee, S.K. Balasingam, H.Y. Jeong, W.G. Hong, H. B. R. Lee, B.H. Kim, Y.
Jun, One-step hydrothermal synthesis of graphene decorated V$_2$O$_5$ nanobelts for
enhanced electrochemical energy storage, Sci. Rep. 5 (2015).
https://doi.org/10.1038/srep08151

[65]   Y. Chen, X. Zhang, D. Zhang, Y. Ma, One-pot hydrothermal synthesis of
ruthenium oxide nanodots on reduced graphene oxide sheets for supercapacitors, J.
Alloy. Comp. 511 (2012) 251-256. https://doi.org/10.1016/j.jallcom.2011.09.045

[66]   G. J. Liu, L. Q. Fan, F. D. Yu, J. H. Wu, L. Liu, Z. Y. Qiu, Q. Liu, Facile one-step
hydrothermal synthesis of reduced graphene oxide/Co$_3$O$_4$ composites for
supercapacitors, J. Mater. Sci. 48 (2013) 8463-8470.
https://doi.org/10.1007/s10853-013-7663-4

[67]   Z. Song, Y. Zhang, W. Liu, S. Zhang, G. Liu, H. Chen, J. Qiu, Hydrothermal
synthesis and electrochemical performance of Co$_3$O$_4$/reduced graphene oxide
nanosheet composites for supercapacitors, Electrochim. Acta 112 (2013) 120-126.
https://doi.org/10.1016/j.electacta.2013.08.155

[68]   Y. Jiang, D. Chen, J. Song, Z. Jiao, Q. Ma, H. Zhang, L. Cheng, B. Zhao, Y. Chu,
A facile hydrothermal synthesis of graphene porous NiO nanocomposite and its
application in electrochemical capacitors, Electrochim. Acta 91 (2013) 173-178.
https://doi.org/10.1016/j.electacta.2012.12.032

[69]   S. Min, C. Zhao, G. Chen, X. Qian, One-pot hydrothermal synthesis of reduced
graphene oxide/Ni(OH)$_2$ films on nickel foam for high performance
supercapacitors, Electrochim. Acta 115 (2014) 155-164.
https://doi.org/10.1016/j.electacta.2013.10.140

[70]   X.A. Chen, X. Chen, F. Zhang, Z. Yang, S. Huang, One-pot hydrothermal
synthesis of reduced graphene oxide/carbon nanotube/α-Ni(OH)$_2$ composites for
high performance electrochemical supercapacitor, J. Power Sourc. 243 (2013) 555-
561. https://doi.org/10.1016/j.jpowsour.2013.04.076

[71]   D. Guan, Z. Gao, W. Yang, J. Wang, Y. Yuan, B. Wang, M. Zhang, L. Liu,
Hydrothermal synthesis of carbon nanotube/cubic Fe$_3$O$_4$ nanocomposite for
enhanced performance supercapacitor electrode material, Mater. Sci. Eng. B 178
(2013) 736-743. https://doi.org/10.1016/j.mseb.2013.03.010

[72]   J. Ji, L.L. Zhang, H. Ji, Y. Li, X. Zhao, X. Bai, X. Fan, F. Zhang, R.S. Ruoff, Nanoporous Ni(OH)$_2$ thin film on 3D ultrathin-graphite foam for asymmetric supercapacitor, ACS Nano 7 (2013) 6237-6243. https://doi.org/10.1021/nn4021955

Chapter 8

Electrochemical Super Capacitors Fabricated by the Layer-by-Layer (LbL) Technique

C. Moganapriya[1], P. Sathish Kumar[2*], Samir Kumar Pal[2], P. Kanagarajan[3], R. Rajasekar[1]

[1] Department of Mechanical Engineering, Kongu Engineering College, Tamil Nadu, India

[2] Department of Mining Engineering, Indian Institute of Technology Kharagpur, West Bengal, India

[3]Department of Automobile Engineering, KSR College of Engineering, Tamil Nadu, India

*sathishiitkgp@gmail.com

Abstract

This chapter provides an overview of current research on electrochemical supercapacitor materials fabricated by the layer-by-layer assembly technique. Emphasis has been given towards the basic principle of supercapacitors and various electrode materials suitable for high-performance supercapacitor applications. The chapter is a detailed discourse regarding the layer-by-layer fabrication technique with their performance and stability. Electrode materials such as metal oxides/hydroxides and conducting polymers exhibit pseudocapacitive behavior whereas carbon-based materials show electrical double layer capacitance. Electrode materials such as, graphene, graphene oxide, carbon nanotubes, polyaniline, and carbon nanofibers, provide high surface area for the layer-by-layer deposition of metal oxides, metal hydroxides, conducting polymers that aid the effective ion diffusion and which in turn enhances the specific capacitance with excellent cyclic stability.

Keywords

Layer-by-Layer, Supercapacitor, Energy Storage, Graphene

Contents

1. Introduction

Energy storage technology is the main aspect in garnering kinetic energy. In recent years there has been a great demand for eco-friendly energy storage systems with high performance. Supercapacitors, widely identified as ultra-capacitors, are the energy storing devices that store and release energy by the principle of nanoscopic charge separation between a carbon electrode and an electrolyte at the electrochemical interface [1].

Supercapacitors show extremely high power density, better lifecycle, and minimum charge separation compared to conventional capacitors. Fascinating features of supercapacitor have created a great interest towards the application in the areas of heavy electric vehicles, consumer electronics, and industrial power management. Fuel cells and batteries store energy by chemical processes, while supercapacitors employ electrostatic charge separation between high surface area electrodes and electrolyte ions for garnering energy [2,3]. Supercapacitors can act as a channel between batteries and ordinary capacitors because of their significant specific energy and power capability. A lot of research work has been performed to increase the efficiency of energy storage capacity of supercapacitors to increase their usefulness in the market and to replace batteries. The development of nanostructured electrode materials created a boom in supercapacitor technologies. The performance of supercapacitor such as power density and energy storage capability mainly depends on the remarkably high surface area and optimal pore dimension of the electrode materials for a pertinent electrolyte solution [4,5].

1.1   Principle of supercapacitors

A supercapacitor is a device used to store energy. It comprises of two electrodes immersed in an electrolyte solution, with a voltage potential across the current collector. The charge propagation between electrodes is prevented by dielectric separator placed in between the two electrodes as shown in Fig. 1 [1].

The principle of supercapacitors can be described in two ways:

(i) Electrical double layer capacitor (EDLC): In this capacitor, the energy is harvested through the ion adsorption technique and the charge transfer process and thus this method is non-Faradic. In EDLC, the electron transfer does not occur between the electrodes and the charge storage is only electrostatic [6].

(ii) Pseudocapacitor: Pseudocapacitor stores energy through fast redox reactions between the electrolyte and electroactive species on the surface of the electrode. Electron transfer induces charge accumulation and the charge transfer in this method is Faradic in nature [6].

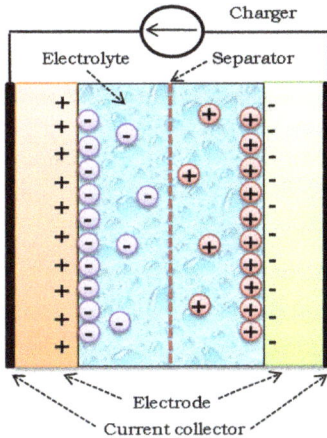

Fig. 1. Supercapacitor cell.

1.2 Layer-by-layer assembly (LbL) techniques

The techniques of thin film formation are essential for achieving high surface area and better sensitivity with low mass devices. Thin film supercapacitor electrodes have been prepared by several synthesis methods, such as sol-gel, dip coating or drop coating [7–10], electrochemical layer deposition [11–15], electrostatic spray deposition [16,17], layer-by-layer (LbL) deposition and physical vapor deposition.

Among the several available deposition methods, LbL deposition method provides ample ways to synthesize different multilayer deposition of desired functions [18–20]. The main valued advantage of LbL deposition is that it permits the synthesis of coatings and depositions of appropriate film morphology, composition, thickness, and functionality onto templates of varying shapes and sizes. Different materials ranging from conventional polyelectrolytes to nanosized objects such as nanoparticles, carbon nanotubes, and copolymer micelles have been assimilated into multilayer films. These materials are incorporated into films through complementary interactions such as electrostatic interactions, hydrogen bonding and covalent bonding [21–26]. Based on the complementary interactions, LbL assembly deposition methods are classified as described in the following paragraph.

1.2.1 Dip assisted, spin assisted and spray assisted

Each deposition method has its distinctive advantages according to the applications. Spin and spray deposited methods are beneficial over dip assisted LbL, since these enable much faster production and can be adapted to many industrial applications. Fig. 2 shows the schematic representation for preparation of a carbon nano-object (i.e., MWNT, GO) based multilayer films by three often used LbL assembly methods (dip, spin, and spray) [18].

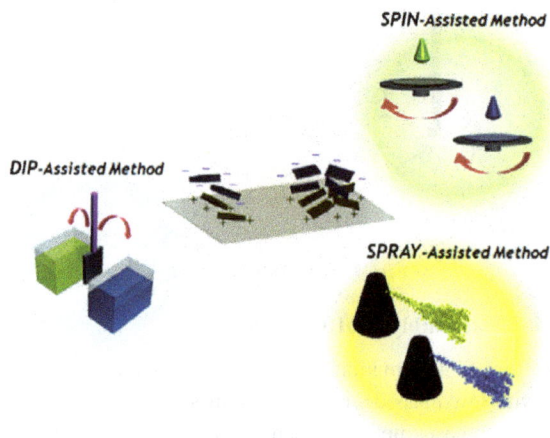

Fig. 2. Schematic representation of multilayer films prepared by layer-by-layer (LbL) assembly deposition methods.

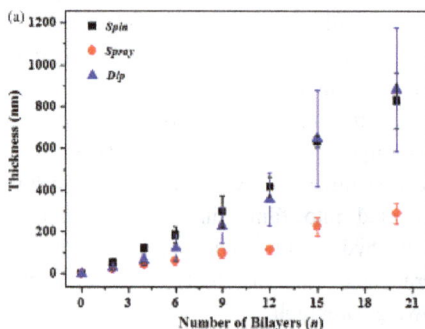

Fig. 3. Growth curve of multilayer films prepared by LbL deposition methods.

The growth process of the multilayer films prepared by each of these methods was measured by ellipsometry experiments in dried state [18] which exhibits different range, as shown in Fig. 3. Each of the methods was revealed to be linear with respect to the number of bilayers after an initial induction period. However, the film thicknesses significantly vary owing to the difference in a number of bilayers [18].

2. Appropriate materials for supercapacitors

Intended at high specific capacitance together with high power and energy density, several carbon-based materials have been considered as electrode materials for supercapacitor applications. The core reason for the advancement in the performance of carbon-based electrodes is owing to the unique combination of physical and chemical properties [1] such as high surface area, better conductivity, good corrosion resistance, excellent thermal stability, controlled pore dimensions, and cost-effectiveness.

Although substantial advancement has been achieved in the last few years in the electrochemical capacitor. There is still a great demand for supercapacitor electrode materials with high performance and outstanding stability.

2.1 Graphene and graphene oxide (GO)

Research on graphene offers great attention due to distinctive characteristics such as high mechanical strength, optical transparency, higher carrier charge mobility, good optical luminescence as well as better high electrical and thermal conductivities [27]. Electrical and optical luminescences drops upon conversion of graphene to graphene oxide. Graphene is hydrophobic as it contains only carbon and hydrogen atoms. However, GO is hydrophilic due to the presence of oxygen along with carbon and hydrogen. The hydrophilic structure of GO is due to the presence of oxidized groups such as alcohols, epoxy, carboxylic acids, and aldehydes. GO can form constant stable aqueous colloids which enable the assemblage of macroscopic assemblies by solution processes owing to its hydrophilic nature. This property is mainly considered for large-scale application of graphene oxide [27–29]. The electronic properties of graphene oxide are widely affected by oxygen content in the microstructure of graphene oxide. Production of graphene oxide with the minimum amount of oxidized groups in different positions can be controlled by oxidation time and degree of oxidation. This class of material shows sufficient electrical conductivity that is best suited for the electrochemical process as a conductive surface [30–32]. The exfoliation of graphite can produce graphene oxide with minimum oxygen content by an electrochemical process that can be employed in several applications [33]. From this perspective, several studies have been performed to explore the usefulness of graphene-based composites.

The graphene-based composites have been used extensively for applications such as optical systems, polymers, supercapacitors, thin films, and electrical conductivity films etc. [34].

2.2   Polyaniline (PANI)

Polyaniline has been one of the most substantial electrode materials for fabrication of pseudocapacitors owing to its excellent specific capacitance and good electrical conductivity [35]. Based on their markable features of graphene and PANI, hybrid materials of these combinations have been developed with synergistic effects which act as promising electrode materials for supercapacitor applications. Untill now, substantial work has been reported on graphene-PANI hybrid materials [36]. From the reported studies, the graphene-PANI composites can be synthesized by

- In-situ chemical polymerization
- Electropolymerization of monomeric molecule of aniline with graphene or GO
- Direct mixing of PANI and graphene

Since graphene has a great tendency to agglomerate and restack, it is difficult for the above techniques to recognize the nano level uniformity in the formation of graphene with PANI in a required structure and composition.

2.3   Metal hydroxides-cobalt based capacitors

Along with carbon-based materials, intensive research work has been carried out on transition metal compounds due to their high specific capacitance, usually in the forms of inexpensive metal oxides and hydroxides [37, 38]. However, widely used in electrocatalysts and secondary batteries, cobalt-based composite materials have been considered as electrode materials for pseudocapacitor. Cobalt-based compound capacitors, containing CoS_x, CoO_x, Co_3O_4, and $Co(OH)_2$, have created research interest because of their high density for storing energy and good stability [37]. Yet, these compounds have poor cycling stability and unsatisfactory high rate capability. Cobalt hydroxides have two polymorphs, which include the metastable a-type and thermodynamically stable b-form of $Co(OH)_2$ [37, 39]. The structure of cobalt hydroxide a-type is poorly and turbostratically crystalline in nature with large interlayer spacing (>0.7 nm) and because of this nature, it is considered to be much superior for electrochemical reactions. Still, the high-rate capability of $Co(OH)_2$ is greatly limited by its low electrical conductivity. The techniques reported in past studies to enhance the electrochemical performance of a-$Co(OH)_2$ can be generally given by:

- Introducing identical intercalated anions leading to a variable interlayer spacing and balancing anions

- Fabricating a micro- and nano-scaled heterogeneous structure by depositing a-Co(OH)$_2$ onto conductive surfaces
- Establishing a porous a-Co(OH)$_2$ film on a current collector and can be used as electrode directly
- Delamination or exfoliating a-Co(OH)$_2$ into ultra-thin nanosheets to improve surface area and active sites

2.4   CNT

CNTs are specifically attracted to their high surface area when bounded via a porous network structure. These specific properties can be used in a wide range of applications such as in electrochemical devices and filtration [40]. In the former case, electrodes can be formed with properties suitable for an efficient supercapacitor applications, compared to activated carbon which is widely used now-a-days. In general, CNT networks have an ability to provide improved electrical conductivity and to be very tunable in pore size distribution. This property is of extreme importance in supercapacitor application, as mesoporous structures with a pore size of 7 nm are good in providing maximum surface area for double layer capacitance [41].

2.5   Metal oxides

In general, metal oxides and carbon materials are the two decisive electrode materials for electrochemical applications. Carbon electrodes exhibit low working voltage, poor specific capacitance, and small energy density and they are not of much use [42]. Amongst different metal oxides, MnO$_2$ has been identified as a favorable supercapacitor electrode material due to its great theoretical capacitance, ample availability, eco-friendly, and reduced toxicity. Though, the major drawback of MnO$_2$ as supercapacitor electrode material is the deprived electrical conductivity which limits its applications. Providentially, it may be enhanced by the introduction of high electrical conductive carbon on the surface of the electrode [42]. The reduced graphene oxide (RGO) with excellent electrical conductivity and the good surface area along with oxygen-containing functional groups can act as a spot for affixing metal ions for consequent growth and nucleation.

2.6   Cellulose nanofibers (CNFs)

In recent years, new recyclable, flexible, and translucent cellulose nanofiber papers may meet the anticipated requirements owing to its outstanding optical transmittance (above 90% invisible region), good mechanical strength, commendable flexibility and the plywood like ordered structure [43, 44]. The presence of carboxylic acid groups, makes

CNFs paper negatively charged in aqueous solution. It is an appropriate condition for layer-by-layer assembly, which reduces the difficulties of surface pretreatment of conventional uncharged substrates and agile for supercapacitor applications.

2.7   Conducting polymers

Conducting polymers such as polyethylene terephthalate (PET), polyethylene naphthalate (PEN), polyimide (PI), polycarbonate (PC), poly(o-methoxy aniline) (POMA), and poly(3-thiophene acetic acid) (PTAA) have been remarkable electrode materials for use as supercapacitors, because of their excellent specific capacitance, minimum environmental impact, and good cycle stability during operations [45].

2.8   Hybrid materials

Hybrid materials made of carbon and metal oxides have been widely studied in energy storage techniques [46]. Wu et al. [47] conveyed that Fe_3O_4 thin electrode films showed high specific capacitances of 7 F/g in aqueous electrolytes and various valence states of metal oxide provide excellent electrical conductivity as a result of hopping of electron in the octahedral site of Fe. Due to its low cost, easiness of preparation, and non-toxicity, Fe_3O_4 has been a suitable material for electrochemical supercapacitor application [46]. Hybrid materials made of Fe_3O_4 and RGO have been reported to be excellent electrode material, in case of electrochemical capacitance. The storage capacitance of an electrochemical supercapacitor is mainly influenced by the microstructure and morphology of supercapacitor electrode materials, which are influenced by the preparation techniques.

3.   Supercapacitors fabricated by LbL assembly

3.1   Reduced GO-PANI composite

Layer-by-layer assembly technique offers smarter and direct approach for developing a graphene-PANI composite, which provides various advantages such as nanometric scale control over the different compositions as well as the characteristics and properties of graphene-PANI composite over the past preparation methods. Furthermore, the electrochemical behavior of the graphene-PANI composite can be simply varied by altering the assembling elements as well as its numbers of cycles.

Recently, the effective fabrication and development of graphene-PANI thin films using the LbL assembly technique have been reported in many studies, which exhibited improved electrochemical storage performance [35]. Three-dimensional graphene-PANI hybrid structures such as hollow structure and hydrogel with aerogel have also been reported [48]. The hollow micro- and nano-structured materials with the shell of nanosize offer advantages in supercapacitors owing to the improved surface area and reduced diffusion length for mass and charge transport.

Liu et al. [49] stated an ingenious method to develop graphene-PANI hollow spheres by enfolding graphene oxide on a hollow sphere of polyaniline through electrostatic interfaces and consequently followed by electrochemical reduction of GO to produce the final hollow graphene-PANI hybrid thin electrode films. Choi et al. [50] described a method to produce three-dimensional graphene-PANI hybrid hollow spheres via self-assembly of graphene oxide and polymethyl methacrylate (PMMA), followed by in-situ polymerization of aniline and elimination of the PMMA base core. The resulted in graphene-PANI hollow sphered structures exhibits improved electrochemical performance including high specificcapacitance, and good cycling stability. With these advantages of LbL technique, it is fairly smarter to fabricate hollow structured graphene-PANI hybrid through the LbL technique.

Jing Luo et al. discussed a novel method to prepare three-dimensional graphene-PANI hybrid hollow spheres (RGO/PANI/HS) using the LbL self-assembly technique. In particular, positively charged polyaniline and negatively charged RGO were used as building particles for alternate layer by layer assembly on the sulfonated polystyrene (PS), consecutively by the removal of the PS. The fabrication procedure for this hybrid material is shown in Fig. 4. The resultant product was characterized by Raman and FTIR spectroscopic studies as well as SEM and TEM measurements, which prove the effective preparation of hollow shaped graphene-PANI hybrid materials [35]. The electrochemical characteristics of the prepared hybrid electrode material was examined by galvanostatic charge-discharge behavior, cyclic voltammetry, and electrochemical impedance spectroscopy.

The step growth of RGO/PANI multilayer deposited on PS spheres can clearly be view from the SEM observations. Fig. 5 shows the typical microscopic images of the uncoated sulfonated PS spheres and RGO/PANI coated PS spheres with 2 and 6 assemblage cycles. From Fig. 5a, it can be viewed that the surface of pure uncoated PS spheres was clean and smooth. Some flocky features were seen on the surface of spheres after the absorption of bilayers of RGO/PANI with two assemblage cycles (Fig. 5b and 5c). By increasing the assemblage cycle to 6, a perceptibly improved surface roughness for RGO/PANI on PS can be witnessed (Fig. 5d, 5e and 5f) which showed that many graphene sheets and PANI particles assemble and cloak on the PS spheres [35].

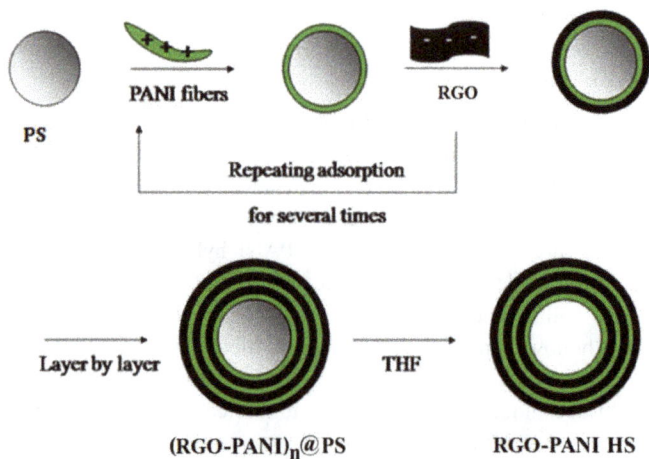

Fig. 4. Step by step procedure for fabrication of graphene-PANI hollow spheres (RGO-PANI-HS).

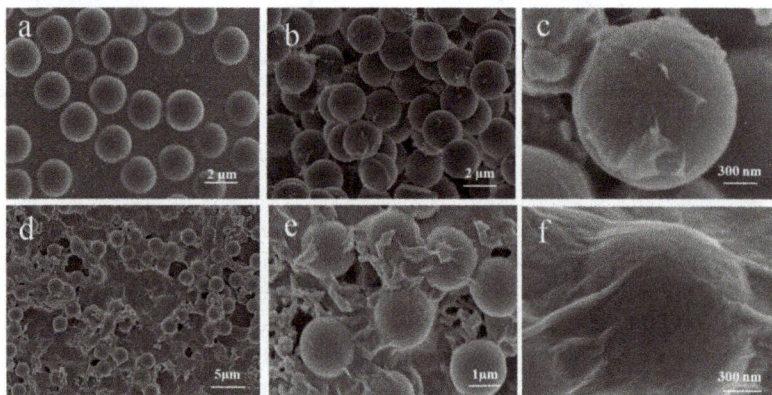

Fig. 5. SEM images of (a) sulfonated PS sphere, RGO/PANI for 2 cycles on PS (b, c) and RGO/PANI for 6 cycles on PS (d, e, f).

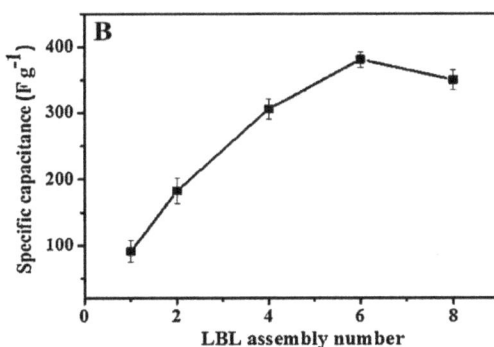

Fig. 6. Specific capacitance for RGO/PANI/HS.

The specific capacitance of prepared RGO/PANI/HS with different assembling bilayers was calculated from the discharge process and shown in Fig. 6. It can be perceived that specific capacitance increases with increasing the bilayers and reached the maximum value at 6, which is reliable with the outcomes of CV values [35].

3.2 GO-propylpyridinium silsesquioxane chloride polymer

Estevam et al. prepared GO with an inorganic polymer, 3-n-propylpyridinium silsesquioxane chloride polymer (SiPy$^+$), through electrochemical exfoliation method. The LbL films were fabricated by applying the prepared composite (SiPy (GO)) as polyelectrolyte with polyanion as nickel tetrasulfophthalo cyanine (NiTsPc). The deposition of NiTsPc^{4-} increased normally with the increase in deposited bilayers. From the inspection of Fig. 7, it is clear that the quantity of NiTsPc^{4-} deposited on the surface of the substrate is always higher to SiPy than SiPy$^+$/NiTsPc^{4-} films [27].

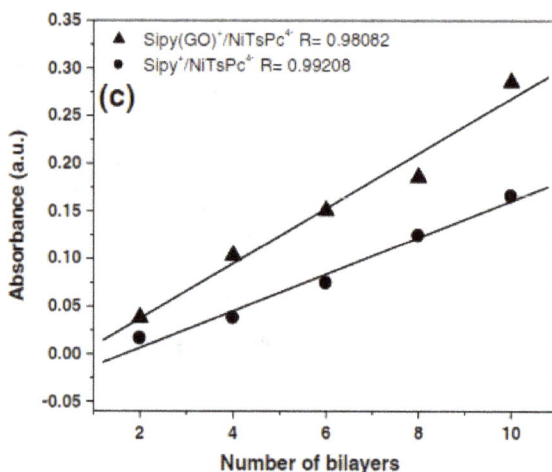

Fig. 7. The relationship between the absorbance and the number of bilayers.

3.3 Reduced GO-MnO$_2$ nanoflowers

Yanhua et al. deposited the MnO$_2$ nanoflowers with reduced graphene oxide composite on the surface of nickel foam as substrate (MnO$_2$NF/RGO on Ni foam) through the LbL technology without any polymer additive, followed by soft chemical reduction as shown in Fig. 8. The prepared layered MnO$_2$NF/RGO composite was uniformly affixed on the Ni foam substrate to form the 3D porous background. The interlayers between them have access to ions channels to enhance the electron transfer and diffusion. This unique construction of the 3D porous structure is beneficial to the improvement of the electrochemical property. The specific capacitance was equal to 246 Fg^{-1} for the current density of 0.5 Ag^{-1}. Subsequently, after 1000 cycles it can reach about 93%, exhibiting good cycle stability [42].

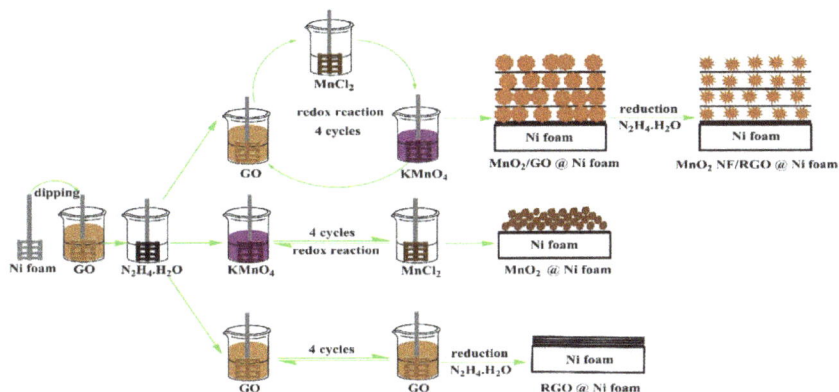

Fig. 8. Preparation of RGO on Ni foam, MnO₂ on Ni foam, MnO2 /GO on Ni foam, MnO₂ NF/RGO on Ni foam electrodes MnO₂ /GO on Ni foam.

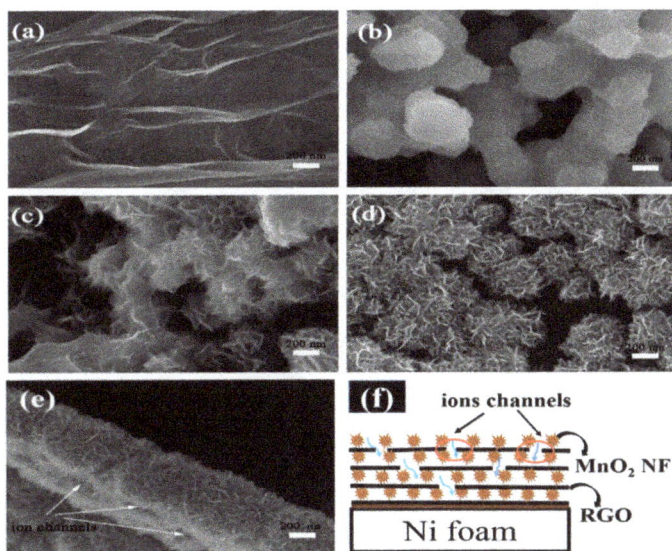

Fig. 9. SEM images of (a) RGO on Ni foam, (b) MnO₂ on Ni foam, (c) MnO₂/GO on Ni foam, (d) RGO/MnO₂ on Ni foam, (e) cross-sectional image of MnO₂NF/RGO on Ni foam (f) schematic representation of MnO₂NF/RGO on Ni foam structure.

249

The morphologies of prepared Ni foam based composites are shown in Fig. 9(a)–9(f). Fig. 9(a) is the SEM image of the RGO on Ni foam subsequently after chemical reduction which is viewed as a wrinkled thin nanosheet. The image of MnO_2 on Ni foam (Fig. 9(b)) shows that MnO_2 nanosheets amassed together to form MnO_2 nanospheres on Ni foam surface whereas the image of MnO_2/GO on Ni foam (Fig. 9(c)) affirms that MnO_2 nanoflowers with an imprecise edge are affixed on the surface of GO nanosheets. After the reduction of GO, the MnO_2 nanoflowers are uniformly attached on the RGO surface and assimilated to form 3D layered structure (Fig. 9(d)). The SEM image of the cross-sectional composite (Fig. 9(e)) exposes the presences of narrow channels within the composite, which can speed up the ion transfer inside the electrode during the discharge process and develop a high surface area on electrode [42].

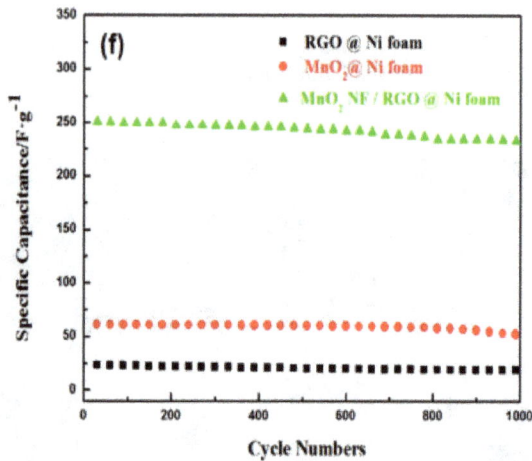

Fig. 10. Cycle stability of composite electrodes.

The cycling stability of supercapacitors is a critical parameter for supercapacitor applications. Fig. 10 demonstrates the specific capacitance retention as a function of the cycle number. The capacitance of RGO electrode reduced by 17.3% of the basic capacitance for the next 1000 cycles, owing to the accumulation of nearby RGO nanosheets. However, the specific capacitance of MnO_2NF/RGO on Ni foam electrode can maintain about 93%, indicating better cycle stability. The specific capacitance of prepared MnO_2NF/RGO on Ni foam substrate is superior to composites reported earlier [42].

3.4 Reduced GO-CNFs- PANI

Poor thermal expansion and high electrolytic absorption of cellulose nanofibers (CNFs) paper is considered to be the prospective substrate material for supercapacitors. CNFs paper avoids surface pretreatment of the LbL method. Xing Wang et al. fabricated LbL assembled multilayer thin electrode film based on CNF paper [43]. In this work, negatively charged GO sheets and poly-3,4 ethylene di-oxy thiophene : polystyrene sulfonate (PEDOT:PSS) particles with positively charged PANI are alternatively deposited onto CNFs paper for preparing multilayer thin film electrodes. Owing to different nanostructures of RGO and PEDOT:PSS, the prepared electrodes showed unique microstructures. The supercapacitor electrode material developed by CNFs/PANI/ RGO (S-PG8) unveils an admirable areal capacitance and better cyclic stability than CNFs/PANI/PEDOT (S-PP8). Recent studies reported a unique method to prepare hybrid electrode materials with CNFs as a substrate with assorted structures that are favorable for imminent supercapacitors [43].

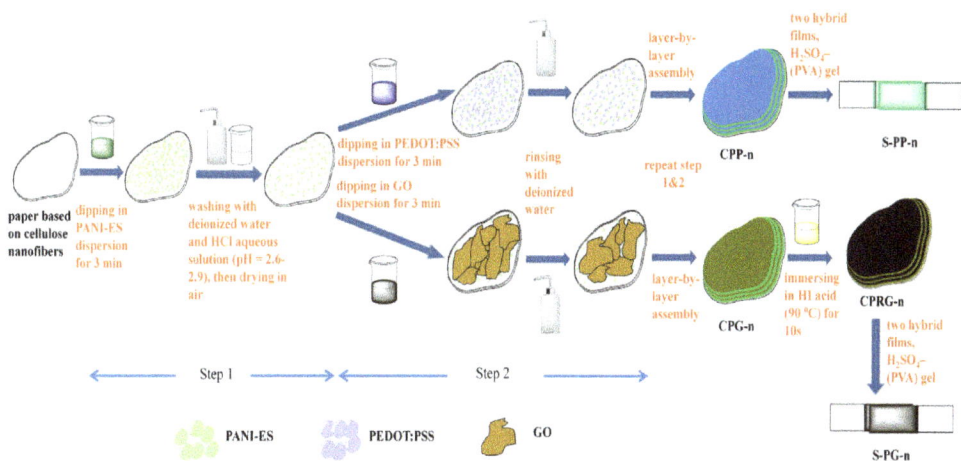

Fig. 11. LbL assembly of CPP-n, CPG-n multilayer films.

Fig. 11 shows the step-by-step procedure of developing thin film electrode material for supercapacitor applications. They are accumulated by the hybrid paper in a stacked configuration, with HE polyvinyl-PVA gel as electrolyte and as separator [43].

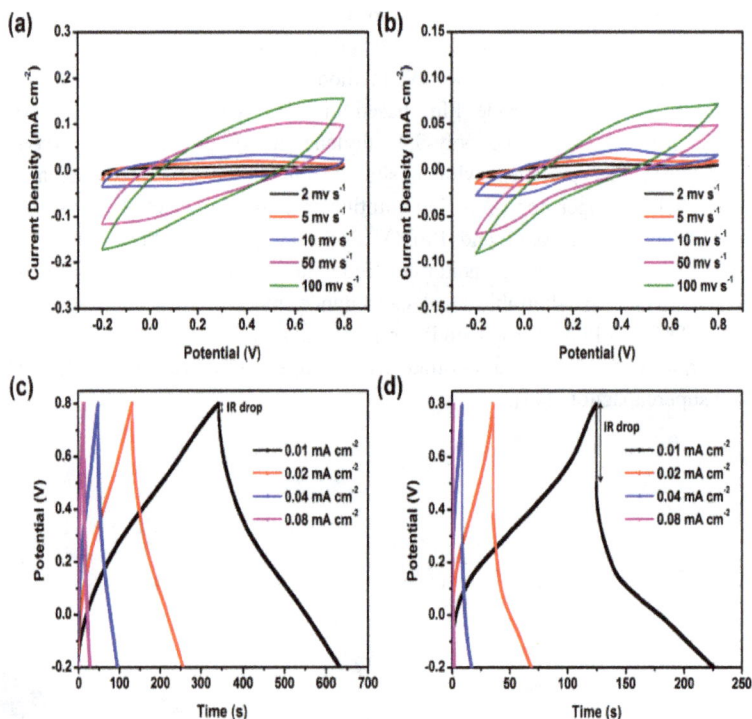

Fig. 12. Characteristic current density/voltage curves of (a) S-PG8, (b) S-PP8 device at various scan rates and the characteristic charge-discharge curves of S-PG8(c), S-PP8(d) devices at various current densities.

Fig. 12a and 12b show the current density/voltage curves of S-PG8 and S-PP8 with a voltage range of 0.2 and 0.8 V for the scan rates varying from 2 to 100 mVs^{-1}. The curves for S-PG8 unveil an analogous rectangular behavior for capacitance with a minor pseudo capacitive redox peaks, whereas the curves for S-PP8 are in some way bias. Fig. 12c and

12d show the galvanostatic charge-discharge curves of S-PG8 and S-PP8. The triangular form of the curves for S-PG8 specifies that the prepared composites have better characteristics for capacitance and hasty current/voltage reaction. But, the curves for charging and discharging are not impeccably linear owing to the existence of pseudocapacitance, which is due to the presence of PANI and oxygen groups of RGO. The inclined triangular shape of curves for S-PP8 predominantly characterize to original pseudocapacitance [43].

3.5 Reduced GO-MnO$_2$-SILAR method

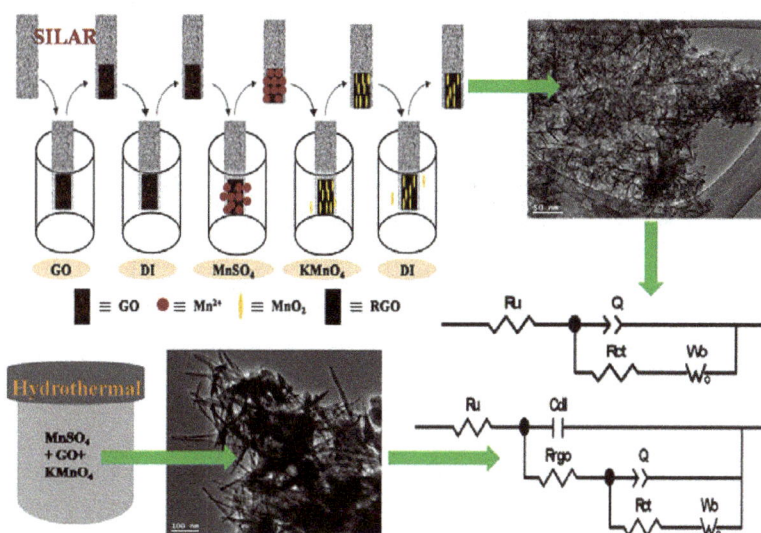

Fig. 13. Successive ionic layer adsorption and reaction (SILAR) method.

Milan et al. demonstrated a superficial, cost-effective, and polymer additive free successive ionic layer adsorption and reaction (SILAR) technique [51] to develop LbL assembly of RGO and MnO$_2$ (MnO$_2$-RGO) on a stainless steel current collector. Fig. 13 shows the step by step procedure of the SILAR technique. The TEM and FESEM images displayed the uniform distribution of RGO and MnO$_2$ SILAR. In the MnO$_2$-RGO SILAR proves enhanced physical and electrochemical properties of the hydrothermally prepared composites. The equivalent circuit of MnO$_2$-RGO, shows the co-occurrence of EDL and constant phase element, indicating a uniform distribution of MnO$_2$ and RGO by the hydrothermal technique. The designed SILAR cell possesses high energy density of ~88 Whkg^{-1}, extensive power density of 23,200 Wkg, and approximately 79% retention in capacitance after 10,000 charge-discharge cycles [51].

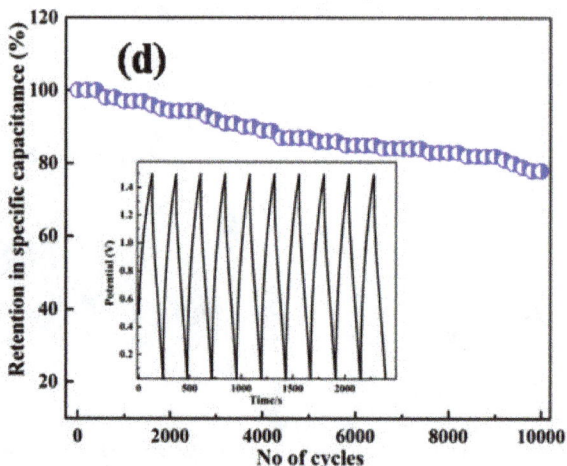

Fig. 14. Charge-discharge cyclic stability of MnO$_2$-RGO-SILAR/TRGO.

It is evidently seen from the Fig. 14, after 1000 cycles SC deviated about 1.5% from its initially existing value. After 5000 charge-discharge cycles, about 86% capacitance was retained, and after 10,000 cycles, it reached 79%. The excellent cyclic stability confirmed the mutable nature of the charge storage electrochemical reaction of the supercapacitor devices [51].

3.6 Reduced GO-magnetite

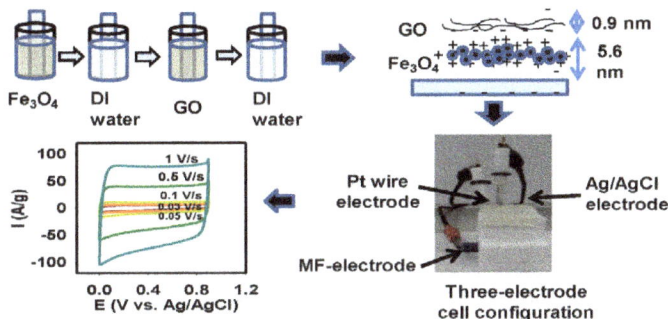

Fig. 15. LbL assembly of magnetite nanoparticles.

Wia Hwa et al. described electrochemical characteristics of LbL assembled multilayer electrode films as shown in Fig. 15. These films consist of metal ion-magnetite nanoparticle (Fe_3O_4 NP) and chemically RGO for supercapacitor applications. Optical ellipsometry was used to determine the thickness of Fe_3O_4/GO bilayer as 6.53 ± 0.17 nm. Every layer was found to be deposited homogeneously and regularly on the substrate. Hydrazine was used to reduce Fe/GO film for fabricating Fe_3O_4/RGO multilayer film. The electrodes developed from 30 bilayers of Fe_3O_4/RGO showed better capacitive characteristics with high specific capacitance (151 F/g) for a current density of 0.9 A/g [46].

3.7 Polymer film

Wania et al. reported supercapacitors developed with the LbL method by using polymers, viz poly(o-methoxy aniline)-POMA and poly(3-thiophene acetic acid)-PTAA. The electrochemical characteristics such as specific capacitance, cyclic retention of POMA/PTAA fabricated supercapacitors were categorized by electrochemical impedance spectroscopy and cyclic voltammetry. The obtained results were related with basic POMA film without coating. The specific capacitance of electrode film surges very linear with a number of bilayers which were not witnessed for POMA films. From the results, it was clearly proven that the self-doping effect between POMA and PTAA altered the characteristics of the films and it can pave the way for effective use of them as an electrode material [45].

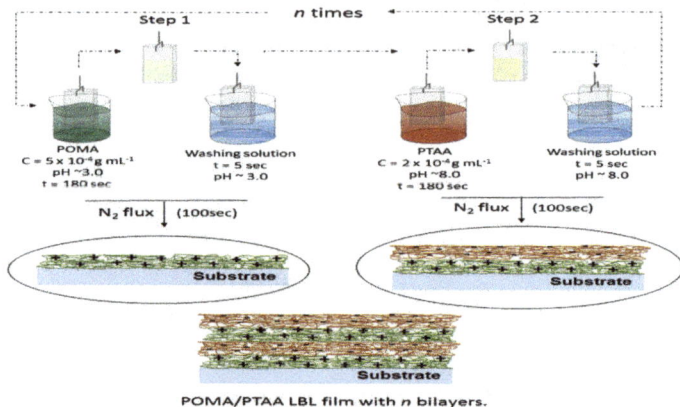

POMA/PTAA LBL film with *n* bilayers.

Fig. 16. Representation of self-assembled electrode film.

Fig. 16 presents a schematic representation of experimental buildup of the layer-by-layer assembled films. From Fig. 17, it is clear that no resistance is offered to the surface of the polymer branch owing to the short connection of each bilayer. Increase in a number of bilayers subsequently increases the active surface area of the film and each bilayer act as a capacitor connected in parallel, which improves the total capacitance of the developed film [45].

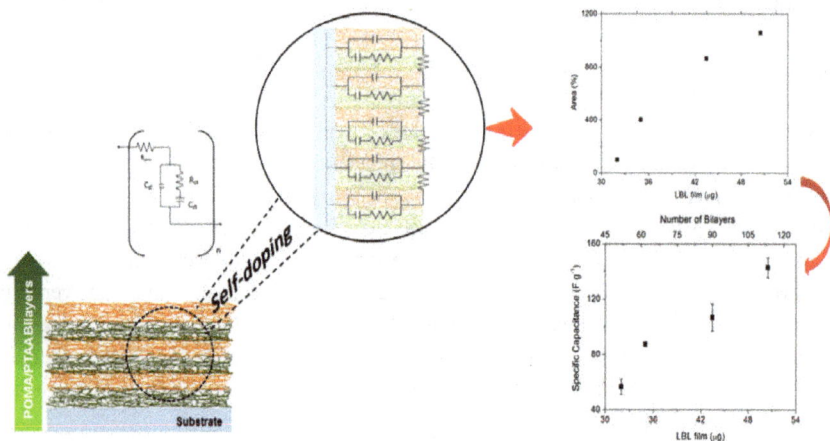

Fig. 17. Schematic demonstration of (a) POMA films and (b) POMA/PTAA films.

Conclusion

Electrochemical supercapacitors have been reported extensively in recent decades as electrochemical energy storage devices. In recent researches, carbon-based nano-sized materials and their allied composites such as reduced GO, conducting polymers, metal oxides, metal hydroxides, carbon nanotubes, graphene sheets, and cellulose nanofibers have been considered broadly as electrode materials for supercapacitor applications for energy storage due to their outstanding properties such as good surface area, excellent electrical conductivity, connected porous structure, better wettability towards the electrolyte, and presence of electrochemically active surfaces. Amongst the several available deposition methods, the layer-by-layer (LbL) assembly method offers numerous ways to synthesize better multilayer coatings with desired functions for supercapacitor applications. A novel group of supercapacitors is likely to supplant batteries in many areas where excellent power, better efficiency, and maximum level of reliability is required. High-performance electrochemical supercapacitors are most auspicious aspirants for imminent energy storage devices.

References

[1]   S. Bose, T. Kuila, A.K. Mishra, R. Rajasekar, N.H. Kim, J.H. Lee, Carbon-based nanostructured materials and their composites as supercapacitor electrodes, J. Mater. Chem. 22 (2012) 767-784. https://doi.org/10.1039/C1JM14468E

[2]   M.D. Stoller, S. Park, Y. Zhu, J. An, R.S. Ruoff, Graphene-based ultracapacitors, Nano Lett. 8 (2008) 3498–3502. https://doi.org/10.1021/nl802558y

[3]   J. Chmiola, G. Yushin, Y. Gogotsi, C. Portet, P. Simon, P.L. Taberna, Anomalous increase in carbon capacitance at pore sizes less than 1 nanometer, Science 313 (2006) 1760–1763. https://doi.org/10.1126/science.1132195

[4]   J.K. Chang, C.H. Huang, W.T. Tsai, M.J. Deng, I.W. Sun, Ideal pseudocapacitive performance of the Mn oxide anodized from the nanostructured and amorphous Mn thin film electrodeposited in BMP–NTf$_2$ ionic liquid, J. Power Sources 179 (2008) 435–440. https://doi.org/10.1016/j.jpowsour.2007.12.084

[5]   L.L. Zhang, R. Zhou, X.S. Zhao, Graphene-based materials as supercapacitor electrodes, J. Mater. Chem. 20 (2010) 5983–5992. https://doi.org/10.1039/c000417k

[6]   H. Pan, J. Li, Y.P. Feng, Carbon nanotubes for supercapacitor, Nanoscale Res. Lett. 5 (2010) 654–668. https://doi.org/10.1007/s11671-009-9508-2

[7]   X. Zhang, W. Yang, D.G. Evans, Layer-by-layer self-assembly of manganese oxide nanosheets/polyethylenimine multilayer films as electrodes for supercapacitors, J. Power Sources 184 (2008) 695–700. https://doi.org/10.1016/j.jpowsour.2008.01.021

[8]   S.C. Pang, M.A. Anderson, T.W. Chapman, Novel electrode materials for thin-film ultracapacitors: Comparison of electrochemical properties of sol-gel-derived and electrodeposited manganese dioxide, J. Electrochem. Soc. 147 (2000) 444-450. https://doi.org/10.1149/1.1393216

[9]   S.C. Pang, M.A. Anderson, Novel electrode materials for electrochemical capacitors: Part II. Material characterization of sol-gel-derived and electrodeposited manganese dioxide thin films, J. Mater. Res. 15 (2000) 2096-2106. https://doi.org/10.1557/JMR.2000.0302

[10]   S.F. Chin, S.C. Pang, M.A. Anderson, Material and electrochemical characterization of tetrapropylammonium manganese oxide thin films as novel electrode materials for electrochemical capacitors, J. Electrochem. Soc. 149 (2002) A379-A384. https://doi.org/10.1149/1.1453406

[11]   J.K. Chang, W.T. Tsai, Material characterization and electrochemical performance of hydrous manganese oxide electrodes for use in electrochemical pseudocapacitors, J. Electrochem. Soc. 150 (2003) A1333-A1338. https://doi.org/10.1149/1.1605744

[12]   M. Nakayama, A. Tanaka, Y. Sato, T. Tonosaki, K. Ogura, Electrodeposition of manganese and molybdenum mixed oxide thin films and their charge storage properties, Langmuir 21 (2005) 5907-5913. https://doi.org/10.1021/la050114u

[13]   N. Nagarajan, H. Humadi, I. Zhitomirsky, Cathodic electrodeposition of MnOx films for electrochemical supercapacitors, Electrochimica. Acta 51 (2006) 3039-3045. https://doi.org/10.1016/j.electacta.2005.08.042

[14]   T. Xue, C.L. Xu, D.D. Zhao, X.H. Li, H.L. Li, Electrodeposition of mesoporous manganese dioxide supercapacitor electrodes through self-assembled triblock copolymer templates, J. Power Sources 164 (2007) 953-958. https://doi.org/10.1016/j.jpowsour.2006.10.100

[15]   M. Nakayama, T. Kanaya, R. Inoue, Anodic deposition of layered manganese oxide into a colloidal crystal template for electrochemical supercapacitor, Electrochem. Commun. 9 (2007) 1154-1158. https://doi.org/10.1016/j.elecom.2007.01.021

[16] Y. Dai, K. Wang, J. Zhao, J. Xie, Manganese oxide film electrodes prepared by electrostatic spray deposition for electrochemical capacitors from the KMnO4 solution, J. Power Sources 161 (2006) 737-742. https://doi.org/10.1016/j.jpowsour.2006.04.098

[17] K.W. Nam, K.B. Kim, Manganese oxide film electrodes prepared by electrostatic spray deposition for electrochemical capacitors, J. Electrochem. Soc. 153 (2006) A81-A88. https://doi.org/10.1149/1.2131821

[18] J. Hong, S.W. Kang, Carbon decorative coatings by dip-, spin-, and spray-assisted layer-by-layer assembly deposition, J. Nanosci. Nanotechnol. 11 (2011) 7771–7776. https://doi.org/10.1166/jnn.2011.4737

[19] M. Olek, J. Ostrander, S. Jurga, H. Mohwald, N. Kotov, K. Kempa, M. Giersig, Layer-by-layer assembled composites from multiwall carbon nanotubes with different morphologies, Nano Lett. 4 (2004) 1889-1895. https://doi.org/10.1021/nl048950w

[20] G. Decher, Fuzzy nanoassemblies: Toward layered polymeric multicomposites, Science 277 (1997) 1232-1237. https://doi.org/10.1126/science.277.5330.1232

[21] P.T. Hammond, Form and function in multilayer assembly: new applications at the nanoscale, Adv. Mater. 16 (2004) 1271-1293. https://doi.org/10.1002/adma.200400760

[22] K.C. Krogman, J.L. Lowery, N.S. Zacharia, G.C. Rutledge, P.T. Hammond, Spraying asymmetry into functional membranes layer-by-layer, Nat. Mater. 8 (2009) 512-518. https://doi.org/10.1038/nmat2430

[23] F.X. Zhang, M.P. Srinivasan, Multilayered gold-nanoparticle/polyimide composite thin film through layer-by-layer assembly, Langmuir 23 (2007) 10102. https://doi.org/10.1021/la0635045

[24] P. Podsiadlo, M. Michel, K. Critchley, S. Srivastava, M. Qin, J.W. Lee, E. Verploegen, A.J. Hart, Y. Qi, N.A. Kotov, Diffusional self-organization in exponential layer-by-layer films with micro- and nanoscale periodicity, Angewandte Chemie Int. Edit. 48 (2009) 7073-7077. https://doi.org/10.1002/anie.200901720

[25] S.S. Shiratori, M.F. Rubner, pH-dependent thickness behavior of sequentially adsorbed layers of weak polyelectrolytes, Macromolecules 33 (2000) 4213-4219. https://doi.org/10.1021/ma991645q

[26] D. Lee, M.F. Rubner, R.E. Cohen, All-nanoparticle thin-film coatings, Nano Lett. 6 (2006) 2305-2312. https://doi.org/10.1021/nl061776m

[27] R.B. Estevam, R.T. Ferreira, A.B.H. Bischof, F.S. DosSantos, C.S. Santos, S.T. Fujiwara, K. Wohnrath, S.R. Lazaro, J.R. Garcia, C.A. Pessoa, Preparation and characterization of LbL films based on graphene oxide nanoparticles interacting with 3-n-propylpyridinium silsesquioxane chloride, Surf. Coat. Tech. 275 (2015) 2–8. https://doi.org/10.1016/j.surfcoat.2015.03.053

[28] T.K. Hong, D.W. Lee, H.J. Choi, H.S. Shin, B.S. Kim, Transparent, flexible conducting hybrid multilayer thin films of multiwalled carbon nanotubes with graphene nanosheets, ACS Nano. 4 (2010) 3861–3868. https://doi.org/10.1021/nn100897g

[29] S. Pei, H.M. Cheng, The reduction of graphene oxide, Carbon 50 (2012) 3210–3228. https://doi.org/10.1016/j.carbon.2011.11.010

[30] M. Jin, H.K. Jeong, T.H. Kim, K.P. So, Y. Cui, W.J. Yu, E.J. Ra, Y.H. Lee, Synthesis and systematic characterization of functionalized graphene sheets generated by thermal exfoliation at low temperature, J. Phys. D: Appl. Phys. 43 (2010) 275402. https://doi.org/10.1088/0022-3727/43/27/275402

[31] H.K. Jeong, M.H. Jin, K.P. So, S.C. Lim, Y.H. Lee, Tailoring the characteristics of graphite oxides by different oxidation times, J. Phys. D: Appl. Phys. 42 (2009) 065418. https://doi.org/10.1088/0022-3727/42/6/065418

[32] Q. Zheng, Z. Li, J. Yang, J.K. Kim, Graphene oxide-based transparent conductive films, Prog. Mater. Sci. 64 (2014) 200–247. https://doi.org/10.1016/j.pmatsci.2014.03.004

[33] K. Parvez, Z.S. Wu, R. Li, Exfoliation of graphite into graphene in aqueous solutions of inorganic salts, J. Am. Chem. Soc. 136 (2014) 6083–6091. https://doi.org/10.1021/ja5017156

[34] D.W. Lee, T.K. Hong, D. Kang, J. Lee, M. Heo, J.Y. Kim, B.S. Kim, H.S. Shin, Highly controllable transparent and conducting thin films using layer-by-layer assembly of oppositely charged reduced graphene oxides, J. Mater. Chem. 21 (2011) 3438–3442. https://doi.org/10.1039/C0JM02270E

[35] J. Luo, Q. Ma, H. Gu, Y. Zheng, X. Liu, Three-dimensional graphene-polyaniline hybrid hollow spheres by layer-by-layer assembly for application in supercapacitor, Electrochimica Acta 173 (2015) 184–192. https://doi.org/10.1016/j.electacta.2015.05.053

[36] E. Coskun, E.A. Zaragoza-Contreras, H.J. Salavagione, Synthesis of sulfonated graphene/polyaniline composites with improved electroactivity, Carbon 50 (2012) 2235–2243. https://doi.org/10.1016/j.carbon.2012.01.041

[37] J.P. Cheng, L. Liu, K.Y. Ma, X. Wang, Q.Q. Li, J.S. Wu, F. Liu, Hybrid nanomaterial of α-Co(OH)$_2$ nanosheets and few-layer graphene as an enhanced electrode material for supercapacitors, J. Colloid Interface Sci. 486 (2017) 344–350. https://doi.org/10.1016/j.jcis.2016.09.064

[38] M. Sevilla, R. Mokaya, Energy storage applications of activated carbons: supercapacitors and hydrogen storage, Energy Environ. Sci. 7 (2014) 1250–1280. https://doi.org/10.1039/C3EE43525C

[39] C. Nethravathi, C.R. Rajamathi, M. Rajamathi, X. Wang, U.K. Gautam, D. Golberg, Y. Bando, Cobalt hydroxide/oxide hexagonal ring–graphene hybrids through chemical etching of metal hydroxide platelets by graphene oxide: energy storage applications, ACS Nano 8 (2014) 2755–2765. https://doi.org/10.1021/nn406480g

[40] T. Bohnenberger, L.D. Rafailovic, C. Weilach, D. Hubmayr, U. Schmid, Thin films from functionalized carbon nanotubes using the layer-by-layer technique, Thin Solid Films 551 (2014) 68–73. https://doi.org/10.1016/j.tsf.2013.11.107

[41] Z. Niu, W. Zhou, J. Chen, G. Feng, H. Li, W. Ma, J. Li, H. Dong, Y. Ren, D. Zhao, S. Xie, Compact-designed supercapacitors using free-standing single-walled carbon nanotube films, Energy Environ. Sci. 4 (2011) 1440-1446. https://doi.org/10.1039/c0ee00261e

[42] Y. Ding, N. Zhang, J. Zhang, X. Wang, J. Jin, X. Zheng, Y. Fang, The additive-free electrode based on the layered MnO2 nanoflowers/reduced, graphene oxide film for high performance supercapacitor, Ceramics Int. 43 (2017) 5374–5381. https://doi.org/10.1016/j.ceramint.2016.10.032

[43] X. Wang, K. Gao, Z. Shao, X.Q. Peng, F. Wang, X. Wu, Layer-by-Layer assembled hybrid multilayer thin film electrodes based on transparent cellulose nanofibers paper for flexible supercapacitors applications, J. Power Sources 249 (2014) 148- 155. https://doi.org/10.1016/j.jpowsour.2013.09.130

[44] T. Saito, S. Kimura, Y. Nishiyama, A. Isogai, Cellulose nanofibers prepared by TEMPO-mediated oxidation of native cellulose, Biomacromolecules 8 (2007) 2485-2491. https://doi.org/10.1021/bm0703970

[45] W.A. Christinelli, R. Gonçalves, E.C. Pereira, A new generation of electrochemical supercapacitors based on layer-by-layer polymer films, J. Power Sources 303 (2016) 73-80. https://doi.org/10.1016/j.jpowsour.2015.10.077

[46] W.H. Khoh, J.D. Hong, Layer-by-layer self-assembled multilayer films composed of graphene/polyaniline bilayers: high-energy electrode materials for supercapacitors, Colloids and Surfaces A: Physicochem. Eng. Aspects 436 (2013) 104–112. https://doi.org/10.1016/j.colsurfa.2013.06.012

[47] N.L. Wu, S.Y. Wang, C.Y. Han, D.S. Wu, L.R. Shiue, Electrochemical capacitor of magnetite in aqueous electrolytes, J. Power Sources 113 (2003) 173–178. https://doi.org/10.1016/S0378-7753(02)00482-2

[48] J. Luo, S. Jiang, H. Zhang, J. Jiang, X. Liu, A novel non-enzymatic glucose sensor based on Cu nanoparticle modified graphene sheets electrode, Analytica Chimica Acta 709 (2012) 47–53. https://doi.org/10.1016/j.aca.2011.10.025

[49] W. Fan, C. Zhang, W.W. Tjiu, K.P. Pallathadka, C. He, T. Liu, Graphene-wrapped polyaniline hollow spheres as novel hybrid electrode materials for supercapacitor applications, ACS Appl. Mater. Inter. 5 (2013) 3382–3391. https://doi.org/10.1021/am4003827

[50] N.B. Trung, T.V. Tam, H.R. Kim, S.H. Hur, E.J. Kim, W.M. Choi, Three-dimensional hollow balls of graphene–polyaniline hybrids for supercapacitor applications, J. Chem. Eng. 255 (2014) 89–96. https://doi.org/10.1016/j.cej.2014.06.028

[51] M. Jana, S. Saha, P. Samanta, N.C. Murmu, N.H. Kim, T. Kuila, J.H. Lee, A successive ionic layer adsorption and reaction (SILAR) method to fabricate a layer-by-layer (LbL) MnO$_2$-reduced graphene oxide assembly for supercapacitor application, J. Power Sources 340 (2017) 380-392. https://doi.org/10.1016/j.jpowsour.2016.11.096

Chapter 9

Graphene-based Composites: Present, Past and Future for Supercapacitors

P. Sathish Kumar[1]*, Samir Kumar Pal[1], T. K. Kannan[2], R. Rajasekar[3]

[1]Department of Mining Engineering, Indian Institute of Technology Kharagpur, Kharagpur, West Bengal – 721302, India

[2]Principal, The Kavery College of Engineering, Salem, Tamil Nadu – 636453, India

[3]Department of Mechanical Engineering, Kongu Engineering College, Erode, Tamil Nadu – 638060, India

*sathishiitkgp@gmail.com

Abstract

Graphene-based materials are promising for applications in supercapacitors and other energy storage devices due to the outstanding properties like highly tunable surface area, outstanding electrical conductivity, good chemical stability and excellent mechanical behavior. This chapter is aimed to encapsulate recent development on graphene-based materials for supercapacitor electrodes. The on-going extensive researches in respect of rationalization of their structures at varying scales and dimensions, development of effective and low-cost synthesis techniques, design and architecture of graphene-based materials, as well as clarification of their electrochemical performance have generated renewed interest in graphene-based composites. The future studies are expected to focus on the overall device performance in energy storage devices and large-scale process in low costs for the promising applications in portable and wearable electronic, transport, electrical and hybrid vehicles.

Keywords

Supercapacitors, Graphene, Thermal Reduction, Energy Materials, Electronic Applications, Electrode Materials

Contents

1.   Introduction and background

In the tide of "new energy fever", among energy storage systems, supercapacitors have aroused special interest to serve as supplementary devices for conventional or advanced batteries to realize economic, efficient, safe, clean and sustainable electricity storage in a broad range of real-life applications. Supercapacitors have gained recent attention due to their high power density, adequate charge/discharge rates, short charge time, light weight, good operational safety and long cycle life performance [1]. Supercapacitors are considered as one of the most promising electrochemical energy storage devices having a potential to complement or eventually replace batteries for energy storage applications, i.e., those for wearable and portable electronics, electrical and hybrid vehicles [1]. Based on the energy storage mechanisms, supercapacitors can be classified into three main categories, i.e., electric double-layer capacitance (EDLC), hybrid supercapacitors and/or pseudocapacitance [2]. They are well suited for applications requiring rapid power delivery and recharging, such as regenerative braking, short-term energy storage, hybrid electric vehicles, large industrial equipment and portable devices. However, commercially available supercapacitors are still based on activated carbon materials and have much less energy density than rechargeable batteries, which severely limit their potential for many applications [3]. The commercialization of hybrid supercapacitors is still lacking due to non-availability of ideal electrode materials of desired nanostructures. However, supercapacitors possess outstanding charge/discharge rate over other types of energy storage devices and hence have practical applications in electric vehicles and large-scale energy storage systems. The capacitance of supercapacitors is primarily determined by the electrode materials which combine a high electrochemical performance with good mechanical properties such as resistance to folding and bending, compactness and lightweight. Flexible energy storage devices have attracted increasing attention for use in foldable electronic equipment (i.e. flexible phones and tablets) and also as components of wearable products (i.e. clothes, bags, etc.) [4].

Ideal electrode materials should possess high surface area accessible to the electrolyte, reasonable electrical conductivity, desirable chemical stability, and low density [2]. An innovative electrode material with high capacity performance is indispensable for developing supercapacitor devices [5]. Considering these parameters and its unique properties, graphene, a well-known monolayer one-atom-thick 2D sheet of carbon has been considered as a promising supercapacitor electrode material. Furthermore, graphene with sp^2 hybridized carbons owns some of the most intriguing properties, i.e., light weight, high electrical and thermal conductivity, highly tunable surface area, strong mechanical strength, and chemical stability. These striking properties enable graphene and graphene-based materials to find applications in high-performance structural nanocomposites, electronics, environmental protection, and energy devices including both energy generation and storage. The combination of favorable physical, mechanical, and chemical properties make graphene-based materials more attractive for electrochemical energy storage and sustainable energy generation, i.e., Li-ion batteries, fuel cells, supercapacitors, and photovoltaic and solar cells. However, the poor mechanical strength of thin graphene films causes difficulty in their handling [4]. Moreover, graphene exhibits unsatisfactory capacitive performance due to unavoidable aggregation or restocking of graphene nanosheets. In order to develop high-performance supercapacitors, various attempts have been made to exploit functionalized graphene and graphene-based composite materials. For conducting polymers, polyaniline (PANI) has been considered to be one of the most important electrode materials for supercapacitor due to its excellent capacity for energy storage, easy synthesis, high conductivity, and low cost. Poly (N-acetylaniline) (PAANI), a substituted PANI conducting polymer has been previously investigated in the field of biosensor fabrication and supporting matrix for catalyst [5]. To the best of our knowledge, no report is available on the synthesis of PAANI/graphene composite and its use in electrochemical performances. Furthermore, free-standing graphene films of flexibility and foldability are more suitable for the requirement in energy storage devices [6].

In isolated single-layer graphene, carbon atoms are connected by conjugated π bond which forms a sheet with high electrical conductivity. However, in bulk materials of multi-layer graphene, the sheets have strong tendency to aggregate due to the strong π-π interaction between the sheets. The aggregation of graphene makes the space between the sheets too narrow to be accessed by the electrolyte [2]. However, the practical capacitive behavior of pure graphene is inferior to the anticipated value due to the serious agglomeration during both the preparation and application processes. Therefore, boosting the overall electrochemical performance of graphene-based materials still remains a great challenge. Graphene-based materials have been extensively investigated as conducting network to support the redox reactions of transition metal oxides, hydroxides and conducting polymers. As regards to 3D graphene, despite its decent capacitance and cyclability when used solely as electrodes, it is more useful to host the high-capacitance phases, where its macro-mesostructure and high electric conductivity can be utilized more effectively.

As an indispensable component in electronics, commercial micro-supercapacitors are disadvantageous due to their cuboid geometry and limited capacity. In comparison, film-like micro-supercapacitors are superior in a miniaturized system integration since these can be folded to fit in restricted spaces while maintaining a high level of volumetric energy density [7]. The prime negative and positive electrodes of graphene-based nano-architectures (such as reduced graphene oxide (rGO), porous graphene, graphene quantum dots (GQD), graphene films, graphene aerogels (GAs), graphene foams (GF), various hybrids of graphene/CNT, graphene/metal oxides, and graphene/conducting polymers) in terms of the major performance parameters including high voltage, high capacitance, high power, and high energy devices as well as new-type device geometry of planar and all-solid-state devices have illustrated the uniqueness and superiority of graphene for supercapacitors. Recently, the emergence of the planar supercapacitor is regarded as an important member of the family of miniaturized energy storage devices. As compared to the conventional supercapacitors which have sandwich structures, a planar layout can render the diffusion length independent to the electrode thickness and thereby it allows more active material loading per unit area while maintaining excellent rate performance and capacitance. More importantly, owing to the planar layout and elimination of separator, the total thickness of the supercapacitor device significantly decreases, showing great potential in microelectronic applications. Most recently, there have been remarkable advances regarding the fabrication of planar supercapacitors. For instance, John Chmiola et al. [8] reported a planar supercapacitor based on carbide-derived carbon originated form sputtered TiC film with chlorination process, showing a high volumetric capacitance. David Pech et al. [9] deposited the onion-like carbon on integrated gold current collector for realizing an ultrafast discharge rate, three times higher than conventional supercapacitors. Wu et al. [10] fabricated a graphene-based planar supercapacitor on arbitrary substrates, which exhibited ultrahigh rate performance and excellent energy density. However, a critical step toward this technology transfer to industry is to elaborate the detailed fabrication and packaging technology of the state-of-the-art planar micro-supercapacitor at the electronic component level. Besides, there is still a lack of a benchmark study regarding such planar micro-supercapacitor component with the commercial-available counterparts [1].

In the last decade, laser fabrication technology featured with high efficiency, ultrahigh resolution, and low cost has presented great potential in miniaturized device fabrication. Because of the interaction between the laser beam and graphene oxide (GO), laser technology has been used to reduce GO into graphene which has drawn much attention in miniaturized energy storage field. For example, Yonglai Zhang et al. [11] fabricated desired micrometer-sized graphene circuits on GO films by using femtosecond laser nano-writing, which brought about a new horizon for applications of graphene-based materials in electronic micro-devices. EI-Kady et al. [12] reported a scalable fabrication of graphene-based planar supercapacitors by laser writing on GO films using a commercial light scribe DVD burner, showing excellent performance. Zhiwei Peng et al. [13] demonstrated that boron can be doped into graphene by direct laser treatment, which delivered higher areal capacitance. In-light of the superior performance of planar supercapacitor based on GO treated with laser, the combination between high-performance planar supercapacitor and micro-electronics may play more important roles for next generation consumer electronics [1]. A variety of materials such as carbonaceous materials, conducting polymers and transition metal oxides have been examined for possible use as supercapacitor electrodes [14]. Due to their low cost, high theoretical capacitance and environmental friendliness, many metal oxides like MnO_2, Fe_2O_3, and Co_3O_4, etc. have been used in supercapacitor applications. The major drawbacks of using these metal oxides are low electrical conductivity, poor stability, and high self-discharge, the low volumetric capacitance for thick-film electrodes and short cycling life [15]. In addition, textile-based electrode materials have been used for supercapacitors in flexible and wearable electronic devices which exhibited high electrochemical performances compared to textile-based pseudocapacitors. Therefore, the world is facing a great challenge in developing metal oxides/textile composite electrode materials by tuning the electrode structure and composition with high electrochemical performance.

2.   Graphite, graphite oxide and graphene

Graphite is a well-known natural material, which is highly anisotropic in both structure and functional behavior, with the in-plane electrical and thermal conductivities being 1000-fold greater than those in the out-of-plane direction. Similarly, the in-plane strength and modulus of graphite are much greater than those of the out-of-plane due to the different types of bonds in the two directions. Some techniques of producing graphite of different qualities in large scales have been developed. Graphite is thus one of the most widely used materials in various structural, functional, chemical, energy storage, and environmental applications. Mitra et al. [16] developed solid-state electrochemical capacitors by using graphite as electrodes, where the calculated specific capacitances were in the range from 0.74 to 0.98 mFcm^{-2}.

Graphene is the monolayer of graphite, which can be prepared by several techniques. Geim et al. [17] prepared graphene from graphite and demonstrated an experimental method to prepare a single layer of graphite with a thickness in atomic scale, named as graphene. Since then graphene has become popular in various application aspects due to its inherently superior electrical/electronic and optical properties (i.e., tunable band gap, extraordinary electronic transport behavior, excellent thermal conductivity, high mechanical strength, and largely tunable surface area). Also, some chemical and physical techniques for synthesis of graphene have been developed.

Graphene oxides (GOs) are important members of the graphene-graphite family, which are considered as derivatives of graphene. GOs can be readily obtained from graphite. They exhibit the layered structure and unique surface-related properties. Depending on the synthesis techniques, there can be different surface groups in graphene oxides as well as their distributions on the surface. For instance, the oxygen-containing functional groups such as hydroxyl (OH), carboxyl (C=O) and epoxy groups (C-O) locate around the edges of graphene sheets and stabilize the quasi-two-dimensional sheets. Graphene oxides can be readily converted into graphene by different reduction processes. As a precursor for graphene, GO can be easily derived from the oxidation of natural graphite at a large scale and low costs. The reduction of GO is a low-cost technique for producing graphene. The atomic layers of GO generally comprise phenol epoxy and epoxide groups on the basal plane and ionizable carboxylic acid groups around the edges.

For both energy storage and other applications, some research work is dedicated to the synthesis of graphene sheets, where the generally facile approaches can be divided into the following categories: i) epitaxial growth and chemical vapor deposition (CVD) of graphene on substrates, such as SiC and matched metal surfaces; ii) mechanical cleavage of graphite, for instance, using atomic force microscopy (AFM) cantilevers or even adhesive tapes; iii) chemical exfoliation of graphite in organic solvents [18]; iv) gas-phase synthesis of graphene platelets in microwave plasma reactor; v) synthesis of multi-layered graphene by arc-discharge; and vi) reduction from GO-derived from the oxidation of natural graphites in a large scale. Among these production methods, graphene prepared by CVD offers better properties, as a result of large crystal domains, monolayer structure and fewer defects in the graphene sheets, which are beneficial for boosting carrier mobility in electronic applications. In addition to mechanical exfoliation by adhesive tapes, Firsov et al. [19] demonstrated the mechanical cleavage of graphite using atomic force microscopy (AFM) cantilevers. Mechanical exfoliation is, however, associated with a low yield and hence it is unsuitable for the mass production of graphene. Hernadez et al. [20] developed chemical methods to exfoliate graphite in organic solvents. In general, the chemical exfoliation of graphite into GOs, followed by controllable reduction of GOs (with reducing agent such as hydrazine hydrate) into graphene has been treated as the most efficient, and low-cost method. Although individual graphene sheets are often partially agglomerated into particles of approximately 15-25 mm in diameter during the reduction process, the product with high specific surface areas of a few hundred m^2g^{-1} can be obtained, offering an electrode material of energy storage devices for the potential application.

Ruoff et al. [21] pioneered the chemically modified graphene (CMG) as electrode materials. They found that the specific capacitances of 135 and 99 Fg^{-1} could be achieved in aqueous and organic electrolytes, respectively. Ajayan et al. [22] reported a two-step method to fabricate a highly reduced GO via deoxygenation with $NaBH_4$ and dehydration with concentrated sulfuric acid. The comb-like connected sheet structure needs to be protected during the harsh oxidation and reduction processes in order to maintain the performance. To reduce the level of toxicity of the reduction agents, some efforts have been made to prepare graphene with non-toxic agents. Zhu et al. [23] synthesized graphene using sugar as a reduction agent. Zhang et al. [24] employed ascorbic acid. In addition, "green" materials (bovine serum albumin, polyphenols of green tea and even bacterial respiration) were also proposed to produce reduced graphene oxides (rGO). Similarly, "green" methods such as solvothermal reduction, hydrothermal dehydration, catalytic reduction and photocatalytic reduction were reported for the reduction of GO into graphene. In particular, the one-step hydrothermal method has been a versatile and low-cost process to produce the rGO. Shi et al. [25] reported the reduction of GO via the one-step hydrothermal approach. The obtained rGO showed high conductance of 5×10^{-3} Scm^{-1} and suitable specific capacitance of 175 Fg^{-1} in aqueous electrolyte. Nevertheless, it is rather challenging to completely reduce GO as some of the oxygen-containing surface groups are rather difficult to be eliminated.

Graphene can be assembled into several different structures, i.e., the free-standing particles or dots, one-dimensional fibers or yarns, two-dimensional films and three-dimensional foams as well as composites. Recent studies focus on graphene-based electrode materials according to their macrostructural complexity, i.e., zero-dimensional (0-D) (e.g. free-standing graphene dots and particles), one-dimensional (1-D) (e.g. fiber-type and yarn-type structures), two-dimensional (2-D) (e.g. graphenes and graphene-based films), and three-dimensional (3-D) (e.g. graphene foams and composites).

3. Graphene-based composites for supercapacitors

In recent years, graphene and other related materials have found potential industrial applications in electrical, electrochemistry, and electronic devices [26] due to their high power density, long cycle life, and low manufacturing as well as maintenance costs. Along with light weight and high surface area, carbon is a more readily available material in nature. Disadvantages of supercapacitors in energy devices have encouraged researchers to do further research to develop graphene-based supercapacitors with improved energy density. The charge storage in supercapacitors is based on the formation of an interfacial double layer on active materials. Several reports on graphene-based supercapacitors are available with improved specific capacitance using various electrolytes like potassium hydroxide, an organic electrolyte, and ionic liquids. Chenguang et al. [27] used high surface area graphene with curved morphology to fabricate graphene-based supercapacitor with high energy density. The reported value of discharge capacitance was very low and the discharge time was 2 minutes at a certain energy density. Yanwu et al. [28] had done many modifications in graphene synthesis and current collecting method to investigate the capacitance properties of the graphene supercapacitors with the ionic liquids. The obtained results for their devices are very low. In addition, pseudocapacitors with metal oxide and polymer graphene composites with improved performance were also reported by many research groups. The use of metal oxide graphene composites depict improvements on the specific capacitance and energy density, but still, the charging and discharging times were low for high power applications. High surface area and good electrical performances are the main key parameters of graphene synthesis for high power supercapacitors. These several methods are for the synthesis of graphene including Hummer's method, dispersion method, microwave method, and electrochemical method. But still, the synthesized graphene showed low capacitance performance and failed to find many practical applications.

3.1 Energy applications

Due to some of the unique features of non-agglomerated 3D inter-networked morphology, large pore volume, high accessible surface area, controlled inter-sheet connectivity, as well as excellent stability, and mechanical strength, 3DG networks showed remarkable improvement compared to 2D graphene/RGO. The 3D hollow structure with the large accessible surface area, the high in-plane electrical conductivity of graphene and the high activity of the incorporated nanomaterials have led to advances in supercapacitors, lithium-ion batteries, dye-sensitized solar cells, and fuel cells based nanomaterial.

Advanced electrochemical devices i.e., supercapacitors which have much higher capacitance are well suited for energy storage with frequent charge/discharge cycles at high power and short duration. Although supercapacitors are commercially available, further research is needed to improve the supercapacitor performance like mass/volume-specific capacitance, rate capability, and power/energy densities. Nanocarbon-based materials, such as activated carbons, carbon nanotubes, and graphene were used due to their high electrical conductivity, improved surface area, and low cost.

Shi's group reported [25] the preparation of graphene hydrogel/Ni foam (G-Gel/NF) composite electrode by direct deposition and in-situ reduction of GO with subsequent free-drying of the hydrogel. Symmetric cells with two-electrode configuration were constructed to test the performance of G-Gel/NF EDLC. The observed capacitance of G-Gel/NF was 45.6 $mFcm^{-2}$ at a discharge current of 0.67 $mAcm^{-2}$, was higher than that of high-rate EDLCs. The excellent performance of G-Gel/NF EDLCs was due to the high conductivity and electrochemical stability of the graphene gel, the 3D porous microstructure of the G-Gel/NF electrode and the short distances of charge transfer. MnO_2-coated free-standing, flexible, light-weight and highly conductive 3DG networks were also fabricated for supercapacitor electrodes by many researchers. In addition, the 3DG network exhibited a high electrical conductivity and possessed a hollow internal structure. Due to high flexibility and stability of the $3DG/MnO_2$ composite electrodes in the symmetrical supercapacitor, no significant difference was found between the CV scans before and after bending. Good capacitive characteristics for the $3DG/MnO_2$ supercapacitor were realized using the linear voltage versus time profiles, the symmetrical charge/discharge characteristics and a quick I–V response. The cycling performance of a supercapacitor was tested. The outstanding performance of the 3DG-based composites was mainly attributed to the following reasons: (1) the porous structure and high surface area of 3DG; (2) the highly conductive 3DG network, and (3) the good contact between the 3DG and active pseudo-species.

3.2   Past, present, and future

Liyi Li et al. [2] synthesized graphene oxide (GO) and amine molecules using one-step hydrothermal reactions. According to cyclic voltammetry (CV) measurement, the control sample without spacers had Cs value of 194 F/g and rGO-spacers had 612 F/g. The increased interlayer distance was confirmed by using x-ray diffraction (XRD) and morphology of rGO-spacers was characterized using scanning electron microscopy (SEM), transmission electron microscopy (TEM) and atomic force microscopy (AFM). The spacers molecules were found to disperse uniformly in the sheets by energy dispersive spectroscopy mapping. Thermogravimetric analysis (TGA), x-ray photoelectron spectroscopy (XPS), UV-visible spectroscopy (UV-vis), Fourier-transform infrared spectroscopy (FTIR), and Raman spectroscopy were used to find out the covalent bonding between the GO and spacers. Based on all the results, the amide formation reactions and epoxide ring opening reactions between amine spacers and GO were identified. The weight measurement and the atomic ratio information from XPS further revealed different reactivity of the spacers to GO [2]. Engineering molecular structure played a major role in determining the Cs. The major finding of the author was to achieve higher Cs by optimizing the molecular structure of the spacers.

Binghe Xie et al. [7] carried out a benchmark study with well packaged thin film micro-supercapacitor towards commercial micro-supercapacitor and aluminum electrolyte capacitor. The micro-planar supercapacitor not only exhibited 3.75 times of a commercial micro-supercapacitor and 8785 times of an aluminum electrolytic capacitor in volumetric energy density under 1000 mVs^{-1} scan rate, but can also be tailored into diversified shapes, rolled up, and plugged into tiny interstitial spaces inside a device [2]. Therefore, such ultrathin micro-supercapacitor with high volumetric energy density can be integrated into an electronic device system for realizing the superior performance characteristics over current commercial benchmarks.

The literature available on graphene-based materials for their application as supercapacitor electrodes and in other energy storage devices due to their highly tunable surface area, outstanding electrical conductivity, good chemical stability, and excellent mechanical behavior was admirably reviewed by Qingqing Ke and John Wang [1]. The recent developments based on their macrostructural complexity (OD, 1D, 2D and 3D) were nicely discussed. The authors suggested that future studies should focus on the overall device performance in energy storage devices and large-scale process in low costs for the promising applications in portable/wearable electronic, transport, electrical, and hybrid vehicles. The preparation methods, resultant structures, and electrochemical performances desired for application in supercapacitors were encapsulated. The corresponding capacitive mechanisms and the effective ways to achieve high energy storage performance were discussed.

M. Sevilla et al. [29] fabricated free-standing, robust and flexible composites consisting of layered graphene deposited over the porous cellulose tissue. The resultant composites with a thickness of around 60 µm and areal densities in the range of 0.6 to 2.4 mgcm^{-2} showed remarkable performance in both liquid and solid electrolytes.

Shuanghao Zheng et al. [4] provided deep insight into recent advances of graphene-based materials for high voltage and high energy asymmetric supercapacitors (ASCs). The latest advances of different graphene-based nano-architectures, such as reduced graphene oxide, porous graphene, graphene quantum dots, graphene nanoribbons, graphene fibers, graphene films, graphene aerogels, graphene foams, and various hybrids of graphene/ carbon nanotubes, graphene/metal oxides, and graphene/conducting polymers, for ASCs have been extensively discussed. Major performance parameters, including voltage, capacitance, power, and energy devices, as well as new device geometries of planar and all-solid-state devices, have been nicely described in detail, highlighting the uniqueness and superiority of graphene for hybrid energy storage. Finally, future perspectives and challenges of graphene-based ASCs were discussed [4].

Bote Zhao et al. [30] developed nanocomposite electrodes with composite and r-GO at room temperature using hydrous hydrazine and proper ratios of composition and morphology. The composites demonstrated high capacity, reasonably sufficient rate capability, and long cyclic life. The authors also examined p-phenylenediamine-modified rGO, which rendered the hybrid supercapacitors with superior energy densities, and excellent cyclic stability.

Xin Yao and Yanli Zhao [3] reviewed the typical synthetic methods used for the preparation of three-dimensional porous graphene networks and their hybrids with different structures. Lithium-ion batteries and supercapacitors with special emphasis on powering flexible electronics were included to highlight their importance.

The smart and light weight all-solid-state supercapacitors (ASSs) have been considered to be a promising candidate to power wearable and miniaturized electronics. Compared with other nanocarbon materials, three-dimensional porous graphene networks (3DPGNs) showed unique advantages in ASSs electrodes. Free-standing, monolithic 3D graphene can be easily manufactured into electrodes without the use of a current collector, polymer binder or additives and thus complex solution processing, high-pressure handling, sintering, and sputtering techniques are avoided. Moreover, the device performance can be enhanced, because the polymer binders and additives are basically non-conductive. Furthermore, the inner porous structure provides ideal channels to hold the solid-state electrolyte made from usual PVA/H_2SO_4 gel. The porous interior also ensures sufficient access to the electrolytes [3].

Wu et al. [31] have fabricated high-performance ASSs from nitrogen and boron co-doped materials. The as-prepared monolithic composites were easily sliced into smaller pieces and then physically pressed into plates of the required size. Then, two of the plates were subsequently assembled into sandwich-like ASSs, separated by one polymer spacer. The ASSs fabricated from prepared composites not only abridged the size and thickness of the device but also gave a decent performance. A specific capacitance was delivered at a particular scan rate. The high rate capability showed a reversible capacitance obtained at

100 mVs^{-1}. Capacitance retention was almost 100% for 1,000 cycles. The energy density achieved was 8.65 Whr kg^{-1} and the power density was 1,650 Wkg^{-1} [3].

Zhang et al. [3] designed a facile and scalable solution-casting technique to prepare graphene foam (GF) from nickel foam (NF) template and GO, precursor. The GFs decorated with MnO$_2$ and CNTs through electrochemical approaches were used as the positive and negative electrodes respectively, to assemble asymmetric ASSs. As indicated by the electrochemical measurements, the devised asymmetric ASSs were able to function with an output cell voltage of 1.8 V and deliver high energy/power density. The supercapacitor retained 84.4% of its initial capacitance for over 10,000 cycles at a current density of 5 Ag^{-1}. Three ASSs connected in series can power a red LED light. Unlike conventional supercapacitors packed with liquid electrolytes, the asymmetric ASSs using a polymer gel of potassium polyacrylate/KCl as the electrolyte exhibited remarkable stability against mechanical stress.

Santhakumar Kannappan et al. [26] developed graphene-based supercapacitors using a modified Hummer's method and tip sonication for graphene synthesis. The prepared supercapacitors have high stability, improved electrical double layer capacitance and energy density with fast charging and discharging time due to increase in ionic electrolyte accessibility.

Shun Mao et al. [32] reviewed the unique structures and properties of three-dimensional graphene-based composites for energy applications. The authors addressed the major merits of 3D graphene (3DG) as a porous and interconnected network, high electrical conductivity, large accessible surface area, excellent mechanical strength and thermal stability, with the high chemical/electrochemical activities of active materials. Due to unique advantages, the developed composites showed great promise as high-performance electrode materials in various energy storage/conversion devices and related applications in supercapacitors, lithium-ion batteries, dye-sensitized solar cells, and fuel cells.

Jing Li et al. [5] deposited poly (N-acetylaniline) (PAANI) nanowires on the surface of graphene. SEM results demonstrated that PAANI nanowires with 70 nm diameter were uniformly distributed with an increase in surface area and charge transfer reaction. The developed composites exhibited good electrochemical properties and enhanced capacitive behavior higher than PAANI nanowire and graphene supercapacitors. The prepared supercapacitors showed excellent charge/discharge rates, and good cycling stability.

Van Hoa Nguyen et al. [14] developed MnO$_2$ intercalated rGO-Co$_3$O$_4$ composite for supercapacitors applications. The ternary composite exhibited higher specific capacitance and better cycling stability compared to binary composites. The ternary composite approach offers an effective methodology for enhancing the device performance of metal-oxide based supercapacitors for long cycling applications.

Cheng Liang et al. [33] deposited nano-scale $Co(OH)_2$/NG (nano-graphene: NG) composites on the macroporous electrically conductive network (MECN) to serve as the active materials for miniature supercapacitors. The $Co(OH)_2$/NG composite in the channels largely contributed to the capacitance. The electrochemical properties were investigated by cyclic voltammetry (CV), galvanostatic charging-discharging and electrochemical impedance spectroscopy (EIS) in 2 M KOH solution. The $Co(OH)_2$/NG composite exhibited a stable capacitance of 3.522 F/cm^2 (880.5 F/g) and retention ratio as 99% after 5000 cycles. The $Co(OH)_2$/NG/MECN can be scanned faster than $Co(OH)_2$/MECN in CV because of the higher conductivity [33]. The authors introduced nano-graphene into the MECN structure by hydrothermal carbonization of a nickel-coated silicon microchannel plate, a typical MECN structure and thermal processing. Cauliflower-like nanostructured $Co(OH)_2$ was deposited on nano-graphene (NG) nanosheets and the cyclic voltammetry (CV), as well as galvanostatic charging-discharging, were utilized to evaluate the electrochemical properties of the composite in KOH.

Rinaldo Raccichini et al. [34] and Kangjun Xie et al. [35] introduced a new method called "facility method" to develop graphene and $LiNO_3$ composites (GLs) in which the mixture of graphene and $LiNO_3$ was burned. The electrochemical properties of GLs as supercapacitor electrode materials were evaluated. The results showed that GLs possess much high specific capacitance at a high current density and scan rate (3339.7 F/g at 1 mV/s and 80.07 F/g at 20 A/g, respectively) with excellent cycling stability. The achievements of such high-performance GLs composites may open up a new window for energy storage applications and wearable electronics in the future.

Ya-ting Zhang et al. [36] explained that the low-cost production of high-performance functional materials based on graphene has been a challenging task. One of the options for tackling this problem is to develop new processes based on cheap starting materials such as coal. Coal-based graphene precursors were prepared from purified Taixi anthracite by catalytic graphitization combined with an improved Hummers method, mixed with MnO_2 and reduced by low-temperature plasma to make MnO_2/coal-based graphene nanocomposites. The composites were characterized by FT-IR, XRD, TEM, and SEM. The electrochemical performance of the developed composites was evaluated using the CV and galvanostatic charge/discharge technique. Results showed that MnO_2 was evenly deposited on the graphene surface and the specific capacitance of the composite as an electrode in a supercapacitor was much higher than that of coal-based graphene without MnO_2. The capacitance (281.1 F/g) was higher i.e., 261.2% of the value for coal-based graphene.

Haitao Zhang et al. [15] reported that supercapacitors exhibited superior performance with reduced graphene oxide and maghemite (g-Fe_2O_3) composites (GgM). Supercapacitors with average electrode thicknesses of up to 60 mm resulted in volumetric capacitance of 230 F cm^{-3} and an outstanding electrode package density of 1.44 g cm^{-3}. XRD and TEM analysis confirmed that there was no detectable change in phase and an effective inhibition of g-Fe_2O_3 refinement after the cycle-life test.

Guanglin Sun et al. [37] synthesized three-dimensional hierarchical porous carbon/graphene composites (3DHCG) from lyophilized graphene oxide-chitosan (GO-CS) composite hydrogels by the combination of direct carbonization and chemical activation processes. The resulting graphene-based composite has been able to construct a hierarchical porous network with randomly opened macropores and micropores, facilitating fast diffusion to the interior surfaces during rapid charge/discharge process. When compared with the pure carbon, the synthesized 3DHCG composite was found to exhibit remarkably enhanced specific capacitance of 320 Fg^{-1} at a current density of 1 Ag^{-1} in 6 M KOH solution, meanwhile maintaining excellent rate performance (capacity retention of 70% at 20 A/g) and outstanding cycling stability (96% retention after 2000 cycles). The improved electrochemical performances of graphene-based materials represent an alternative promising candidate for the application as supercapacitor electrodes.

Ke-Jing Huang et al. [38] reported that in order to control the formation of $MoSe_2$-graphene composites, graphene and the flexible Ni foam substrate were used in a facile hydrothermal process. The degree of graphene content was found to affect the structure of prepared composites. Graphene nanosheets on the surface and interspace of $MoSe_2$ bars led to the large specific surface area and plenty of macropores. The as-prepared materials exhibited good catalytic activity towards the electrochemical oxidation of dopamine with a linear range of 0.01-10 mM and the detection limit of 1.0 nM. The porous layered $MoSe_2$-graphene composite was found more suitable for applications in high-performance supercapacitors and electrochemical sensors.

Qinxing Xie et al. [39] prepared sandwich-like nitrogen-enriched porous carbon/graphene composites by carbonization of polyacrylonitrile (PAN) nanofiber paper with bulk-doped and/or surface-coated graphene oxide (GO) and followed by activation with KOH at high temperatures. The composites exhibited high specific surface areas in the range of 1957.2–2631.8 m^2g^{-1}, as well as superior energy storage capability as electrodes for supercapacitors. A very high specific capacitance of 381.6 Fg^{-1} at a current density of 0.1

Ag^{-1} was achieved in 6 M KOH aqueous electrolyte for the composites which contained two types of graphene. According to the investigation results, the surface-coated graphene played a more positive effect on improved supercapacitance than the bulk-doped graphene in the composites.

Mei-Xia Guo et al. [40] reported that flexible supercapacitors with textile-based electrode materials hindered their application due to low electrochemical capacitance. The hydrothermal method was used to deposit the MnO_2 (high theoretical capacitance) onto the conductive graphene/polyester composite fabric. Its morphology was controlled by adjusting the reaction time. The electrochemical performance of MnO_2/graphene/polyester composite electrode materials based on MnO_2 morphology, electrode structure, mechanical bending, and stretching was studied in detail. The developed composite exhibited a specific capacitance of 332 Fg^{-1} at a scan rate of 2 mVs^{-1} and excellent cycling stability. The corresponding solid-state supercapacitor device was stable under mechanical bending and stretching conditions.

Yuvaraj Haldorai et al. [41] fabricated titanium nitride/reduced graphene oxide nanocomposite (TiN/rGO) by a two-step process. The resulting TiN particles had a mean diameter of less than 10 nm and were densely decorated onto the rGO surface. The TiN/rGO composite was used as a support matrix to anchor platinum (Pt) nanoparticles by the polyol method to fabricate a Pt@TiN/rGO ternary hybrid catalyst for methanol oxidation. An increase in the methanol oxidation current density was observed for Pt@TiN/rGO when compared to Pt/rGO and Pt/Vulcan, confirming that the inclusion of TiN along with rGO improved the electrocatalytic activity. The electrochemical surface area was significantly higher for the Pt@TiN/ rGO catalyst (84.5 m^{2}g^{-1}) compared to Pt/rGO (51.7 m^{2}g^{-1}) and Pt/Vulcan (33.7 m^{2}g^{-1}), highlighting the importance of TiN. The Pt@TiN/rGO hybrid showed excellent electrocatalytic activity, long-term stability, and better carbon monoxide tolerance for the electrooxidation of methanol when compared to more traditional catalysts, namely Pt/rGO and Pt/Vulcan with same Pt content. Conversely, the TiN/rGO composite (without Pt) showed higher capacitance of 415 Fg^{-1} and a long cycle life, with 7.0% capacitance loss after 10,000 cycles. The capacitance was as high as 275 Fg^{-1} at a current density of 5 A.

In the past, substantial research efforts have been carried out regarding the synthesis of graphene/Pt nanostructures for methanol oxidation [42]. The graphene support improved the catalytic activity in direct-methanol fuel cells (DMFCs) compared to other carbonaceous supports. Xia et al. [43] reported that Pt@graphene nanocomposites exhibited high electrocatalytic performance and good stability towards methanol oxidation when compared to Pt alone. Similarly, Li et al. [44] and Dong et al. [45] reported higher electro-catalytic efficiencies of Pt@graphene electro-catalysts for electrooxidation of methanol when compared to CNT and carbon black supports. Other researchers have synthesized metal oxide/graphene/Pt composites with enhanced electrocatalytic activity for methanol oxidation [41]. Zeng et al. [46] prepared mesoporous carbon/WO_3 hybrid as a support for Pt electrocatalyst for methanol oxidation. Xia et al. [47] synthesized titanium dioxide (TiO_2)/reduced graphene oxide (rGO) composite decorated with Pt nanoparticles as an electrocatalyst support for methanol oxidation. Yu et al. [48] fabricated cerium oxide/rGO/Pt nanostructure that showed significant electrocatalytic activity for DMFCs.

Tao Yu et al. [42] developed an interesting method for the synthesis of micro-spherical polyaniline/graphene (PANI/G-MS) composites using sheet-like polyaniline/graphene oxide (PANI/GO) composites as raw materials via spray-drying and chemical reduction process, in which the granulated polyaniline (PANI) was in-situ grown on the surface of graphene oxide. In case of PANI/G-MS composites, PANI was uniformly coated on the surface of graphene to create a high conductive network for accelerating electronic transmission in the composites electrode for supercapacitors. Moreover, PANI/G-MS composites generated numerous channels among their spherical particles during random stacking of sheet-like PANI/GO composites via the spray-drying process. Due to the special structure, the electrochemical capacitances of the as-synthesized PANI/G-MS composite were 596.2 and 447.5 Fg^{-1} at current densities of 0.5 and 20 Ag^{-1}, respectively, indicating superior rate capability. Additionally, after 1500 cycles at a current density of 2 Ag^{-1}, about 83.7% of the initial capacitance was retained.

As a typical representative of conductive polymers, polyaniline (PANI) has enjoyed some advantages for future practical applications, such as high theoretical capacity, ease of synthesis and low cost [42]. However, when employed in supercapacitors, it suffered from some shortcomings of relatively low electrical conductivity and volumetric deformation that arose from the doping/dedoping of dopants during charging and discharging processes. These drawbacks affect the actual capacitance and the cycling stability of PANI and then seriously restrict the use of PANI in supercapacitors. To conquer these obstacles, significant attentions have been focused on integrating PANI with high conductive carbon materials to obtain high-performance composite electrode materials. Many carbon materials, for example, porous carbon, carbon aerogel, carbon nanotubes and carbon nanofibers have been utilized to hybridize with PANI. Recently, as a special monolayer nanostructure of carbon atoms network, graphene demonstrated superiority towards the above-mentioned carbon materials, e.g. excellent electrical and thermal properties, as well as high mechanical strength [42]. These outstanding properties made graphene as an extremely promising material for electrochemical energy storage, especially supercapacitors [42]. Therefore, a strategy for combining PANI with graphene would be an attractive method to overcome the drawbacks of PANI. A lot of research work concerning various structures of graphene/PANI composites has been carried out.

Comparing to the lamellar structure, three-dimensional (3D) structures (3D porous structures, sandwiched structure, tremella-like, spheres and hollow spheres) have exhibited much more improved electrochemical performance. For instance, Stanciu et al. [49] employed PANI nanoparticles as a spacer to obtain sandwiched PANI/graphene nanocomposites and realized high specific surface area (891 m^2g^{-1}) and specific capacitance (257 Fg^{-1} at 0.1 Ag^{-1}) [42]. Among those complicated structures, the spherical structure was highly conductive to improve the performance of electrode materials as well. Liu et al. [50] have synthesized tremella-like PANI/graphene spherical composites by self-assembly of graphene nanosheets. Owing to the 3D porous spherical architecture, the specific capacity of 497 Fg^{-1} at 0.5 Ag^{-1} was achieved [42]. Moreover, Cao and his co-workers fabricated PANI/graphene nanocomposites through in-situ polymerization using graphene microspheres as substrate. The final product exhibited an improved electrochemical capacitive performance [42]. However, due to the existence of concentration gradient of aniline between the surface and interior in graphene microspheres, the polymerization process occurred on the surface of graphene microspheres. In order to fully utilize the synergistic effect, graphene and PANI must be well in contact with each other. Therefore, a rational sphere structure of PANI/graphene composites with good contact between two components can enhance the performance of electrode materials.

Shuying Kong et al. [51] used carbon materials, especially graphene nanosheets (GNS) and/or multi-walled carbon nanotube (MWNT) as electrode materials for supercapacitor due to their advantages of the higher specific surface area and electronic conductivity. However, the relatively low specific capacitance and energy density are hampering their large applications. On the contrary, MnO_2 exhibits higher energy density but poor electrical conductivity. In order to obtain high-performance supercapacitor electrode, by combining the advantages of these materials, researchers have designed a facile two-step strategy to prepare 3D MnO_2-GNS-MWNT-Ni foam (MnO_2-GM-Ni) electrode. First, GNS and MWNT were wrapped on the surface of Ni foam (GM-Ni) via a "dip & dry" method using an organic dye as a co-dispersant. Then, by using this 3D GM-Ni as a substrate, MnO_2 nanoflakes were in-situ supported on the surface of GNS and MWNT through a hydrothermal reaction. The specific capacitance of MnO_2-GM-Ni electrode was 470.5 Fg^{-1} at 1 Ag^{-1}. Furthermore, the authors have successfully fabricated an asymmetric supercapacitor using MnO_2-GM- Ni, and GM-Ni as the positive and negative electrodes, respectively. The MnO_2-GM-Ni//GM-Ni asymmetric supercapacitor exhibited maximum energy density of about 35.3 $Whkg^{-1}$ at a power density of around 426 Wkg^{-1} and with favorable cycling performance (83.8% capacitance retention) after 5000 cycles. These results showed manageable and high-performance which offer a promising future for practical applications.

Balsydulu Singuand Kuk Ro Yoon [52] proposed a simple one-step method for the synthesis of reduced graphene oxide-manganese oxide (rGO-MnO$_2$) nanocomposite using graphene oxide (GO) and KMnO$_4$ in the presence of sulfuric acid. The crystal structure, morphology, thermal, pore size, and other physical properties of the rGO-MnO$_2$ nanocomposite were systematically analyzed by XRD, TGA, XPS, FE-SEM, TEM, and Brunauer-Emmett-Teller techniques. XPS analysis confirmed the synthesis of exfoliated GO and rGO-MnO$_2$ nanocomposite. The rGO-MnO$_2$ nanocomposite exhibited highest values of specific capacitance, energy, and power density as 290 Fg^{-1}, 25.7 Whkg^{-1}, and 8008.7 Wkg^{-1}, respectively, in 1M Na$_2$SO$_4$ electrolyte along with a high retention (87.5%) of capacitance after 5000 cycles.

Yangshuai Liu et al. [53] prepared manganese dioxide nanotubes and activated carbon coated multiwalled carbon nanotubes (AC-MWCNT) composites by hydrothermal methods. MnO$_2$-graphene positive electrodes with good dispersion of individual components were fabricated using poly[1-[4-(3-carboxy-4-hydroxyphenylazo) benzenesulfo-namido]-1,2-ethanediyl, sodium salt] (PAZO) as a co-dispersant. The unique structure of PAZO, containing chelating aromatic monomers allowed efficient adsorption of this polyelectrolyte on MnO$_2$ and graphene, a prerequisite for their efficient electrosteric co-dispersion. The MnO$_2$-graphene electrodes with the active mass loading of 30 mgcm^{-2} showed capacitance of 3.3 Fcm^{-2} at a scan rate of 2 m s^{-1}. The capacitance retention of 64% was achieved with an increase of scan rate from 2 to 100 mVs^{-1}. The use of AC-MWCNT with thick AC coating facilitated the fabrication of negative electrodes, which closely matched with the capacitive performance of the positive electrodes. The asymmetric supercapacitors containing MnO$_2$-graphene positive electrodes and AC-MWCNT negative electrodes showed capacitance of 1.42 F cm^{-2} at a scan rate of 2 mVs^{-1} and the capacitance retention of 52% in the scan rate range of 2–100 mVs^{-1}.

Kuo Song et al. [6] reported that flexible supercapacitors based on paper-like electrodes have attracted significant interest because of the increasing demands in the energy storage systems. As promising binder-free electrode materials in the supercapacitors, graphene-based films have been developed for enhancing their performance in energy storage systems by insetting "spacers" in-between nanosheets to prevent inevitable aggregations. A facile and versatile strategy was presented for fabricating graphene-based composite films by introducing activated carbonized cotton fibers to regulate the chemical composition, surface area, and pore size distribution.

Yru Jiang et al. [54] stated that designing and optimizing the electrode materials and studying the electrochemical performance or cycle life of the supercapacitor under different working conditions were crucial for their practical applications. The authors proposed a rational design of 3D-graphene/$CoMoO_4$ nanoplates on the basis of a facile two-step hydrothermal method. Owing to the high electron transfer rate of graphene and the high activity of the $CoMoO_4$ nanoplates, the three-dimensional electrode architectures achieved remarkable electrochemical performances with high areal specific capacitance and superior cycling stability. The all-solid-state asymmetric supercapacitor composed of 3D-graphene/$CoMoO_4$ and activated carbon (AC) exhibited a specific capacitance of 109 F/g at 0.2 A/g and an excellent cycling stability with only 12.1% of the initial specific capacitance after 3000 cycles at 2 A/g. The effects of temperature and charge-discharge current densities on the charge storage capacity of the supercapacitor were also investigated in detail for practical applications.

Qinxing Xie et al. [18] prepared polystyrene (PS) foam derived nitrogen-enriched porous carbon/graphene (AC/Gr) composites using ammonia, ethylene diamine, and melamine as nitrogen sources. The as-prepared AC/Gr composites and graphene-free PS-derived porous carbon (PSAC) exhibited moderate specific surface areas. Compared to PSAC, AC/Gr composites demonstrated significantly improved energy storage capability. High gravimetric capacitance (339 Fg^{-1}) and volumetric capacitance (365 Fcm^{-3}) were achieved at a current density of 0.05 Ag^{-1} in 6 M KOH aqueous electrolyte. The assembled aqueous symmetric supercapacitors were capable of delivering both high energy density as well as power density.

4. Conclusion

Graphene-based composites are facing a big problem for practical applications because of their unavailability in large-scale and low-cost production techniques. In addition, to find the optimum composition of graphene-based composites for specific applications is also a great challenge for the researchers in this field. Electronic, electrochemical and structural properties are important to assess the performance of the graphene-based composites in energy devices. It is expected many more energy storage/conversion devices and environmental applications from graphene-based composites will emerge in the near future.

This chapter encapsulates present progress and future scope of development on graphene-based materials for supercapacitor electrodes based on various properties and particularly on their structural complexity: zero-dimensional (0D) (e.g. free-standing graphene dots and particles), one-dimensional (1D) (e.g. fiber-type and yarn-type structures), two-dimensional (2D) (e.g. nanocomposites films or papers), and three-dimensional (3D) (e.g.

graphene-based foams and hydrogels). The rapid development of flexible electronics requires flexible and deformable energy storage devices. Therefore, future studies focus on the development of mechanical flexibility of graphene-based materials for supercapacitors and other energy storage devices. The multifunctional or self-powered hybrid systems will be of considerable interests for future development. Recent pioneer work on the combination of flexible supercapacitors with other electronic and energy devices (i.e., solar cells, Li-ion batteries, electrochromic devices, and nanogenerators) has been very well presented. It is hoped that the integration of graphene-based supercapacitors with such devices will be of considerable values but still faces challenges for future research.

References

[1]   Q. Ke, J. Wang, Graphene-based materials for supercapacitor electrodes-A review, J. Materiomics 2 (2016) 37-54. https://doi.org/10.1016/j.jmat.2016.01.001

[2]   L. Li, B. Song, L. Maurer, Z. Lin, G. Lian, C.C. Tuan, K.S. Moon, C.P. Wong, Molecular engineering of aromatic amine spacers for high-performance graphene-based supercapacitors, Nano Energy 21 (2016) 276-294. https://doi.org/10.1016/j.nanoen.2016.01.028

[3]   X. Yao, Y. Zhao, Three-dimensional porous graphene networks and hybrids for lithium-ion batteries and supercapacitors, Chem. 2 (2017) 171–200. https://doi.org/10.1016/j.chempr.2017.01.010

[4]   S. Zheng , Z.S. Wu, S. Wang, H. Xiao, F. Zhou, C. Sun, X. Bao, H.M. Cheng, Graphene-based materials for high-voltage and high-energy asymmetric supercapacitors, Energy Stor. Mater. 6 (2017) 70-97. https://doi.org/10.1016/j.ensm.2016.10.003

[5]   J. Li, H. Xie, Y. Li, Enhanced electrochemical performance of poly(N-acetylaniline)/graphene composites as electrode materials for supercapacitors, Mater. Lett. 124 (2014) 215-218. https://doi.org/10.1016/j.matlet.2014.03.060

[6]   K. Song, H. Ni, L.Z. Fan, Flexible graphene-based composite films for supercapacitors with tunable areal capacitance, Electrochim. Acta 235 (2017) 233-241. https://doi.org/10.1016/j.electacta.2017.03.065

[7]   B. Xie, Y.Wang, W. Lai, W. Lin, Z. Lin, Z. Zhang, P. Zou, Y. Xu, S. Zhou, C. Yan, F. Kang, C.P. Wong, Laser-processed graphene based micro-supercapacitors for ultrathin, rollable, compact and designable energy storage components, Nano Energy 26 (2016) 276–285. https://doi.org/10.1016/j.nanoen.2016.04.045

[8]   J. Chmiola, C. Largeot, P.L. Taberna, P. Simon, Y. Gogotsi, Monolithic carbide derived carbon films for micro-supercapacitors, Sci. 328 (2010) 480-483. https://doi.org/10.1126/science.1184126

[9]   D. Pech, M. Brunet, H. Durou, P.Huang, V. Mochalin, Y. Gogotsi, P.L. Taberna, P. Simon, Ultrahigh power micrometer sized supercapacitors based on onion like carbon, Nat. Nanotechnol. 5 (2010) 651-654. https://doi.org/10.1038/nnano.2010.162

[10]   Z.S. Wu, K. Parvez, X. Feng, K. Mullen, Graphene-based in-plane micro-supercapacitors with high power and energy densities, Nat. Commun. 4 (2013) 2487. https://doi.org/10.1038/ncomms3487

[11]   Y. Zhang, L. Guo, S. Wei, Y. He, H. Xia, Q. Chen, H.B. Sun, F.S. Xiao, Direct imprinting of microcircuits on graphene oxides film by femtosecond laser reduction, Nano Today 5 (2010) 15-20. https://doi.org/10.1016/j.nantod.2009.12.009

[12]   M.F. El-Kady, R.B. Kaner, Scalable fabrication of high-power graphene micro-supercapacitors for flexible and on-chip energy storage, Nat. Commun. 4 (2013) 1475. https://doi.org/10.1038/ncomms2446

[13]   Z. Peng, R. Ye, J.A. Mann, D. Zakhidov, Y. Li, P.R. Smalley, J. Lin, J.M. Tour, Flexible boron doped laser induced graphene micro supercapacitors, ACS Nano 9 (2015) 5868-5875. https://doi.org/10.1021/acsnano.5b00436

[14]   V.H. Nguyen, V.C. Tran, D. Kharismadewi, J.J. Shim, Ultralong MnO_2 nanowires intercalated graphene/Co_3O_4 composites for asymmetric supercapacitors, Mater. Lett. 147 (2015) 123-127. https://doi.org/10.1016/j.matlet.2015.01.139

[15]   H. Zhang, X. Zhang, H. Lin, K. Wang, X. Sun, N. Xu, C. Li, Y. Ma, Graphene and maghemite composites based supercapacitors delivering high volumetric capacitance and extraordinary cycling stability, Electrochim. Acta 156 (2015) 70-76. https://doi.org/10.1016/j.electacta.2015.01.041

[16]   S. Mitra, S. Sampath, Electrochemical capacitors based on exfoliated graphite electrodes batteries, fuel cells, and energy conversion, Electrochem. Solid-State Lett. 7 (2004) 264-268. https://doi.org/10.1149/1.1773752

[17]   X. Wang, L.J. Zhi, K. Mullen, Transparent, conductive graphene electrodes for dye-sensitized solar cells, Nano Lett. 8 (2008) 323-327. https://doi.org/10.1021/nl072838r

[18]   Q. Xie, G. Chen, R. Bao, Y. Zhang, S. Wu, Polystyrene foam derived nitrogen-enriched porous carbon/graphene composites with high volumetric capacitances for aqueous supercapacitors, Micropor. Mesopor. Mater. 239 (2017) 130-137. https://doi.org/10.1016/j.micromeso.2016.10.007

[19]   K.S. Novoselov, A.K. Geim, S.V. Morozov, D. Jiang, Y. Zhang, S.V. Dubonos, Electric field effect in atomically thin carbon films, Sci. 306 (2004) 666-669. https://doi.org/10.1126/science.1102896

[20]   Y. Hernandez, V. Nicolosi, M. Lotya, F.M. Blighe, Z. Sun, S. De, High-yield production of graphene by liquid-phase exfoliation of graphite, Nat. Nanotechnol. 3 (2008) 563-568. https://doi.org/10.1038/nnano.2008.215

[21]   M.D. Stoller, S. Park, Y. Zhu, J. An, R.S. Ruoff, Graphene-based ultracapacitors, Nano Lett. 8 (2008) 3498-3502. https://doi.org/10.1021/nl802558y

[22]   W. Gao, L.B. Alemany, L.J. Ci, P.M. Ajayan, New insights into the structure and reduction of graphite oxide, Nat. Chem. 1 (2009) 403-408. https://doi.org/10.1038/nchem.281

[23]   C. Zhu, S. Guo, Y. Fang, S. Dong, Reducing sugar: new functional molecules or the green synthesis of graphene nanosheets, ACS Nano 4 (2010) 2429-2437. https://doi.org/10.1021/nn1002387

[24]   J. Zhang, H. Yang, G. Shen, P. Cheng, J. Zhang, S. Guo, Reduction of graphene oxide via L-ascorbic acid, Chem. Commun. 46 (2010) 1112-1114. https://doi.org/10.1039/B917705A

[25]   W. Shi, J. Zhu, D.H. Sim, Y.Y. Tay, Z. Lu, X. Zhang, Y. Sharma, M. Srinivasan, H. Zhang, H.H. Hng, Q. Yan, Achieving high specific charge capacitances in Fe_3O_4/reduced graphene oxide nanocomposites, J. Mater. Chem. 21 (2011) 3422-3427. https://doi.org/10.1039/c0jm03175e

[26]   S. Kannappan, K. Kaliyappan, R.K. Manian, A.S. Pandian, H. Yang, Y.S. Lee, J.H. Jang, W. Lu, Graphene based supercapacitors with improved specific capacitance and fast charging time at high current density, https://arxiv.org/abs/1311.1548.

[27]   C. Liu, Z. Yu, D. Neff, A. Zhamu, B.Z. Jang, Graphene based supercapacitor with an ultrahigh energy density, Nano Lett. 10 (2010) 4863-4868. https://doi.org/10.1021/nl102661q

[28]   Y. Zhu, S. Murali, M.D. Stoller, K.J. Ganesh, W. Cai, P.J. Ferreira, A. Pirkle, R.M. Wallace, K.A. Cychosz, M. Thommes, D. Su, E.A. Stach, R.S. Ruoff,

Carbon-based supercapacitors produced by activation of graphene, Sci. 332 (2011) 1537-1541. https://doi.org/10.1126/science.1200770

[29]   M. Sevilla, G.A. Ferrero, A.B. Fuertes, Graphene-cellulose tissue composites for high power supercapacitors, Energy Stor. Mater. 5 (2016) 33–42. https://doi.org/10.1016/j.ensm.2016.05.008

[30]   B. Zhao, D.C. Chen, X. Xiong, B. Song, R. Hu, Q. Zhang, B.H. Rainwater, G.H. Waller, D. Zhen, Y. Ding, Y. Chen, C. Qu, D. Dang, C.P. Wong, M. Liu, A high-energy, long cycle-life hybrid supercapacitor based on graphene composite electrodes, Energy Stor. Mater. 7 (2017) 32–39. https://doi.org/10.1016/j.ensm.2016.11.010

[31]   Z.S. Wu, A. Winter, L. Chen, Y. Sun, A. Turchanin, X. Feng, K. Mullen, Three-dimensional nitrogen and boron co-doped graphene for high-performance all solid-state supercapacitors, Adv. Mater. 24 (2012) 5130–5135. https://doi.org/10.1002/adma.201201948

[32]   S. Mao, G. Lu, J. Chen, Three-dimensional graphene-based composites for energy applications, Nanoscale 7 (2015) 6924–6943. https://doi.org/10.1039/C4NR06609J

[33]   C. Liang, L. Chen, D. Wu, C.Z.S. Xu, Y. Zhu, D. Xiong, P. Yang, L. Wang, P.K. Chu, Hybrid $Co(OH)_2$/nano-graphene/Ni nano-composites on silicon microchannel plates for miniature supercapacitors, Mater. Lett. 172 (2016) 40-43. https://doi.org/10.1016/j.matlet.2016.02.132

[34]   R. Raccichini, A. Varzi, S. Passerini, B. Scrosati, The role of graphene for electrochemical energy storage, Nat. Mater. 14 (2015) 271-279. https://doi.org/10.1038/nmat4170

[35]   K. Xie, J. Yang, Q. Zhang, H. Guo, S. Hu, Z. Zeng, X. Fang, Q. Xu, J. Huang, W. Qi, Burning the mixture of graphene and lithium nitride for high-performance supercapacitor electrodes, Mater. Lett. 195 (2017) 201-204. https://doi.org/10.1016/j.matlet.2017.02.129

[36]   Y.T. Zhang, J.K. Li, G.Y. Liu, J.T. Cai, A. Zhou, J.S. Qiu, Preparation of MnO_2/coal-based graphene composites for supercapacitors, New Carbon Mater. 31 (2016) 545-549.

[37]   G. Sun, B. Li, J. Ran, X. Shen, H. Tong, Three-dimensional hierarchical porous carbon/graphene composites derived from graphene oxide-chitosan hydrogels for

high performance supercapacitors, Electrochim. Acta 171 (2015) 13-22.
https://doi.org/10.1016/j.electacta.2015.05.009

[38]   K.J. Huang, J.Z. Zhang, J.L. Cai, Preparation of porous layered molybdenum
       selenide-graphene composites on Ni foam for high-performance supercapacitor
       and electrochemical sensing, Electrochim. Acta 180 (2015) 770-777.
       https://doi.org/10.1016/j.electacta.2015.09.016

[39]   Q. Xie, S. Zhou, A. Zheng, C. Xie, C. Yin, S. Wu, Y. Zhang, P. Zhao, Sandwich-
       like nitrogen-enriched porous carbon/graphene composites as electrodes for
       aqueous symmetric supercapacitors with high energy density, Electrochim. Acta
       189 (2016) 22-31. https://doi.org/10.1016/j.electacta.2015.12.087

[40]   M.X. Guo, S.W. Bian, F. Shao, S. Liu, Y.H. Peng, Hydrothermal synthesis and
       electrochemical performance of MnO2/graphene/polyester composite electrode
       materials for flexible supercapacitors, Electrochim. Acta 209 (2016) 486-497.
       https://doi.org/10.1016/j.electacta.2016.05.082

[41]   Y. Haldorai, D.A. Salas, C.S. Rak, Y.S. Huh, Y.K. Han, W. Voit, Platinized
       titanium nitride/graphene ternary hybrids for direct methanol fuel cells and
       titanium nitride/graphene composites for high performance supercapacitors,
       Electrochim. Acta 220 (2016) 465-474.
       https://doi.org/10.1016/j.electacta.2016.10.130

[42]   T. Yu, P. Zhu, Y. Xiang, H. Chen, S. Kang, H. Luo, S. Guan, Synthesis of
       microspherical polyaniline/graphene composites and their application in
       supercapacitors, Electrochim. Acta 222 (2016) 12-19.
       https://doi.org/10.1016/j.electacta.2016.11.033

[43]   Y.G. Zhou, J.J. Chen, F.B. Wang, Z.H. Sheng, X.H. Xia, A facile approach to the
       synthesis of highly electroactive Pt nanoparticles on graphene as an anode catalyst
       for direct methanol fuel cells, Chem. Commun. 46 (2010) 5951–5953.
       https://doi.org/10.1039/c0cc00394h

[44]   Y. Li, W. Gao, L. Ci, C. Wang, P.M. Ajayan, Catalytic performance of Pt
       nanoparticles on reduced graphene oxide for methanol electro-oxidation, Carbon
       48 (2010) 1124–1130. https://doi.org/10.1016/j.carbon.2009.11.034

[45]   L. Dong, R.R.S. Gari, Z. Li, M.M. Craig, S. Hou, Graphene-supported platinum
       and platinum–ruthenium nanoparticles with high electrocatalytic activity for
       methanol and ethanol oxidation, Carbon 48 (2010) 781–787.
       https://doi.org/10.1016/j.carbon.2009.10.027

[46]   J. Zeng, C. Francia, C. Gerbaldi, V. Baglio, S. Specchia, A.S. Arico, P. Spinelli, Hybrid ordered mesoporous carbons doped with tungsten trioxide as supports for Pt electrocatalysts for methanol oxidation reaction, Electrochim. Acta 94 (2013) 80–91. https://doi.org/10.1016/j.electacta.2013.01.139

[47]   B.Y. Xia, H.B. Wu, J.S. Chen, Z. Wang, X. Wang, X.W. (David) Lou, Formation of Pt–TiO_2–rGO_3-phase junctions with significantly enhanced electro-activity for methanol oxidation, Phy. Chem. 14 (2012) 473–476.

[48]   X. Yu, L. Kuai, B. Geng, CeO_2/rGO/Pt sandwich nanostructure: rGO-enhanced electron transmission between metal oxide and metal nanoparticles for anodic methanol oxidation of direct methanol fuel cells, Nanoscale 4 (2012) 5738–5743. https://doi.org/10.1039/c2nr31765f

[49]   Z.F. Li, H.Y. Zhang, Q. Liu, L.L. Sun, L. Stanciu, J. Xie, Fabrication of high-surface area graphene/polyaniline nanocomposites and their application in supercapacitors, ACS Appl. Mater. Inter. 5 (2013) 2685-2691. https://doi.org/10.1021/am4001634

[50]   H.Y. Liu, W. Zhang, H.H. Song, X.H. Chen, J.S. Zhou, Z.K. Ma, Tremella-like graphene/polyaniline spherical electrode material for supercapacitors, Electrochim. Acta 146 (2014) 511-517. https://doi.org/10.1016/j.electacta.2014.09.083

[51]   S. Kong, K. Cheng, T. Ouyang, Y. Gao, K. Ye, G. Wang, D. Cao, Facile dip coating processed 3D MnO_2-graphene nanosheets/MWNT-Ni foam composites for electrochemical supercapacitors, Electrochim. Acta 226 (2017) 29-39. https://doi.org/10.1016/j.electacta.2016.12.158

[52]   B.S. Singu, K.R. Yoon, Synthesis and characterization of MnO_2-decorated graphene for supercapacitors, Electrochim. Acta 231 (2017) 749-758. https://doi.org/10.1016/j.electacta.2017.01.182

[53]   Y. Liu, K. Shi, I. Zhitomirsky, Asymmetric supercapacitor, based on composite MnO_2-graphene and N-doped activated carbon coated carbon nanotube electrodes, Electrochim. Acta 233 (2017) 142-150. https://doi.org/10.1016/j.electacta.2017.03.028

[54]   Y. Jiang, X. Zheng, X. Yan, Y. Li, X. Zhao, Y. Zhang, 3D architecture of a graphene/$CoMoO_4$ composite for asymmetric supercapacitors usable at various temperatures, J. Colloid Interf. Sci. 493 (2017) 42-50. https://doi.org/10.1016/j.jcis.2017.01.009

Keyword Index

About the Editors

Dr. Inamuddin is currently working as Assistant Professor in the Chemistry Department, Faculty of Science, King Abdulaziz University, Jeddah, Saudi Arabia. He is a permanent faculty member (Assistant Professor) at the Department of Applied Chemistry, Aligarh Muslim University, Aligarh, India. He obtained Master of Science degree in Organic Chemistry from Chaudhary Charan Singh (CCS) University, Meerut, India, in 2002. He received his Master of Philosophy and Doctor of Philosophy degrees in Applied Chemistry from Aligarh Muslim University (AMU) in 2004 and 2007, respectively. He has extensive research experience in multidisciplinary fields of Analytical Chemistry, Materials Chemistry, and Electrochemistry and, more specifically, Renewable Energy and Environment. He has worked on different research projects as project fellow and senior research fellow funded by University Grants Commission (UGC), Government of India, and Council of Scientific and Industrial Research (CSIR), Government of India. He has received Fast Track Young Scientist Award from the Department of Science and Technology, India, to work in the area of bending actuators and artificial muscles. He has completed four major research projects sanctioned by University Grant Commission, Department of Science and Technology, Council of Scientific and Industrial Research, and Council of Science and Technology, India. He has published 85 research articles in international journals of repute and seventeen book chapters in knowledge-based book editions published by renowned international publishers. He has published nineteen edited books with Springer, United Kingdom, Nova Science Publishers, Inc. U.S.A., CRC Press Taylor & Francis Asia Pacific, Trans Tech Publications Ltd., Switzerland and Materials Science Forum, U.S.A. He is the member of various editorial boards of the journals and also serving as associate editor for a journal Environmental Chemistry Letter, Springer Nature. He has attended as well as chaired sessions in various international and national conferences. He has worked as a Postdoctoral Fellow, leading a research team at the Creative Research Initiative Center for Bio-Artificial Muscle, Hanyang University, South Korea, in the field of renewable energy, especially biofuel cells. He has also worked as a Postdoctoral Fellow at the Center of Research Excellence in Renewable Energy, King Fahd University of Petroleum and Minerals, Saudi Arabia, in the field of polymer electrolyte membrane fuel cells and computational fluid dynamics of polymer electrolyte membrane fuel cells. He is a life member of the Journal of the Indian Chemical Society. His research interest includes ion exchange materials, a sensor for heavy metal ions, biofuel cells, supercapacitors and bending actuators.

Dr. Mohammad Faraz Ahmer is presently working as Assistant Professor in the Department of Electrical Engineering, Mewat Engineering College, Nuh Haryana, India, since 2012 after working as Guest Faculty in University Polytechnic, Aligarh Muslim University Aligarh, India, during 2009-2011. He completed M.Tech. (2009) and Bachelor of Engineering (2007) degrees in Electrical Engineering from Aligarh Muslim University, Aligarh in the first division. He obtained a Ph.D. degree in 2016 on his thesis entitled "Studies on Electrochemical Capacitor Electrodes". He has published six research papers in reputed scientific journals. His scientific interests include **electrospun nano-composites** and supercapacitors. He has presented his work at several conferences. He is actively engaged in searching of new methodologies involving the development of organic composite materials for energy storage systems.

Prof. Abdullah M. Asiri is the Head of the Chemistry Department at King Abdulaziz University since October 2009 and he is the founder and the Director of the Center of Excellence for Advanced Materials Research (CEAMR) since 2010 till date. He is the Professor of Organic Photochemistry. He graduated from King Abdulaziz University (KAU) with B.Sc. in Chemistry in 1990 and a Ph.D. from University of Wales, College of Cardiff, U.K. in 1995. His research interest covers color chemistry, synthesis of novel photochromic and thermochromic systems, synthesis of novel coloring matters and dyeing of textiles, materials chemistry, nanochemistry and nanotechnology, polymers and plastics. Prof. Asiri is the principal supervisors of more than 20 M.Sc. and six Ph.D. theses. He is the main author of ten books of different chemistry disciplines. Prof. Asiri is the Editor-in-Chief of King Abdulaziz University Journal of Science. A major achievement of Prof. Asiri is the discovery of tribochromic compounds, a new class of compounds which change from slightly or colorless to deep colored when subjected to small pressure or when grind. This discovery was introduced to the scientific community as a new terminology published by IUPAC in 2000. This discovery was awarded a patent from European Patent office and from UK patent. Prof. Asiri involved in many committees at the KAU level and on the national level. He took a major role in the advanced materials committee working for KACST to identify the national plan for science and technology in 2007. Prof. Asiri played a major role in advancing the chemistry education and research in KAU. He has been awarded the best researchers from KAU for the past five years. He also awarded the Young Scientist Award from the Saudi Chemical Society in 2009 and also the first prize for the distinction in science from the Saudi Chemical Society in 2012. He also received a recognition certificate from the American Chemical Society (Gulf region Chapter) for the advancement of chemical science in the Kingdome. He received a Scopus certificate for the most publishing scientist in Saudi Arabia in chemistry in 2008. He is also a member of the editorial board of various journals of international repute. He is the Vice- President of Saudi Chemical Society (Western Province Branch). He holds four USA patents, more than one thousand publications in international journals, several book chapters and edited books.

Dr. Sadaf Zaidi is an Associate Professor of Chemical Engineering at the Department of Chemical Engineering, Zakir Hussain College of Engineering, Aligarh Muslim University, Aligarh, India. He has a rich and varied experience of more than two and a half decades of teaching and guiding students of the B. Tech and M. Tech. (Process Modeling and Simulation) programs in Chemical Engineering at his department. He is an active researcher and is currently supervising five Ph.D. scholars. He has written 32 technical publications including those published in journals of repute like Chemical Engineering Science (Elsevier), Chemical Engineering Research and Design (Elsevier), Waste Management and Research (Sage Publication), Process Safety and Environment Protection Journal (Elsevier Publication) and chapters in knowledge-based editions published by renowned publishers like Springer Verlag and Nova Science Publishers Inc. etc. He has also contributed/presented 39 papers in conferences and attended 47 workshops, conferences, seminars and training courses. His research interests are in the fields of soft computing, boiling heat transfer (thermosiphon reboilers), energy technology, particularly supercapacitors, petroleum processing, nanocomposites and solid waste management. Dr. Zaidi received his B.Sc. Engg., M.Tech. and Ph.D. degrees in Chemical Engineering from the Aligarh Muslim University, Aligarh, India. He is a life member of the Indian Institute of Chemical Engineers, India.

www.ingramcontent.com/pod-product-compliance
Lightning Source LLC
Chambersburg PA
CBHW071331210326
41597CB00015B/1418